THE REVOLVER, 1818–1865

By A. W. F. Taylerson

THE REVOLVER 1865-1888

THE REVOLVER

1818–1865

By

A. W. F. TAYLERSON, R. A. N. ANDREWS
and J. FRITH

CROWN PUBLISHERS, INC.
NEW YORK

Published in the
United States of America
by Crown Publishers, Inc.

Library of Congress Catalog
Card Number: 68–29115

Printed in Great Britain by Cox & Wyman Limited, London, Fakenham and Reading

CONTENTS

CONTENTS

LIST OF PLATES

LIST OF FIGURES

FOREWORD

A QUARTER of a century has passed since the late J. N. George published his trail-blazing *English Pistols and Revolvers*, and during those years collector interest in British revolving arms has reached proportions that would have delighted him.

This book builds upon the foundations laid down by J. N. George, and we trust that it will prove to be as valuable and useful to the average collector as its predecessor.

Before turning to our subject, the sense in which certain words are used should be made clear. Revolving arms are not basically complicated devices, and a grasp of their principles is open to anyone who can understand the workings of a bicycle, but some of their component parts and lock-mechanisms are named in one manner by British collectors, and in another by American enthusiasts. This fact has confused many new-comers to the field, and warrants attention at the beginning of our book.

Throughout these pages, lock-mechanisms (i.e. those parts of the revolver which together rotate its cylinder, and discharge the arm) are described either as "single-action", or "double-action", or "self-cocking", or "hesitating".

The hammer of single-action lock-mechanisms can only be cocked, for firing, by drawing it back with the thumb; the hammer of a self-cocking lock-mechanism, as the name suggests, can only be raised and dropped by pulling the trigger. Double-action lock-mechanisms are so constructed that the firer may work them single-action *or* self-cocking, as need or fancy takes him; many Americans use the term "double-action" to describe self-cocking lock-mechanisms, but we prefer to use that term in the context mentioned above, as it has been used in England for more than a century. Hesitating lock-mechanisms are modified self-cocking designs, in which pressure on the trigger raises the hammer to full-cock (where it is automatically held), and a second pull on the trigger discharges the arm.

As to those components making up a complete weapon, our terms of description are largely self-explanatory. However, some parts do have different names on either side of the Atlantic, and we therefore confirm use here of the following terminology, in three specific cases:

1. *Pawl.* This is the lever in a lock-mechanism by which the cylinder or barrels are rotated.

2. *Arbor.* This is the axle upon which barrels or cylinder rotate.

3. *Cylinder-bolt.* This is the lock-mechanism component by which the chambers of a revolver cylinder are successively bolted in alignment with both barrel and hammer, for discharge. In multi-barrelled "pepperbox" firearms (see Chapter 3), this device bolts the barrel cluster.

Generally, we have stated the calibre or bore size of muzzle-loading weapons described in this book, by reference to the latter measurement, as (for examples) "38 bore" or "54 bore". This form of description was in general use by British revolver manufacturers of our period, and meant simply that a stated number of round lead balls exactly fitting the barrel would weigh one pound. There are, of course, equivalent decimal parts of an inch which can be stated instead of the bore size, but we have only used such measurements (as, for examples, "·36" or "·31" of an inch) in relation to American arms. In the United States, this decimal calibration found favour quite early, and that fact has been recognized in our pages. When dealing with breech-loading revolvers, however, we have (whether British or American) described their calibres in the decimal calibration by which their metallic cartridges were (and still are) described or ordered.

Finally, we have to acknowledge the kindly and unstinted assistance that we have received (from every quarter) in the writing and compiling of this book.

Our first debt here was to John Darwent, whose unselfish help made possible the chapters dealing with Adams and Tranter revolvers, and who contributed so many of the Plates.

Hard upon his heels, trod both John Bell (whose magnificent collection also figures large in these pages), and Douglas Fryer; the latter helped both as a friend, and as representing Messrs. Wallis & Wallis, the famous Lewes arms auctioneers. We are indebted to the latter for Plates 1 (*lower*), 2 (*top*), and 4 (*top*).

Through the courtesy of A. N. Kennard Esq., B.A., F.S.A., access was arranged to the magnificent revolver collection in the Armouries, H.M. Tower of London, from which we secured Plates 4 (*lower*); 5 (*lower*); 10; 16 (*top*); 26; and 34 (*lower*). It is our pleasure to acknowledge here the source assisting us to secure these Plates.

For the rest, we can only take refuge in the anonymity of an alphabetical listing, for it would be impossible adequately to single out further the individuals who have so unselfishly contributed their time, their collections and their encouragement.

De Witt Bailey
A. H. Baker (Fig. 17)
P. A. Bedford
R. Bedford } (Plate 30; Plate 34, *top*)
Eric G. Bewley, F.C.I.S.; *Messrs. Webley & Scott Ltd.*
H. L. Blackmore, F.S.A., F.G.A.: *President of the Arms & Armour Society of Great Britain*
C. Blair, B.A., F.S.A., *Victoria and Albert Museum*
A. G. Briggs
E. Buchanan
Institution of Civil Engineers, London
Percy Clare, M.B.E., A.L.A.; *Chief Librarian, Central Library, Metropolitan Borough of Bermondsey*
I. McW. Davidson (Plate 7, *bottom*)
Major W. C. Dowell
F. J. Drew (Plate 1, *top*)
R. Duncombe; *Kelly's Directories Ltd.* (Figures 40; 42; Plates 38; 39, *top*)
R. K. Dunham
T. W. Eagle, *Herbert Jenkins Limited*
The Guardians of the Birmingham Proof House
Staff of the Guildhall Library, London
S. A. Hampton, *Central Library, Metropolitan Borough of Bermondsey*
S. B. Haw
James J. Heslin; *The New York Historical Society, U.S.A.*
Miss R. M. Hill
B. Hinchley, B.A.
A. N. Kennard, B.A., F.S.A.; *The Armouries, H.M. Tower of London*
J. Kesterton (Plate 2 *lower*; Plate 3)
Miss W. S. Kerr
Miss L. Kerr
Lt.-Col. R. C. Kuhn, Ill., U.S.A. (Plate 15)
J. C. Lawson
Paul S. Lederer, Silver Spring, Maryland
P. S. McCarthy (Plate 32)
Miss G. McKinstry, Christ Church, Barbados
J. L. Mitchell, Georgia, U.S.A.
W. Keith Neal, F.S.A. (Scot.)
D. Nie
Staff of the Patent Office Library, London

Mrs. P. M. Pendrill
Roger C. Peterson, Maine, U.S.A. (Plate 15)
Staff of the Long Room, Public Record Office, London
W. Reid, F.S.A. (Scot.); *The Armouries, H.M. Tower of London*
A. V. S. de Reuck, M.Sc., F.R.Met.S.
Raymond L. J. Riling, Philadelphia, U.S.A. (Plate 35)
F. Rogers
Rutherford D. Rogers; *The New York Public Library, N.Y., U.S.A.*
Robert C. Sale; *State Librarian, Connecticut State Library, U.S.A.*
R. Scurfield
Royal United Services Museum, London
The Curator and Staff, Pattern Room, Royal Small Arms Factory, Enfield
Mrs. F. O. Taylerson, B.Sc. (Econ.)
Major E. W. Taylerson (Plate 18, *lower*)
Dr. M. Techy; *Directeur du Musée d'Armes à Liège, Belgium* (Plate 25, *top*)
E. J. Turner (Plate 11, *top*)
F. W. Wilkinson
Larry Wilson; *Wadsworth Atheneum, Connecticut, U.S.A.*
V. H. Woods, F.L.A.; *Reference Library, Birmingham*

We also wish to express our appreciation for the professional efforts of Messrs. Phelps & Marchant (of Reigate), and to Malcolm Pendrill (of Malcolm Pendrill Limited, Reigate); the latter photographed the bulk of those weapons illustrated.

CHAPTER I

FLINTLOCK REVOLVERS

A GLOBAL history of revolving arms can be traced from the close of the reign of Henry VIII (1509–47), but there is little evidence to suggest that the British gun trade, as a whole, contributed significantly to any dossier, prior to the third quarter of the eighteenth century. In Europe, on the other hand, revolving guns and pistols reached surprising levels of sophistication and distribution early in the seventeenth century, and were made with match-, wheel- and flintlocks; such arms were probably imported into England, as playthings for the wealthy, but native gunmakers rarely attempted either to imitate or to compete. Flint-ignition was almost exclusively used here, when emulating the revolving arms made abroad, and what little enterprise was exhibited in this field was too often that of emigré craftsmen from the Continent.

However, small numbers of such British revolving arms were eventually made to both of the basic designs described later in these pages, and can be encountered either as multi-barrelled weapons, or as "cylinder" guns and pistols with single barrels, and separate revolving charge-carriers. Commonly, both designs were hand-rotated after each shot, but exceptional pieces were made to be rotated mechanically.

Due to the weight of metal necessarily concentrated in their barrels, multi-barrelled arms were usually made as pistols of comparatively small calibre, at least where more than two barrels were embodied. The British gun trade in general left this field alone until about 1780, at which time a number of multi-barrelled pistols did begin to appear; however, it was still uncommon for any but a handful of makers to apply the revolving principle to such arms, even twenty years later. Instead, a cluster of barrels might be fired simultaneously in a volley, or discharged in succession by using ingeniously compartmented priming-pans for the purpose.

That there was a modest demand for multi-barrelled revolving arms prior to 1780 is proven by the existence of small-calibre, all-metal arms imported here from Belgium (of the type shown on Plate 1 (*top*)), and by the products of one or two exceptional British gunmakers, but it was not apparently on a scale to invite competition for the market, prior to that date.

It cannot have been any lack of ability amongst British gunmakers which caused this situation to arise.

As early as 1680, Nicholas Paris (a Warwickshire gunmaker) was producing superb six-barrelled, hand-rotated flint-lock guns (one of which has survived in the Keith Neal collection), in which the barrels were not brazed into a bundle, but painstakingly assembled about a rotating wooden core. Paris dispensed with the separate pan-covers and steels that bristled about most Continental equivalents, and fitted one steel, which embodied a priming-powder magazine within it. This refinement gave an air of grace and efficiency to his arms, which was lacking amongst many similar weapons of that period.

A later worker in the same field was J. G. Kolbe (formerly of Suhl) who worked in London between 1730 and 1737; a pair of multi-barrelled, mechanically-rotated flint-lock pistols, by this maker, are still preserved in the Armouries at Windsor Castle. Kolbe accomplished barrel-rotating by using a gear-train, to transmit motion to the barrels as the hammer was cocked.

A better-known flint-lock design appeared some fifty years after Kolbe's pistols, and did achieve modest success. Popularly, Henry Nock (a leading London gunsmith) is accepted as the inventor of those sturdy little pistols, but the names of so many gunsmiths or retailers appear upon surviving specimens that it seems more likely that a member of the Birmingham gun trade was responsible for their manufacture.

As may be seen from a specimen shown on Plate 1, these were single-action, hand-rotated arms, measuring about 9½ in. overall, and (although stoutly built) of modest calibre.

Seven barrels were screwed separately on to powder-chambers formed in a single massive breeching, from which projected a spindle that could be inserted into the lock-case of the pistol. With its spindle once inserted, this breeching was rotatably held by a thumbscrew, and could be manually rotated against the thrust of a flat spring, which bore upon a ratchet cut around the spindle-head. The purpose of this spring and ratchet was to hold each barrel, by turn, in alignment with the flint and steel used to discharge it, and a second thumbscrew could be turned in (or out) to adjust the tension of that spring upon the ratchet. Both thumbscrews had to be appropriately adjusted, before barrels and breeching could either be removed from the pistol, or inserted into it.

Six of the barrels were arranged about the circumference of a seventh, and each had its own flash-pan cut in the breeching. One of these pans also served the central barrel, and caused a double-discharge at any selected moment in the firing cycle.

Around the rotatable breeching was a fixed gun-metal sleeve (attached to the lock-case), which prevented spillage of priming-powder from the pans, and also served to mount a hinged steel and pan-cover. This latter was knocked forward at each discharge, and had to be shut down again (after cocking the hammer), before rotating the barrels for another shot.

As may be seen from Plate 1, the hammer was centrally hung, and provided with a sliding safety-bolt to reinforce the conventional half-cock bent. Loading was slow, since each barrel had to be unscrewed from the breeching (by use of a suitable muzzle-key), powder and ball placed in the breech-cup, and the barrel then replaced; priming was effected by filling the main pan in the gun-metal sleeve, and then rotating the barrels to allow a charge of priming powder to drop into each pan in the breeching as it passed. These acts accomplished, however, the weapon must have been both effective and easy to use; there was little that could go wrong with it mechanically, and the seven balls (normally of about 120-bore calibre) would be adequate to settle any normal dispute. It is an indication of the satisfaction given by these pistols, that some of them were converted to use percussion caps, after the first quarter of the nineteenth century had passed.

A six-barrelled coaching carbine of this general type was made by Henry Nock (who had earlier produced multi-barrelled "volley" guns for the British Navy), but here the screw-off barrels were replaced by a brazed bundle. In a healthy 59-bore calibre, and weighing 9 lb. 13 oz., this weapon must have been a comforting fellow-traveller; in view of the scarcity of examples, today, it seems unlikely that many were made. The specimen encountered used a magazine primer, in exactly the same manner as that followed by Nicholas Paris (during the previous century), and one or two other gunmakers also apparently used such primers during the first decade of the nineteenth century, for multi-barrelled pistols; such arms were, however, exceptional by that date.

The reasons for this general lack of interest in multi-barrelled revolving arms, amongst British gunmakers, are not known. Clearly, however, it did not stem from any inability to handle the techniques of their manufacture, for a number of such makers successfully tackled the cylinder-gun design next considered, and this type of weapon was quite as taxing (in the matters of both materials and workmanship) as the multi-barrelled configuration.

One of the earliest of such arms is a six-chambered, mechanically-rotated pistol now preserved at the Armouries, H.M. Tower of London; made almost exclusively of brass, this weapon has a snaphaunce flint-lock, with single hinged steel serving pans in the cylinder which have sliding covers that are knocked open, in turn, as the hammer falls. Although bulky, this pistol appears perfectly viable as a design, and it is to be regretted that the maker is still unknown; dating from about the same period as Paris's multi-barrelled arms, this weapon has been tentatively ascribed to J. Dafte, a London gunmaker who produced manually-rotated carbines of the cylinder type.

James Gorgo (a Huguenot gunmaker) also worked in London in the 1690's, and produced both three-and four-shot hand-rotated cylinder pistols, of true flint-lock type. Like the Paris arms, a primer magazine was embodied in the single steel, but the chambers of Gorgo's cylinders were not aligned with the barrel; on discharge, each bullet was funnelled into the bore *via* a generous leed formed at the breech.

During the late seventeenth and eighteenth centuries, a number of manually-rotated flint-lock cylinder guns and pistols were also made to embody multiple steels and pan-covers, earlier criticized in relation to multi-barrelled arms. A four-shot 20-bore musket of this type is shown on Plate 2, and the ease with which a pan-cover could be inadvertently knocked open (and the priming-charge lost) can be readily seen.

Thereafter, and until after 1800, British interest in cylinder arms appears generally to have lapsed. This is not to say that interest in repeating or multi-shot firearms had been abandoned, but that other systems for accomplishing a similar end were under study. The multi-barrelled "volley" arms of Henry Nock have already been noticed, and similar pistols and wild-fowling guns were made by William Ketland and other Birmingham makers. Henry Mortimer (and some other London craftsmen) approached the problem by way of magazine guns or pistols, which fed loose powder and ball successively to the barrel, through the agency of a lever-operated rotating breech-block.

In 1818, however, a really serious attempt was at last made to produce mechanically-rotated flint-lock cylinder firearms that were not merely a curiosity or plaything for rich men. On November 24th, 1818, Elisha Haydon Collier (an American) secured English Patent No. 4315 for a "firearm combining single barrel with several chambers to obtain succession of discharges from one loading", which introduced a series of guns and pistols that are now keenly sought by collectors.

Collier's patent was filed as "partly communicated" (see Chapter II), and he never denied his collaboration with two other men in the design of his arms. Captain Artemus Wheeler (of Concord, Mass.,) originally conceived a hand-rotated flint-lock cylinder musket, and this was briefly studied by the American Navy in 1821; Wheeler was apparently related, by marriage, to a Boston merchant named Cornelius Coolidge, and had been helped by Collier in constructing his first arms. Using Wheeler's basic idea, and Coolidge's money, Collier then came to the United Kingdom and patented an improved, mechanically-rotated, version of the original Wheeler firearm.

Wheeler and Coolidge have always remained rather shadowy figures, but it is known that Collier gave Coolidge a copy of the Specification to his own English Patent No. 4315 of 1818, and that the latter secured a French patent in such terms, for a period of five years from August 5th, 1819. It seems likely that Coolidge accomplished French manufacture of mechanically-rotated arms on the Collier principle, but no specimens have been identified in any collection to date.

The remarkable feature of Collier's patent was not that it envisaged a revolving flint-lock cylinder arm requiring only to be cocked (and the pan-cover shut) between each shot, for earlier arms have already been mentioned as capable of such operation. It was, in fact, the method of cylinder-rotation, and the highly efficient system for indexing barrel and cylinder on discharge, which made Collier's design so noteworthy.

Rotation was effected by a coiled spring, attached to the cylinder, which was put into tension by first hand-rotating the cylinder in a direction contrary to that in which it turned when firing.

A male cone was formed at the breech end of the barrel, and mated successively with a female countersink cut in the mouth of each chamber to lock barrel and chamber firmly in alignment. A helical spring (distinct from the coiled rotating-spring earlier mentioned) thrust cylinder and barrel into engagement at all times, except when it was overcome (in a manner to be described) to permit rotation of the cylinder.

In addition to these features, Collier's Specification also claimed a most effective priming-powder magazine, formed as part of the steel and pan-cover. A ratchet-and-pawl, or a linkage (see Plate 2), accomplished successive deposits of priming-powder into the flash-pan, by turning a feed-plug mounted transversely across the bottom of the steel, each time that the pan-cover was shut down for discharge.

Once the chambers of a Collier firearm had been loaded, and the priming magazine filled, the sequence of operation was simple. First the cylinder was drawn back from engagement with the barrel-cone, then rotated in an anti-clockwise direction until the coiled rotation-spring was in tension, and finally allowed to snap forward again, on to the barrel-cone. The arm was now ready for use and the act of cocking thereafter permitted the cylinder to rotate automatically, bringing the chambers successively into line with the bore of the barrel.

This intermittent rotation was accomplished by means of a small retracting hook linked to the hammer, which normally engaged a circular notched skirt mounted on the rear face of the cylinder. As the cock was drawn back, this hook moved rearwards, pulling back the cylinder with it, and as soon as the cylinder was clear of engagement with the barrel, the compressed rotating spring spun the latter on its spindle. As the next chamber came into line with the barrel, this retracting hook found a suitable notch in the skirt and disengaged from the cylinder, whereupon the helical spring thrust the latter forward, and the rear end of the barrel snapped into the mouth of that chamber.

The helical spring alone was insufficient to hold cylinder to barrel against the recoil of discharge, but this problem was solved by means of a small sliding bolt, cammed forward by the fall of the cock to butt against the rear end of the cylinder and lock the latter firmly forward; this bolt also acted as a safety device by preventing the complete fall of the cock, unless barrel and cylinder were correctly engaged.

It would appear that Collier experienced great difficulty with the system of mechanical rotation that he had patented, for no arms are known to survive embodying it, and only a few show evidence that the rotation device was once fitted and subsequently removed. This modification was probably introduced between 1823 and 1824, and was referred to in a lawsuit (*Samuel Colt* v. *The Massachussetts Arms Company*, tried June 30th, 1851. Circuit Court of the U.S. District of Massachussetts) when Collier was shown a gun of his manufacture and asked:

Ques. Is that a gun of your manufacture?

Ans. Yes, Sir, it works upon that principle, except that there is one difference in it; the spiral spring, the one which revolved it, is left out; the helical spring is there; I considered this an improvement. I think the gun was made in 1825; it is No. 103.

Ques. Did you make the 102 previous ones like this?

Ans. I had a right to make improvements. I left it out. This contains the spring for shoving forward; it rotates by hand, the only thing left out is the spiral spring.

Ques. Is there any process in that gun by which the cylinder can be turned, except by the application of the hand charge by charge?

Ans. No, Sir, the act of cocking only releases it.

Ques. Look at these guns; do they all turn by hand or by machinery?

Ans. They all turn by hand; this is No. 137; I considered it an improvement to leave out the spring.

Ques. Now state whether the guns which you sent to India were made on this plan.

Ans. The first guns had the spring, perhaps 30 or 40 of them; I had the whole principle in them.

Ques. Will you say that you ever made a practical working gun which operated upon that principle?

Ans. I have said that I made a little model here on that principle. The gunmakers of London witnessed the operation of some which were made in that way; they were fired in the hunting grounds there. I suppose I made a hundred in London, on the rotating principle or from 50 to 100. I improved the gun as I went on, and left out the spring because I thought it was useless. I wanted to get rid of all superfluous parts, that is I considered the gun better without it.

It should be made clear that Collier only deleted the coiled rotation-spring; his helical spring, for mating chamber and barrel, was never discarded, and it remained necessary to pull back the cylinder (against the tension of this spring) before rotating it by hand for the next shot. Some Collier arms had a circular shield attached to the barrel by a spring clip, and this covered all the chamber-mouths to prevent charges from being shaken out in transit. This shield is shown detached in Plate 2, and may sometimes have caused multiple discharge, by deflecting the flash, on firing, into neighbouring chambers.

After obtaining his English Patent of 1818, Collier tried to interest the British government in his invention. On November 19th, 1819 the Duke of Wellington's Secretary, Lord Fitz Roy Somerset, sent a letter to Lt.-Gen. John Ramsey, Commandant of the Garrison at Woolwich, instructing him to convene a Committee of senior Artillery officers to examine and report on the Collier gun. Collier himself was present at the examination. Eight days later, Lt.-Gen. Ramsey submitted the following report:

Woolwich,
27th November 1819.

My Lord,

In reference to your Lordship's letter of the 19th instant, communicating the Master-General's orders that I should convene a Committee of Colonels and Field Officers to consider and report upon an Invention for an improvement in Fire Arms to be submitted to them by Mr. E. H. Collier.

I have the honour to report for the information of his Grace that Mr. Collier has attended the Committee and produced a Pistol with seven revolving barrels, so contrived as to allow a like number of rounds to be fired without stopping to load or prime between the rounds. A rotary motion given to the Barrels by Machinery within the locks brings each of them successively under the pan and the priming is communicated from a Magazine which is fixed behind the hammer. The principle of this is not new, as Sir W. Robe showed to the Committee a very old Swedish Firelock of similar construction.

The Committee are of the opinion that Mr. Collier's improved mode of applying this principle in construction of his Pistol is very ingenious but at the same time consider it by far too complicated and expensive to be applicable to the Public Service.

It may not be improper to remark that Mr. Collier stated himself to be an American; and informed the Committee that he first obtained a Patent for his Invention in America and afterwards in England; and that he seemed desirous of selling to the latter Government.

I have the honour to be etc.,

John Ramsey,

Col. and Lt.-General.

Lord Fitz Roy Somerset.

It will be noted that this letter gave no indication that the pistol was actually tested, but did suggest that a mechanically-rotated arm was shown to the Committee, who understandably rejected it for use in "the Public Service".

Oddly enough; it would seem that the findings of the Committee were not forwarded to Collier and, tired of waiting, he wrote to the Treasury inquiring whether they wished to purchase his patent. The Treasury had no knowledge of the matter and wrote to the Secretary of the Board of Ordnance (on March 19th, 1820) enclosing Collier's letter and seeking information; both letters were sent to the Duke of Wellington (the Master-General of Ordnance) who subsequently gave instructions that a copy of Ramsey's report should be sent to the Secretary, with a note saying "that the Board would inform the Lords of the Treasury that they cannot advise them to purchase Mr. Collier's arm".

Collier was undoubtedly upset over the outcome, but four years elapsed before he succeeded in obtaining a further examination of his arm. The correspondence covering this event has perished, and it is fortunate that two of Collier's broadsheets still exist to provide some information about it.

On March 24th, 1824 (the broadsheets reveal), Collier once again appeared before a Committee of Artillery Officers at Woolwich; it is not clear how he obtained this hearing, after his first rebuff, but he may have been assisted by a Lieutenant-Colonel J. D'Arcy, to whom he had loaned one of his rifles. In returning the arm, D'Arcy wrote:

Suffolk Street,
November 15th.

Dear Sir,

I return your gun with much satisfaction having had an opportunity of exhibiting it at Chatham on 12th inst. to His Royal Highness the Duke of Clarence, who expressed his approbation of it in the strongest terms, and recommended it should be shown to His Royal Highness the Duke of York, as worthy the attention of Government. Among the officers present were Admiral Sir John Gore, Major-Gen. D'Arcy, Colonel Christie and many others of rank, who all confirmed the approbation bestowed upon it by His Royal Highness and agreed in the superiority of the construction for the service of Light Cavalry, Riflemen, Marines in Tops or Boats and Travellers in general.

With my best wishes for your success,
I remain,
Yours very truly,
J. D'Arcy.

It is unfortunate that D'Arcy omitted to add the year of dating his letter, but this may have been 1823, and the Woolwich trial of March, 1824, may therefore have come about at the instance of the Duke of Clarence.

The details of this Woolwich trial are set out clearly enough in a letter sent by Fitz Roy Somerset (later Lord Raglan, and Commander-in-Chief of the British Expedition to the Crimea) to Collier:

Office of Ordnance,
30th March, 1824.

Sir,

With reference to your letter of 27th instant, I have to acquaint you that the Select Committee of Artillery Officers appointed to examine your invention state—That on your appearance at Woolwich on 24th instant you selected a Rifle Piece from among those you brought down: and that with a charge of two drams you loaded and fired from the shoulder one hundred rounds in twenty-nine minutes at a target distant one hundred yards, putting seventy-one balls through the target and lodging one in it. . . . You then fired five rounds at two hundred yards and put the balls through the target—and there was not a missfire in the 105 rounds: That the shot chambers continued to revolve freely without clogging—That Detonating Caps were used with this Rifle, which, as well as the other Firearms shown by you, appear to be well finished.

I am, Sir,
Your most obedient Servant,
Fitz Roy Somerset.

Whilst this letter does not refer to the point in terms as clear as Ramsey's letter of 1819, it does suggest that Collier again put forward a mechanically-rotated arm; there seems little point in Lord Fitz Roy Somerset's reference to the cylinder as having rotated "freely and without clogging", unless this were so.

Although this letter did not categorically reject the Collier design, as useless for service use, the inventor could not draw encouragement from it, and made no further attempt to win official adoption for his arms. Thereafter, his efforts were devoted to the exploitation of a civilian market.

To supply this market, carbines, pistols, rifles and shotguns were produced in some numbers, but it is actually doubtful if Collier made any of them, apart from pattern arms; although a man of considerable ingenuity, he was not a trained gunmaker, and sensibly sought the help of others. Writing in 1858, John Deane stated in his *Manual of the History, etc., of Firearms* (London, 1858) that "the solid chambered cylinders were bored for him by the present Mr. Evans, Engineer of 104 Wardour Street, Soho, in 1821–2", and Evans himself had already recorded such a claim, in the *Mechanics Magazine* of November 1st, 1851. Writing on behalf of his firm (John Evans & Son, Engine Lathe and Tool Manufacturers), he stated that between February and September, of 1822, his firm had made the barrels of 200 muskets and 200 pistols for Collier's business at Fountain Court, Strand. It is not clear whether Evans meant cylinders, when he spoke of "barrels", but we are inclined to accept that he did make both of these items for the inventor.

As to the remaining parts of the Collier weapons, opinion remains divided as to those responsible for making-up and completing the main components made by Evans. J. N. George (*English Guns and Rifles*, Plantersville, 1947) suggested W. & J. Rigby, who were Dublin and London gunmakers of great skill, but there is a Collier broadsheet in existence to suggest other craftsmen as more directly interested. The broadsheet contained a testimonial for Collier arms, signed by S. Nock, W. A. Beckwith, and T. Mortimer; the gun trade has seldom been noted for disinterested gestures in favour of competitors, and it therefore seems at least possible that these men were boosting something in which they had a real interest.

Whatever their source, Collier arms were produced. The rarest arm today is probably a six-shot coaching blunderbuss, with a hand-rotated bronze cylinder and 14-in. barrel; this arm had a folding bayonet attached to the barrel, and the ramrod was concealed in the gunstock. A known specimen is marked "E. H. Collier. Patent No. 9", to suggest that it was amongst the earliest arms made. Rifles are also rare today, but a fine 30-bore 5-shot specimen is shown on Plate 4; serially numbered 112, the arm has a 27-in. barrel with conventional full-length ramrod beneath it, and is also hand-rotated.

Shotguns are the Collier arms which have survived in greatest numbers, the 20-bore calibre being apparently popular, with pistols ranking second in scale of survival. The latter were usually smooth-bored weapons, with 6-in. barrels in 44-bore calibre; an unusual 60-bore arm, with 4-in. barrel, is held at the Tower of London, and is serially-numbered "1". Both types of arm were, of course, hand-rotated.

Another rarity (from the J. Kesterton Collection) is the rifled carbine shown on Plate 3, with a detachable butt; the 27-in. barrel is in ·54-in. calibre, and the arm is unnumbered. Like the blunderbuss, this beautiful hand-rotated weapon may be unique.

It is apparent from surviving Collier broadsheets (and from Lord Fitz Roy Somerset's letter of March 30th, 1824) that customers could order either percussion or flint-lock arms, at choice, quite early in Collier's career. His earliest flint-lock arms (like the blunderbuss and the revolver serially numbered "1") had hammers mounted internally on their lockplates, and plain round cylinders; later flint-lock weapons (like those shown in Plates 3 and 4) had external hammers with fluted cylinders, and this design was followed in the percussion arms.

It would appear that some of Collier's flint-lock arms were converted to use percussion caps, during their service life, but it is difficult to suggest any feature that will immediately identify such a conversion, by comparison with percussion arms initially made as such.

Collier's first business address was given, in his broadsheets, as at No. 6, Herbert's Passage, Beaufort Buildings, Strand. Here he sought orders for his "Patent Feu-de-Joie Firearms", but in the second half of 1824 directories showed him at No. 54 Strand; it is possible, of course, that both addresses were for the same premises.

In 1825, the directories gave his trading style as "Collier & Co., Gunmakers", but without details of any associates. During the Massachussetts Arms Co. case, he stated that £10,000 worth of rifles and shotguns were once shipped to India, via the house of Bazette, Farquhar, Crawford & Co., and this shipment may have dated in that year. Another firm (Collier claimed) advanced 22 guineas on each weapon that he made for them, and sold some of the pieces for 60 guineas each.

Between 1825 and 1827, he moved to No. 3, North Piazza, Royal Exchange, possibly as a subtenant in part of the premises. As he did not pay rates for the premises in 1828, it would seem that he ceased business at the close of 1827 or early in the following year. After closing down his gun business, he took up Civil Engineering until 1850, when he returned to America.

Mr. S. B. Haw has studied Collier's career in detail, and we gratefully acknowledge his help in reciting testimonials (taken from Collier broadsheets), as to the satisfaction given by these ingenious arms.

6 Euston Square
25th December, 1824.

Sir,

I am delighted with your Pistol. I was in the country yesterday and took the opportunity of trying it, and the result was that in one-third of a minute, that is to say 20 seconds, I fired the five barrels and put four balls out of five through the mark. This, for the very first time I had ever fired one of your Pistols, I thought excellent, but my friend's admiration was unbounded and he requested me to order one for him immediately. Therefore you will have the goodness to prepare one exactly like mine and send it as soon as possible to Miles Stringer, Esq. of Effingham Hill, Surrey, with two or three dozen balls ready made. If you have no objection, I should be glad to exchange my single pistol for a brace. I shall then consider my house and myself well fortified. Let them be sent home immediately, as if a brother Captain, who is on the point of sailing, is not already provided I shall present him with mine and get another pair for myself.

Yours obedient Servant,
William Manning.

(Manning was captain of the East Indiaman *Waterloo*)

Another letter came from a Member of a Volunteer Rifle Corps, who stated:

127 Mount Street,
Grosvenor Square,
December 4, 1826.

Sir,

The Rifle you were kind enough to lend me I took the opportunity of trying yesterday. The result was that I beat every other effort of mine, although I belong to a Rifle Corps and have been in practice for at least 17 years. I fired at a painted man target at 200 yards distant, the five men were standing abreast and in the astonishing space of 15 seconds

I hit each man. I must have one of these Rifles and will call on you in a few days to give some directions how I wish it to be finished.

Meanwhile, I am your obedient servant,

Joseph Pellingell, Jun.

The significance of these letters lies in the status of those who penned them. A century earlier, such weapons would have been the recreation of noblemen, and although Collier's designs never achieved the widespread use for which he had hoped, they were in truth the first revolving cylinder arms available to a man of moderate means.

CHAPTER II

LETTERS PATENT

NO summary of British patent law will be attempted here, but one or two facets of the subject have to be considered, in order that later remarks or suggestions, about specific weapons, may become intelligible to the reader.

The development of every British revolver made after 1851 was affected, in greater or lesser degree, by Robert Adams's English Patent No. 13,527 of that year, and it is almost impossible to grasp the significance of serially-numbered arms, unless the reader has, however briefly, been informed as to the British system for protecting and encouraging inventors.

"Letters patent" were originally used by English monarchs for granting privileges to individual favourites, and were open letters (of the "To whom it may concern . . ." variety) commanding any reader to honour or permit exercise of the privilege stated in them. An early use for such letters was to reward men introducing new manufactures into England, from abroad, by conferring an exclusive right to produce such manufactures, under royal protection, for a certain period. In course of time, this practice was extended to benefit native inventors of anything absolutely new, and by the reign of James I (Statute of Monopolies, 21 James I c. 3) statutory recognition was given to the granting of such rights for either class of originator.

Originally, men seeking letters patent appear to have petitioned the monarch through his Lord Chancellor, but as government became increasingly sophisticated, a number of other Crown officers became involved in any application for such protection. The procedure for securing letters patent was almost all customary, before October 1st, 1852, but was eventually put onto a proper statutory basis by the Patent Law Amendment Act, 1852 (15 & 16 Vict. c. 83).

Effectively, two important principles were involved in the concept of British patent law, whether an applicant was introducing some new manufacture from abroad, or had invented something novel himself.

Firstly, it was essential that the idea to be patented was new, and unknown within the kingdom. The point here was that if an idea was already known, then to grant letters patent would deprive others of a right hitherto enjoyed, namely, that of using the invention; if, on the other hand, a device was genuinely novel, then no person was deprived of an existing right to enjoy it, by such grant of letters patent.

Secondly, it was grasped that the Crown and its subjects would suffer, if control of new inventions remained always under the control of their respective inventor or introducer. Accordingly, a bargain was struck, between State and individual, as to the terms upon which exclusive rights to manufacture, use and sell a patented novelty would be enforced by the Crown, and enjoyed by the inventor. Those exclusive rights granted by the letters patent were cancelled, if the inventor did not provide the Crown with such a description as would enable people to use an invention, once protection had expired.

Prior to October 1st, 1852, letters patent normally gave a fourteen-year period of protection, although other terms were occasionally granted; after the Patent Law Amendment Act, this protection was given for a maximum period of fourteen years, but would lapse at either the third or seventh year, if renewal fees were not then paid. There was also a difference as to the point in time from which protection ran; under the old law, protection operated from date of sealing by the Lord Chancellor's office, but this legislation of 1852 gave protection from the date of application for letters patent. This was an important change, because pre-1852 applicants had sometimes lost the "novelty" of their idea before sealing, through theft of it (or by chance invention or importation at other hands), whereupon grant of letters patent was refused.

One further change in the law, by this Patent Law Amendment Act, was to create British letters patent; prior to October 1st, 1852, an inventor wishing to secure protection throughout the United Kingdom must take out separate letters patent for England (with Wales and Berwick-on-Tweed), Ireland, Scotland, and the Channel Islands, at a not inconsiderable cost of £393. Under the new law, initial cost was reduced to £25 (with a third-year renewal fee of £50, and another of £100 at the seventh year), for letters patent covering the whole country.

It will be appreciated, from the foregoing, that "letters patent" were necessarily personal documents, and their survival rate has not proved high. This is unfortunate, since they sometimes contained special terms and conditions relating to the subject of their grant, which information could be valuable to the modern student; examples of such conditions are that protection was to run only during the life of a certain foreign patent, or (before 1852) that the term of protection might be more, or less, than fourteen years.

However, as earlier mentioned, it was each inventor's responsibility to provide details of his invention, from which any citizen might discover how to make or use it, after the inventor's monopoly had ended. This information was filed at the Patent Office (or with the High Court, before 1852) and was called a "Complete Specification"; under the old law, a patentee had to file this specification not less than two months, nor more than six, after the date of sealing for his letters patent, dependent upon whether he was taking protection for England only, or for any permutation of countries. After 1852, the period was fixed at six months.

All Complete Specifications are still preserved at the Patent Office, and available for inspection; they comprise an enormous hoard of information for the collector of revolvers, and copies can be purchased for a few shillings.

In studying these Specifications, however, it is as well to understand the reasoning by which they were drafted. William Carpmael (*The Law of Patents for Inventions*, 6th edn., London, 1860) was a well-known patent agent in London, and stated two primary requirements for a perfect specification.

"In preparing a specification of an invention" (Carpmael wrote) "there are two things to be carefully attended to:

"Firstly—That it shall be so clear that a workman, or other qualified person, shall be able to realize the invention by following the description of it in the specification.

"Secondly—That the invention claimed as new shall not be greater than what is indicated by the title of the patent and the provisional specification."

It should also be explained that after 1852 an application for letters patent was accompanied by a Provisional Specification, which broadly described the invention sought to be protected. Upon acceptance of the application, provisional patent protection was secured upon the idea outlined by this Provisional Specification, and the inventor could safely discuss and finalize his invention with other men, but without destroying any vital "novelty" in it.

The true subject-matter of an invention once having been established, and Letters Patent issued to protect it, a Complete Specification was drawn up and filed, within the period mentioned above. In normal course, this was done by a professional patent agent, who guided the inventor through all legal formalities, and many of these men drafted

Complete Specifications that were models of clarity, and which met those two principal requirements suggested by Carpmael. However, some agents were more devious, and filed Complete Specifications that were deliberately misleading, either by inference or by omission; in the former case, wording might infer that the inventor was claiming a wider area of invention than was actually so, whilst in the latter case some vital detail would be omitted, in order that (it was hoped) an inventor could still have an advantage over others when his protection officially ended.

Accordingly, in studying Complete Specifications to revolver patents, the reader's attention should be directed to specific claims appearing at the end of these documents; the first of these is likeliest to reveal exactly what was patented, whilst those later in the list tend to be increasingly more hopeful than factual.

In an Appendix to this book, a list of all English and British letters patent relating to revolving firearms is set out; an opportunity has also been taken to give those dates at which these patents either expired or became void, and names of patentees are given, together (where this has not been dealt with in preceding chapters) with brief notes of the matter patented.

Upon scanning this Appendix, it will be noticed that protection was sometimes secured by one man, upon a design "communicated" to him by another, and this was a practice often adopted by inventors domiciled overseas. A British patent agent would file application for protection of their invention, but stating it as being "a communication" to him; after guiding the application to a successful conclusion, the agent would then assign (i.e. transfer) the letters patent to his principal. After October, 1859, the inventor's name had to be stated, but prior to that date (as the Appendix shows) an inventor's name was seldom revealed in respect of a communicated invention.

To conclude this chapter, a few paragraphs are devoted to methods used for extracting a commercial return from letters patent, and to changes in title to them.

In the first instance, a patentee might lack financial means or manufacturing capacity to produce what he had patented. An obvious course, here, was to assign or sell his patent to somebody interested, either for a cash sum or for a royalty to be paid on every article made under the patent; an example of such an arrangement will be found in Chapter VIII, as having occurred in respect of F. B. E. Beaumont's British Patent No. 374/1855.

Another patentee might find that his invention could best be developed or used in association with another man's patent. William

Tranter certainly had the means to produce revolvers under his own British Patent No. 212/1853, but was shrewd enough to accept that his double-trigger lock-mechanism (see Fig. 44) could most advantageously be used with the one-piece barrel and frame protected by Robert Adams's English Patent No. 13,527/1851. Here an appropriate course was to obtain a licence to manufacture, use and sell Adams's patented frame, subject to payment of an appropriate royalty.

In general, a well-to-do or experienced patentee preferred to keep and work his own patent, but to licence others to use it where demand was such that it outstripped his own personal production. Letters patent usually authorized such licensing, and there were a number of terms on which this could be arranged; if, for example, the patent was established and successful, licensees might pay a lump-sum premium to use it, as well as paying a royalty for each article made in accordance with its claims.

In form, a patent licence was merely a sealed agreement, which set out those terms by which the patentee (or owner of the patent) authorized his licensee to make, use, sell, or put into practice his invention. Such an agreement might be only for a portion of the patent life, and could be limited to a particular part of the United Kingdom; it might also be an exclusive licence, or a common one where other licences existed or were envisaged.

We have never been fortunate enough to study a licence agreement covering the manufacture and sale of revolvers, but although that part of the "Register of Proprietors" (required under s. XXXV of the Act of 1852) which was supposed to contain details of all patent licences was destroyed many years ago, we would expect such an agreement to have imposed some or all of the following terms, upon the licensee:

1. District or area licensed.
2. Period of the licence, or, alternatively, quantities to be produced under the patent (including a minimum number).
3. The royalty to be paid on each weapon produced, decreasing in amount, perhaps, as quantities climbed. If the alternative under 2 above had not been inserted, there might be provision for a minimum annual payment in this respect.
4. A requirement that each weapon be marked with either the patentee's or licensee's name and that each be numbered. If a number of licences were to be granted, or had already been created, there would be a block allocation of serial-numbers to be used by the licensee, and a letter prefix or suffix to his serial-numbers might be required.

5. A requirement to meet certain quality standards in manufacture, or to permit inspection and rejection by someone acting for the patentee.
6. A requirement that the licensee would manufacture no other type of revolver in competition with the patented weapon during the term of the licence.

The various terms and conditions that could be written into such an agreement are obviously legion, but we think this list is sufficient to indicate the likely scope of any licensing arrangement during our period.

There were also more complicated arrangements which might affect the ownership or use of a patent. The patentee might sell shares in it, or lease it to someone else, or sell the rights to it in (say) Scotland, or mortgage it, or assign it in trust like a piece of land. Similarly, he might die or become bankrupt, and in either case a change in ownership would occur.

There is no point in examining these various events which could affect title to patents, if the point has been grasped that rights to them could vest in administrators, beneficiaries under a will, creditors, executors, or trustees in bankruptcy, in addition to the patentee or his licensees. In the cases of our better-known British percussion revolvers, there is little evidence that the relevant patents were subjected to this sort of intervention; in other cases, however, a collector is well advised to bear such possibilities in mind, where markings or serial numbering appears to vary between weapons, and no obvious reason can be seen for any change in a system previously used. Later remarks (in Chapter XVII) upon James Webley's "Longspur" revolvers can be studied on this point.

CHAPTER III

PERCUSSION "PEPPERBOX" PISTOLS

PERCUSSION caps helped to make possible a truly compact multi-barrelled pistol, suitable for carrying in the pocket, and such arms have principally survived, in modern collections, as revolving weapons. However, a few designers preferred to rotate the striker (rather than the cluster of barrels) and pistols of that type also enjoyed some success. Both varieties of the weapon are generally called "pepperboxes" today (having a fair similarity to pepper-shakers in appearance), although contemporary reports often spoke of the revolving arms as "revolvers", "self-revolving pistols", or "repeating pistols". The majority of them had a self-cocking lock-mechanism, but small numbers of single- and modified double-action pistols were also made, some of the former having hand-rotated barrels.

One interesting and unusual example of an early British pepperbox was the "Budding" weapon shown on Plate 4; this was a hand-rotated five-barrelled pistol of about 120-bore calibre, in which horizontal nipples were entirely enclosed within the lock-case. To load, the barrels were removed from their arbor, the nipples capped, the weapon charged with powder and ball (using the arbor as a ramrod), and then reassembled. The lock had no trigger in a conventional sense, that component being replaced by a cocking-spur on the single-action under-hammer, which slid to-and-fro in a groove on the underside of the lock-case and was powered by a spiral spring. This cocking-spur was pressed into a notch at the side of the groove, after sliding back the hammer, and it was only necessary to press the spur slightly to the left, for it to become disengaged from the notch and thus free the hammer to strike a nipple in line with it. Another model had a trigger and trigger-guard, of conventional type, in conjunction with the "door-bolt" hammer, and was slightly longer than the first-mentioned version; both models were made of brass and were also Birmingham-proved.

The late Major R. H. S. Brown, of Nevern Place, S.W.5., once stated his belief that Budding may have been an Irishman, whilst the late J. N. George suggested that he might have been a clock- or scale-maker, in business between 1825 and 1830. The mystery of the maker's identity was finally solved, through the efforts of W. Keith Neal, in an article published by *Guns Review*; Vol. 2, No. 3; May, 1962. Mr. Neal there illustrated a cased specimen, with a case-lid label stating that the weapon was "Manufactured by E. Budding, Thrup Mill, Near Stroud, Gloucestershire", and his researches established that one Edward Budding had indeed occupied a mill near Stroud, and had also invented a lawn-mower, which was duly patented as English Patent No. 5990, of August 31st, 1830. As an unexpected dividend, the pistol case yielded a small, handwritten card, which stated that the weapon had once belonged to a "Col. Hewett, about 1830". Although we have heard others quarrel with the opinion that it is safe to date the weapon from 1830 on the basis of this evidence, we prefer to accept Mr. Neal's view; Edward Budding was a "machinist" by trade, and these ingenious weapons would have lent themselves admirably to manufacture by such a craftsman. As Mr. Neal also pointed out, however, it remains to be established whether Budding actually invented this pistol, for all specimens known to us were merely signed "Budding, Maker", on top of their lock-cases.

Budding's was not the only single-action, hand-rotated pepperbox to appear in England around that date (although weapons of that period are extremely rare), and the reliability of more conventional arms on this system ensured their use for a surprising length of time.

Initially, these single-action pepperboxes were little more than adapted versions of the double-barrelled hand-rotated pocket pistol, sometimes even having two hammers, but, during the 1840's, a few manufacturers were at pains to produce more compact arms, really suitable for the pocket. In this category, five- or six-shot pistols, normally with centre hammers and folding triggers, were offered by various provincial gunsmiths, and it is significant of the popularity enjoyed by such arms, that a patent relating to them was secured as late as May 7th, 1853; from Birmingham, John Crofts (a foreman) and John Hall (a gunmaker) took out British Patent No. 577 for a spring catch used to index the barrels of hand-rotated pepperboxes beneath the hammer.

Another single-action pepperbox can be associated with the names of William and John Rigby, of Dublin, one of whose broadsheets (dating after 1846) claimed its invention, and whose names appear upon a number of specimens. In these arms, a fixed cluster of from three to six small-calibre barrels was served by a single centre hammer, carrying a hand-rotated striker-plate upon its nose, which could be adjusted to strike each nipple in turn; folding triggers were almost invariably fitted, and the barrels were generally screwed off, one by one, for loading. J. Dunlap's admirable *American, British and Continental Pepperbox Firearms* (Recorder-Sunset Press, 1964) illustrated ten of these pistols, and it is curious to note that only two of them were actually signed by William and John Rigby.

Overseas, hand-rotated, single-action pepperboxes rapidly became obsolete. Serviceable arms of this type appeared in America around 1836, and later excellent mechanically-rotated arms were made in quantity by Stocking & Co. (of Worcester, Mass), Allen & Thurber, and Bacon & Co. (both of Norwich, Conn.) up to the outbreak of the American Civil War; some of these American pepperboxes undoubtedly reached England, in the hands of travellers, but there is no evidence of their wholesale importation.

Conversely, the single-action, mechanically-rotated British pepperbox is today extremely rare. J. N. George (*op. cit.*) illustrated and described such a weapon as made by Joseph Lang, a London gunmaker, and H. J. Dunlap (*op. cit.*) illustrated two more six-barrelled pistols, retailed by provincial gunsmiths; something is said of the Lang weapon, in Chapter XIV, and it is enough here to emphasize that we are unable to accept the view, argued by these authors, that such arms antedated the self-cocking, bar-hammer pepperboxes next described, and were made in the mid-1830'.

FIG. 1. *Lock-mechanism for self-cocking bar-hammer pepperboxes.*

Specimens of the more common bar-hammer, self-cocking pepperbox are shown on Plate 5, and appear in even the most modest of collections, today, yet opinion remains divided both as to when the type was first introduced into Great Britain, and as to a source for the ingenious lock mechanism fitted to the majority of such arms; see Fig. 1.

Whatever the means of introduction, or the date for it, this design was so successful that during some fifteen years these weapons became completely identified with the term "revolver", in British lay

PLATE 1. HAND-ROTATED PEPPERBOX
Four-barrelled, manually-rotated, all metal flintlock pepperbox pistol. Continental manufacture, but sold in Britain, during the late eighteenth century, in some numbers. Sometimes called a "Segallas-type" pistol. *F. J. Drew Collection.*

PLATE 1. HAND-ROTATED PEPPERBOX
Seven-barrelled, manually-rotated flintlock pepperbox pistol, signed "H. Nock, London". Early nineteenth century. *Wallis & Wallis, Lewes.*

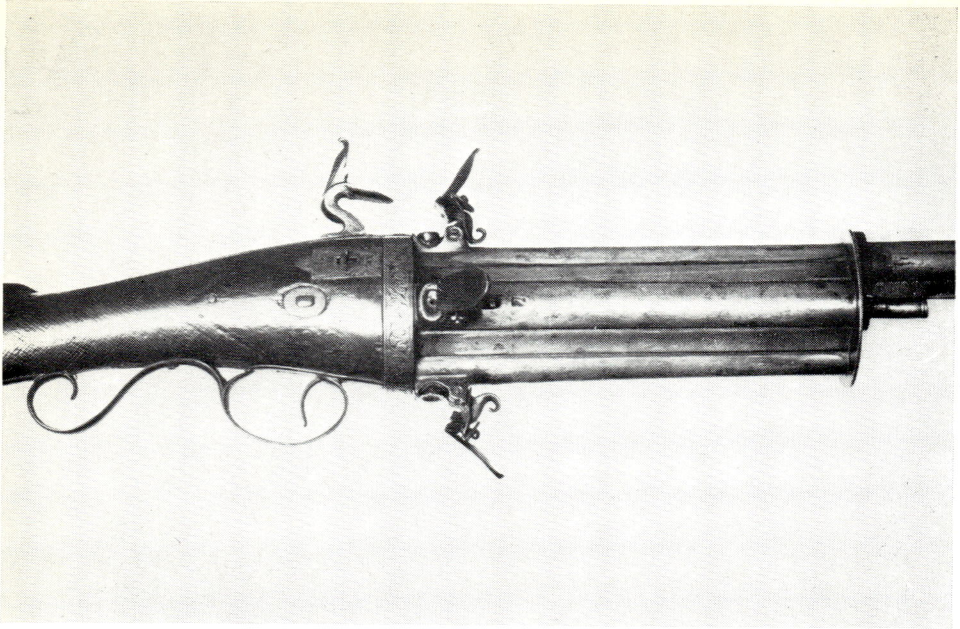

PLATE 2. HAND-ROTATED CYLINDER ARMS
Four-chambered, manually-rotated eighteenth-century flintlock musket, believed to be of English manufacture, but unsigned. 20 bore, with $22\frac{1}{2}''$ barrel, and measuring $35''$ overall. Cylinder is released for rotation by pressing up the trigger-guard, and an additional manual cylinder-bolt can be seen just below the cock. *Wallis & Wallis, Lewes.*

PLATE 2. HAND-ROTATED CYLINDER ARMS
Detail of Collier pistol carbine shown as Plate 3, showing the magazine primer in the steel. Note plate for covering the loaded chambers. *J. Kesterton Collection.*

PLATE 3. Collier rifled carbine pistol (from Plate 2) with detachable butt. Marked "E. H. Collier". Overall length 48 inches. *J. Kesterton Collection.*

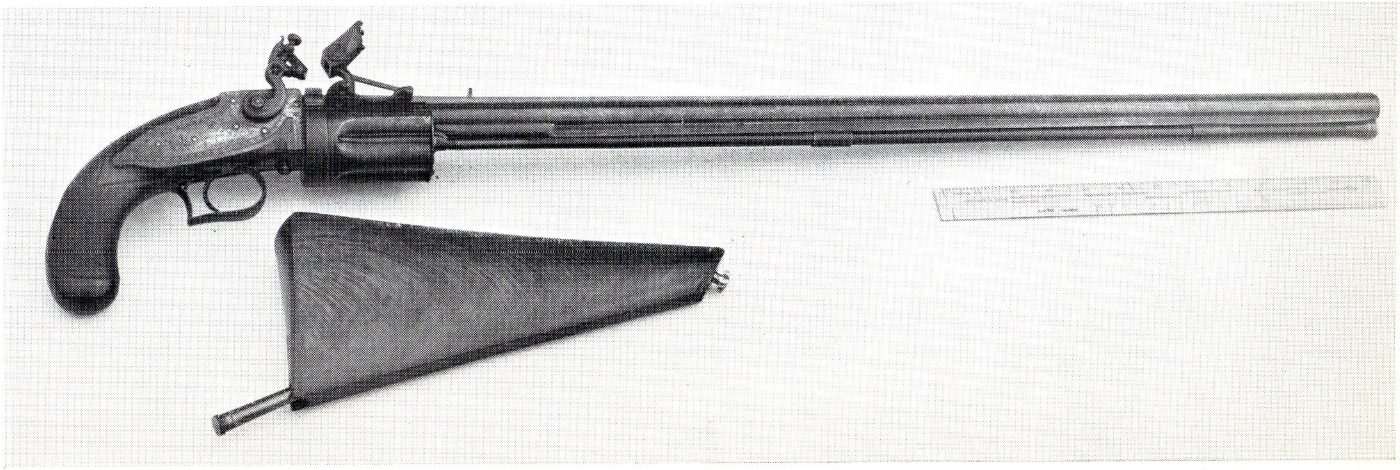

PLATE 3. The same, with butt detached. 27 inch barrel, in about 40 bore, rifled with 9 lands.

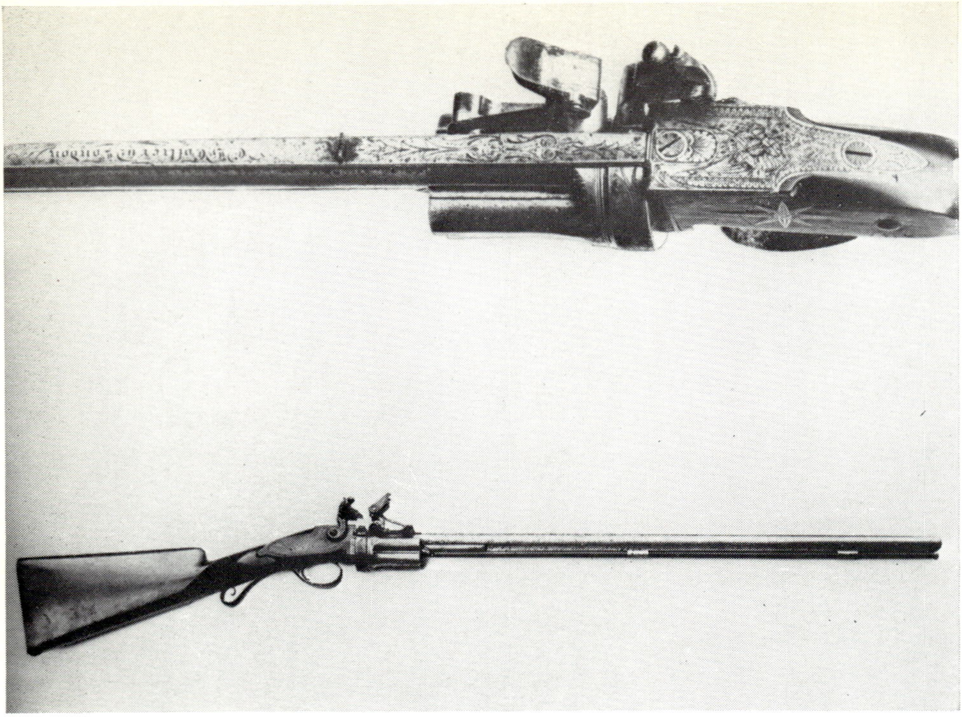

PLATE 4. HAND-ROTATED CYLINDER ARMS
Five-shot, hand-rotated flintlock rifle, in 30 bore calibre. Signed "E. H. Collier, London", and serially-numbered 112. *Wallis & Wallis, Lewes.*

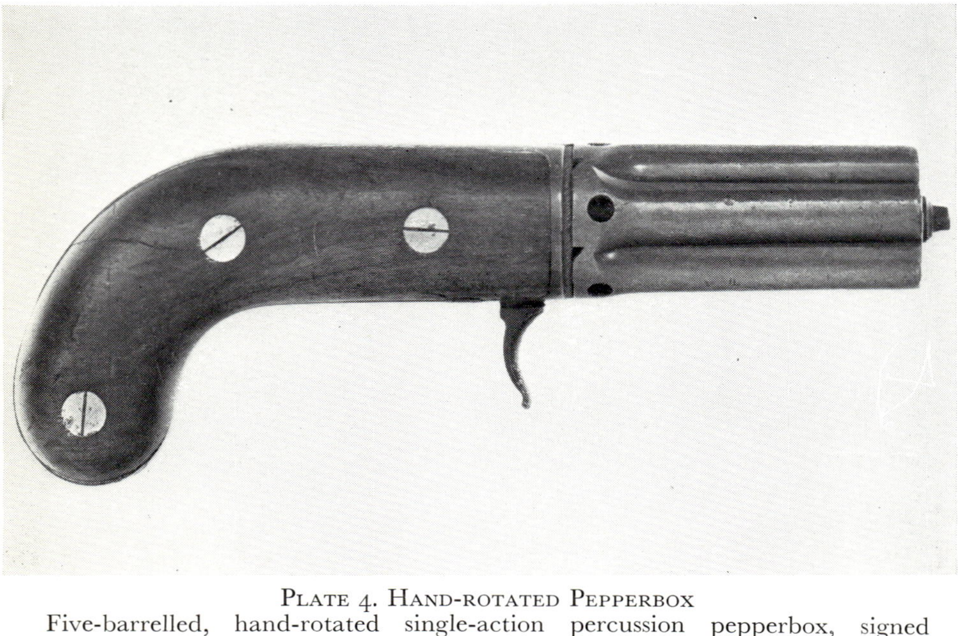

PLATE 4. HAND-ROTATED PEPPERBOX
Five-barrelled, hand-rotated single-action percussion pepperbox, signed "Budding, Maker". *Tower Collection.*
Crown Copyright. By permission of the Master of the Armouries.

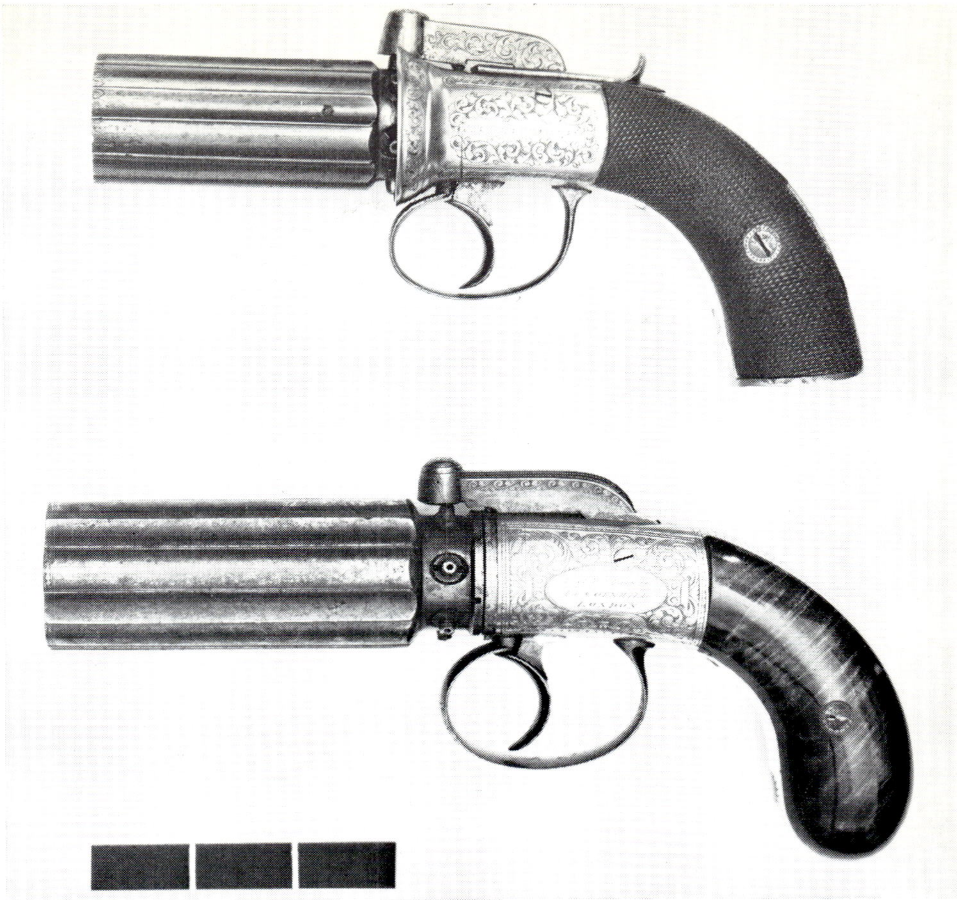

Plate 5. Self-cocking Percussion Pepperbox Pistols

Two bar-hammer self-cocking percussion pepperboxes. The upper pistol (retailed by "Baker, Hereford") was made in Birmingham during the 1850s, and has six barrels in 88 bore calibre. The lower arm is marked "E. & W. Bond, 45 Cornhill, London, Improved Revolving Pistol", but was probably also made in Birmingham, for the retailers. It has five barrels in 64 bore calibre, and is of earlier date than the "Baker" weapon. The address given for the retailers, and the absence of any capping shield, suggests that it was made during the life of Stocker & Bentley's Eng. Pat. 8024/1839 (*see Figure 2*). *Andrews Collection.*

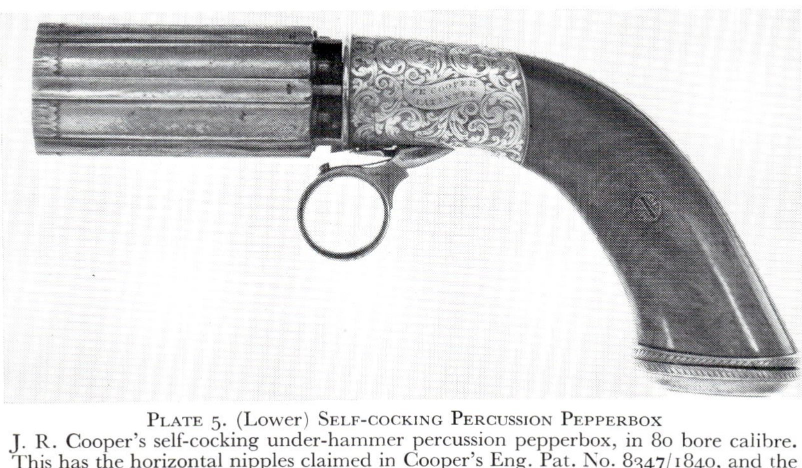

Plate 5. (Lower) Self-cocking Percussion Pepperbox

J. R. Cooper's self-cocking under-hammer percussion pepperbox, in 80 bore calibre. This has the horizontal nipples claimed in Cooper's Eng. Pat. No. 8347/1840, and the "Mariette" lock-mechanism shown in Figure 6. The quality of this specimen is better than that normally encountered. *Tower Collection.*

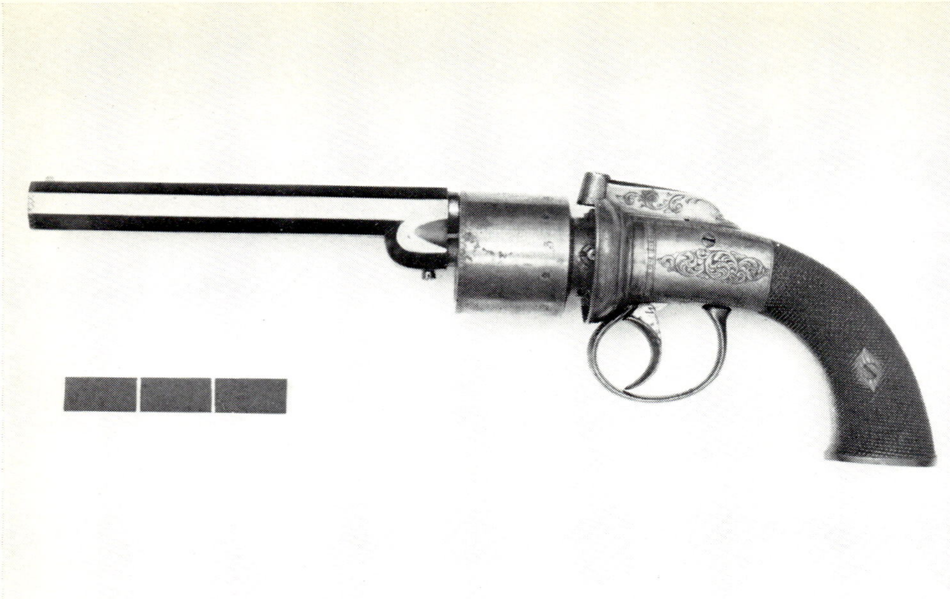

Plate 6. Transition-type Revolver
Bar-hammer, self-cocking, percussion revolver of transition-type, in 58 bore calibre. No maker's or retailer's name. *Andrews Collection.*

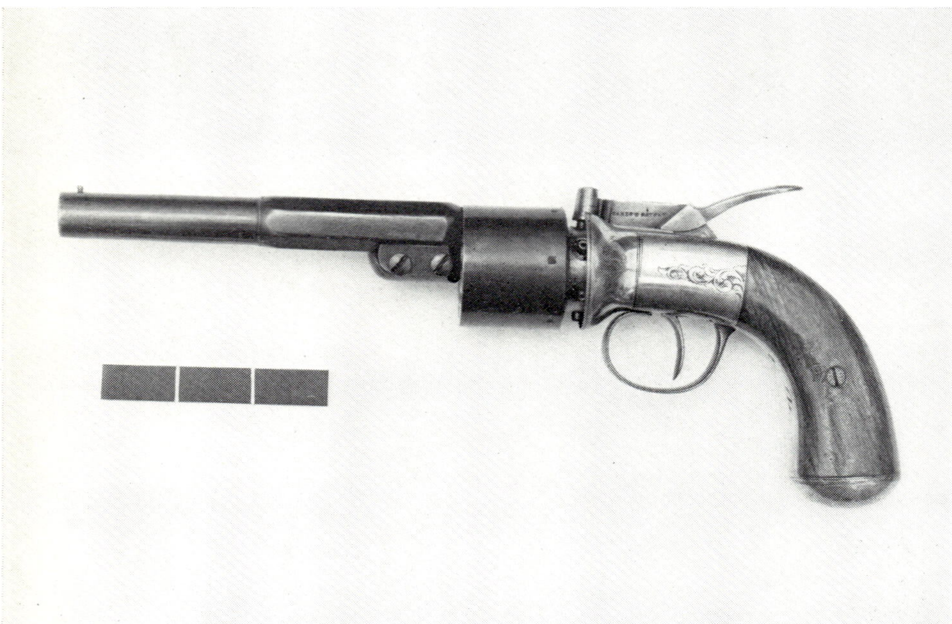

Plate 6. T. K. Baker's Single-action Percussion Revolver
54 bore revolver. Numbered 1496. Marked "Baker's Patent" and also "Registered April 24th, 1852". *Andrews Collection.*

PLATE 7. ROBERT ADAMS' SELF-COCKING PERCUSSION REVOLVER
Five-chambered self-cocking percussion revolver, retailed by "Deane, Adams & Deane", under Robert Adams' Eng. Pat. 13,527/1851. A very early model in 38 bore calibre, serially-numbered 56. *Derwent Collection.*

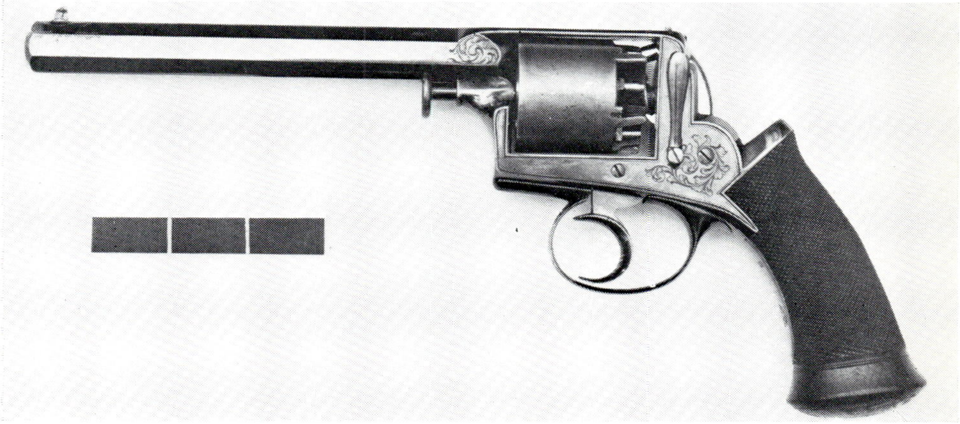

54 bore version of Robert Adams' self-cocking percussion revolver, Model of 1851. Retailed by "Deane, Adams & Deane", and serially-numbered 4727(R). An early model in this calibre. *Andrews Collection.*

120 bore pocket model of the weapon illustrated in upper Plates. Presentation-engraved, and retailed by "Deane, Adams & Deane". Serially-numbered 4008 (with "R" on lower butt-strap), and an extremely early version of Adams' self-cocking revolver in this calibre. *Davidson Collection.*

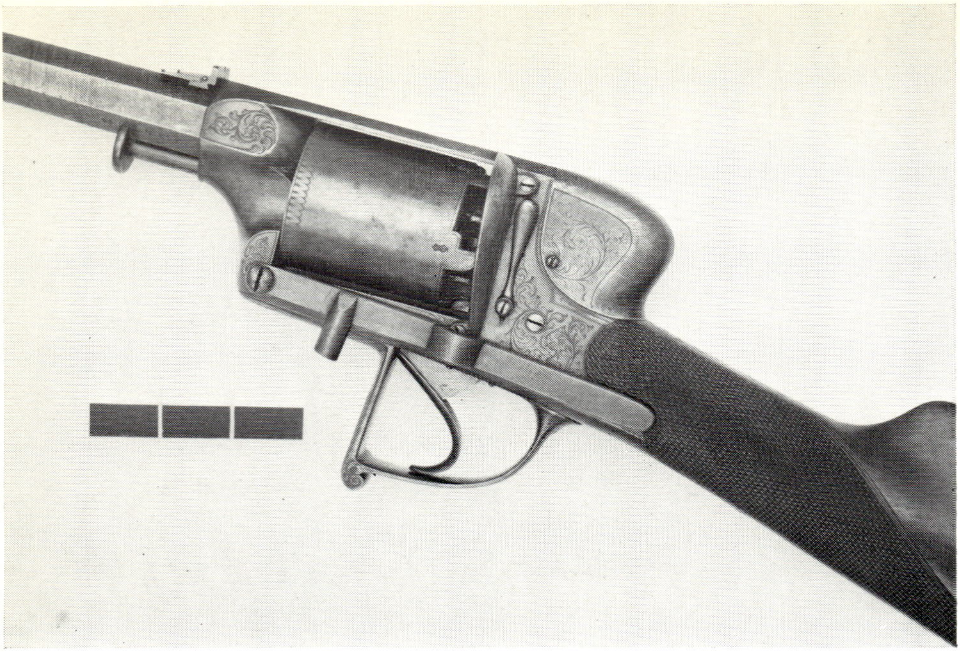

PLATE 8. ROBERT ADAMS' SELF-COCKING REVOLVING ARMS
(*Top*) Five-chambered self-cocking percussion revolving rifle, in 24 bore calibre, retailed by "Deane Adams & Deane". Serially-numbered 11,526R, and fitted with detachable hammer shroud, the rammer (after Adams' Br. Pat. 2645/1854) may have been added after sale. *J. B. Bell Collection.*

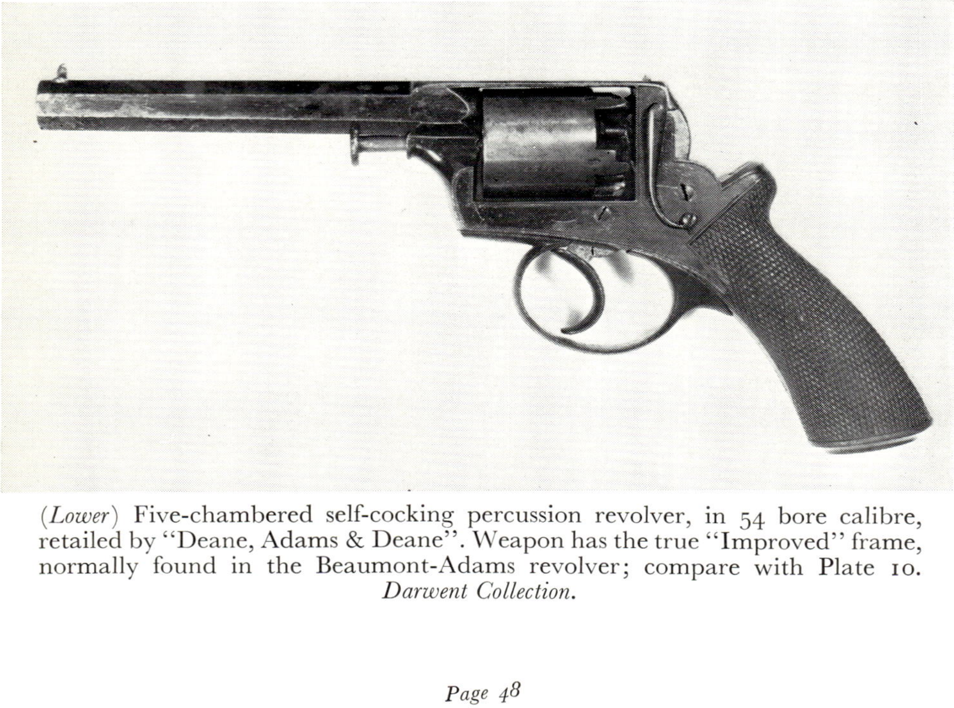

(*Lower*) Five-chambered self-cocking percussion revolver, in 54 bore calibre, retailed by "Deane, Adams & Deane". Weapon has the true "Improved" frame, normally found in the Beaumont-Adams revolver; compare with Plate 10. *Darwent Collection.*

terminology, almost to the exclusion of the now commonly accepted configuration, of a single fixed barrel and a short revolving cylinder.

It seems at least probable, however, that the lock-mechanism was copied from an American design, for which Ethan Allen (of Grafton, Mass.) was granted a U.S. patent for "Locks for firearms", on November 11th, 1837. A bar-hammer, self-cocking, lock-mechanism was used in a series of single-shot pistols made by Allen (and marked "Patented 1837"), who later became a noted pepperbox manufacturer. By inference, a strong claim can thus be urged for an Allen parentage of at least

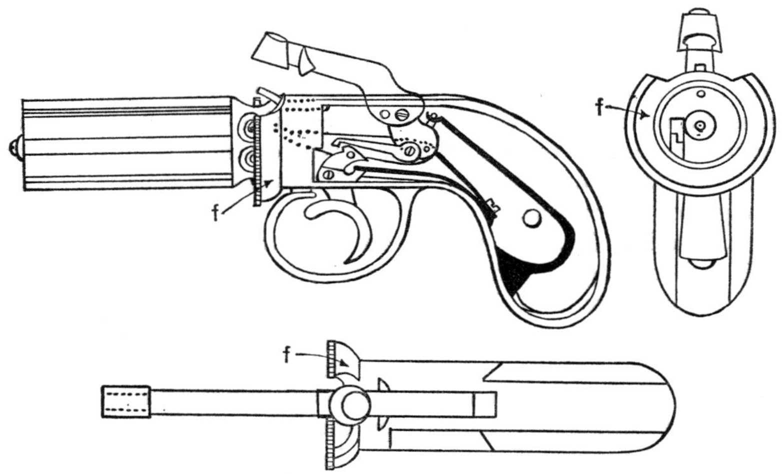

FIG. 2. *Drawing from Stocker & Bentley's Patent of 1839.*
Note that the self-cocking pepperbox design was stabilized, even then.

part of the bar-hammer lock-mechanism used in the British arms. Where, or how, the thrust-lever to revolve a cluster of barrels was allied to the basic lock-mechanism, is not known, but clearly the self-cocking lock-mechanism used in these bar-hammer pepperboxes was common property amongst the British gun trade manufacturers, but could not be patented in Great Britain; as early as April 9th, 1839 (see Fig. 2), Stocker & Bentley's English Patent No. 8024 had to disclaim most of the prominent design-features for the bar-hammer self-cocking pepperbox weapon. A study of Fig. 2 will disclose the circular nipple-guard that was the subject of this protection, and also show that the weapon thus improved was even then recognizable in the form familiar to modern collectors.

A sketch of the components used in the bar-hammer pepperbox lock mechanism is shown as Fig. 1, from this it can be seen that the trigger (C) has a notch at the top, which engages with a hooked member (B) that is hinged to the hammer (A). When the trigger (which moves against a spring, as shown) is pulled back, the hooked member (B) is pulled forward, raising the hammer (A) about its pivot until a point is reached when (B) is disengaged from (C) and the hammer falls to strike a nipple on the barrel waiting to receive it. The rotation of the cylinder is achieved by attaching a thrust-lever (D) loosely to the side of the lock-case, by a screw, and actuating this lever by a transverse pin protruding at the right of the trigger (C). Thus as the trigger is pressed backwards, the thrust-lever is pushed upwards by that pin, and the tip of it bears against a ratchet cut at the back of the barrels, to rotate them through part of a revolution.

The barrels were drilled from a cylindrical block of metal, usually iron, and a central hole was bored through this block to take an arbor on which the barrels rotated; normally, the external walls of the barrel cluster were also planed with a series of grooves that effectively reduced the weight of the barrels.

H. J. Dunlap (*op. cit.*) set up an elaborate classification for the British self-cocking, bar-hammer, pepperbox, and recognized three groups, which he tentatively dated as pre-1837 (Group 1), *circa* 1840–45 (Group 2), and 1850–60 (Group 3); the arguments put forward by this author in favour of his groupings are ingenious, but we are unable to accept his basis for them. Dunlap teaches that a declining standard of quality characterizes his three Groups, the coarse construction of the late Group 3 demonstrating an exhaustion in the demand for pepperboxes, and presumably in favour of conventional revolvers.

We cannot quarrel with the view that demand eventually tailed off, but equally we cannot accept the three Groups as relevant to the dating of these pepperbox arms. In our view (backed by the proof figures and dates of attempted improvements to the basic design later given), the different qualities to which Dunlap points merely recognized that demand required varying qualities of weapon throughout the service life of the self-cocking design, and that the Birmingham gun trade was equipped to provide pepperboxes in price-ranges acceptable to almost any buyer. The fact that a pistol of mediocre quality carried the name of a leading London gunmaker does not argue, to us, that this was the only quality of weapons available to him, but only that a good tradesman carries all qualities of goods suitable for his class of customer. We are prepared to accept that very small quantities of such pepperboxes were made before 1840, but regard the bulk of them (in all qualities) as dating between 1849, and 1855.

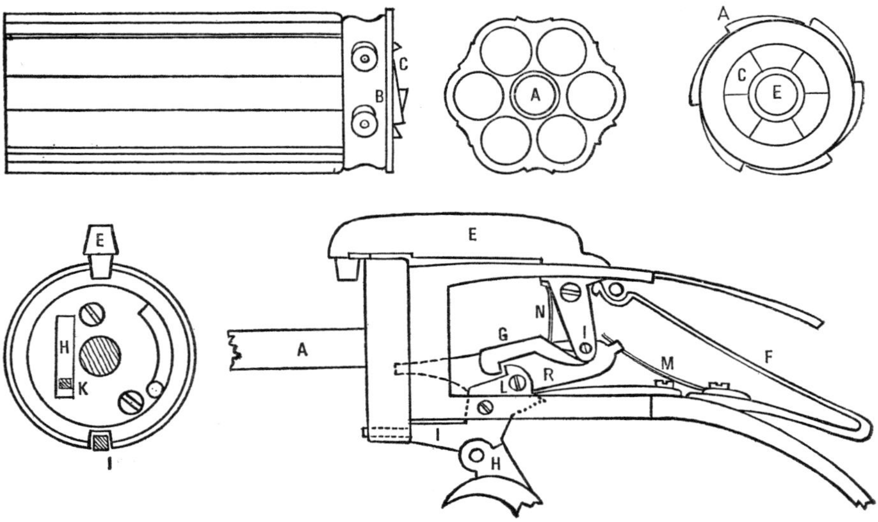

FIG. 3. *British Registered Design No. 86. of November 7th, 1843.*
J. R. Cooper's "Many Barrelled Revolving Pistol and Gun".

Arms embodying the lock-mechanism shown in Fig. 1 were made in every quality, in calibres as small as 140 bore, and as large as 20 bore, and it was even applied to light carbines; belt-hooks were sometimes

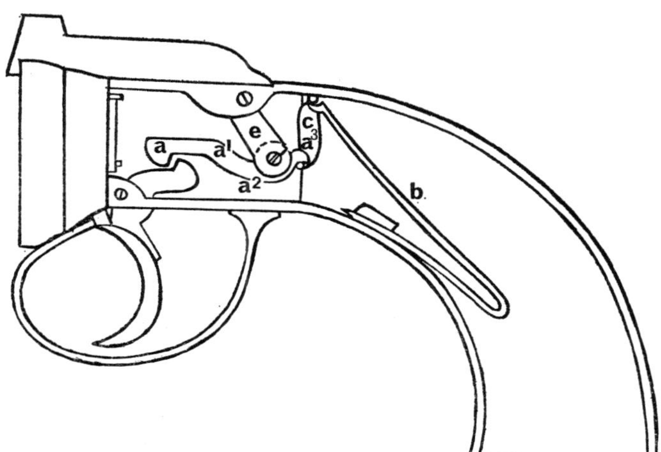

FIG. 4. *British Registered Design No. 2054, of October 13th, 1849*
William Tranter's "Design for a Lever-Catch for Pistol locks".

This design substituted one spring, for the three action-springs normally employed in a bar-hammer self-cocking pepperbox lock-mechanism.

fitted, and elaborate casings were provided with the more expensive pistols. German silver or brass lock-cases are found in some of the better weapons, which were often elaborately engraved, and also provided with a cap-box in the one-piece butt; cheaper pistols often followed the American practice of casting the lock-case with integral butt-straps, and attaching to them a pair of wooden butt-cheeks, secured by a cross-pin.

Naturally enough, attempts to improve either the basic mechanism or the overall design of these bar-hammer, self-cocking arms were made, and three such ideas are shown in Figs. 3, 4, and 5; we have never secured specimens of weapons embodying these improvements (although H. J. Dunlap (*op. cit.*) did show specimens of the Cooper weapon shown on Fig. 3), and we have been equally unsuccessful in tracing any pistol made after Thomas Hunt's British Patent No. 125 of October 1st, 1852; this inventor (a Gunmaker, of Leman Street, London) sought to perfect a sliding "needle-lock" powered by "springs having rollers . . . or spiral springs . . .", to replace the bar-hammer lock design.

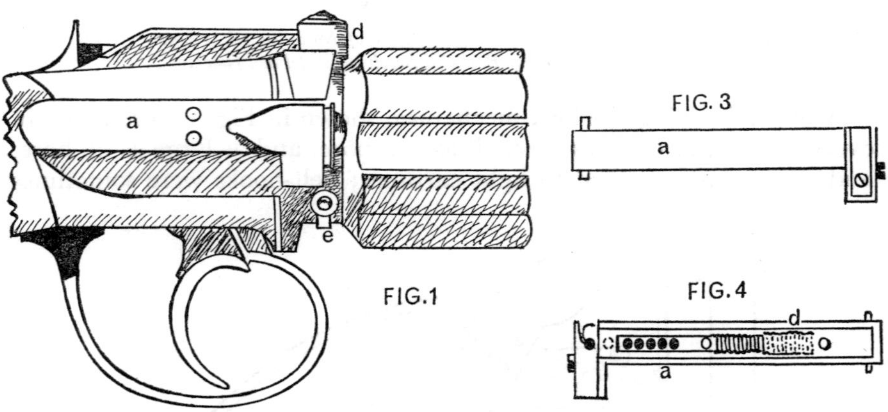

FIG. 5. *B. Cogswell's Registered Design No. 1378, of March 2nd, 1848—"Revolving and self-priming pistol."*

Fortunately, for collectors, one weapon demonstrating a radical departure from the standard British bar-hammer lock is not rare; see Plate 5 (*lower*).

This pistol is known as the "Cooper" pepperbox, and was a well-made but normally rather roughly-finished weapon, with a few lines of decorative engraving on the lock-case. The six barrels were usually between 3½ in. and 4 in. long and in about 80-bore calibre, whilst the nipples were unusual in being both shielded by partitions and horizontally aligned with their respective bores. All specimens examined have been marked "J. R. COOPER'S PATENT", but no British patent relevant to the lock-mechanism used therein can be traced; certainly, Cooper's rare Registered Design bar-hammer pepperbox (shown on Fig. 3) was neither forerunner to, nor descendant of, the pistol now considered.

The weapon appears to have been first manufactured in Great Britain after 1840, but to have been copied from an earlier Belgian "Mariette" pepperbox, without using the separate "screw-on" barrels of that arm. An almost identical pistol was retailed, in America, by Messrs. Blunt & Syms, of New York, and in various calibres; opinion remains divided as to whether or not these weapons were imported into America, either by Cooper (who had a New York office, around 1850) or by another importer. H. J. Dunlap (*op. cit.*) inclined to the view that Blunt & Syms arranged local manufacture of barrels, for imported lock-cases.

Deane, in his "*Manual of the Science & History of Firearms*" (London, 1858) wrote of the parent "Mariette" pistol that ". . . the rotation is continuous, it does not require to be re-cocked, the tumbler having no half-bent, nor full-cock notch, but merely a bent of escape, the release from which is instantaneous . . . each chamber of the breech has behind a well-imbedded nipple that presents itself to the action of a hammer under the lock and fulfilling the office of cock".

In the conventional bar-hammer self-cocking pepperbox, all nipples were necessarily screwed in vertically to the axis of the barrels; Cooper's pistols (with those by Blunt & Syms or Mariette) were undoubtedly superior in this respect, with their nipples screwed directly and horizontally into the breech of each barrel. This arrangement gave a very direct flame at ignition of the cap, and reduced the chance of a misfire; since the hammer was situated to strike the lowest of the nipples, a more compact pistol resulted, and an attempt could be made to aim along the barrels.

A sketch of this lock mechanism is shown as Fig. 6, from which it can be seen that the ring-trigger (A) has attached to it both a hooked

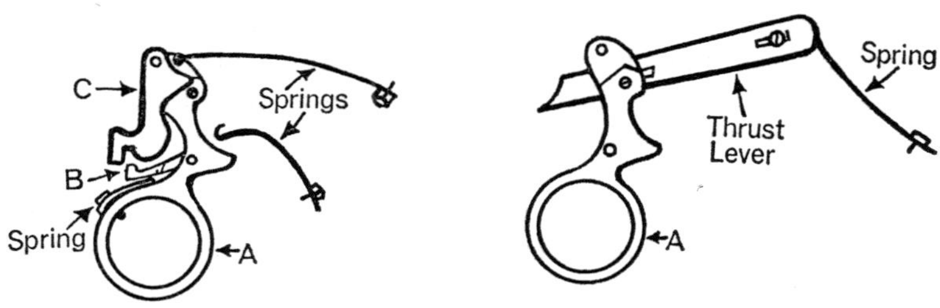

FIG. 6. *"Mariette" or "Cooper" self-cocking pepperbox lock-mechanism.*

member (B) and a rearward-swinging inverted hammer (C). Both ring-trigger and hammer each have a heavy spring acting upon them, and the former has a smaller spring on top of the ring, which causes (B) to engage in a notch in the hammer (C). Loosely attached to the right side of the lock-case, at a level about two-thirds up the ring-trigger, is a thrust-lever similar to that used in the bar-hammer lock-mechanism, upon which a light spring acted. Pulling back the ring-trigger caused this thrust-lever to pivot upwards (by the action of a pin protruding from the right of the trigger), and by engaging the ratchet cut at the rear of the barrel group, to rotate it and bring another nipple into line with the hammer. Continued pressure upon the trigger caused the hammer to be drawn farther and farther back, until the hooked member (B) could no longer hold it, and it was released to strike a nipple. The whole design was an exact duplicate of the lock-mechanism used in Mariette and Blunt & Syms pistols, but this was never patented in Great Britain.

In fact, the source of the "Patent" legend engraved upon Cooper pepperboxes of this type will be found in his English Patent No. 8347, of January 21st, 1840, which covered various improvements to firearms in general, and also related to pepperboxes in particular. The lock-mechanisms claimed therein are rarely found, but the Complete Specification to the Patent also claimed ". . . a mode of constructing pistols which have a series of barrels revolving on an axis and which barrels are fired in succession by a single lock and the improvement consists in so arranging the parts that the nipples may be affixed at the end of the barrels and thus obtain . . . a nearly central fire".

Obvious as this improvement may appear to the modern reader, who is also aware of the earlier existence of Budding's hand-rotated pistol and the Mariette weapon, both embodying horizontal nipples, the Patent Office regarded this particular idea as novel, and a Patent was accordingly granted to embrace it. Cooper was an inveterate patentee, and later obtained English Patent No. 12,781 of September 20th, 1849, in relation to pepperboxes; this envisaged a lock-mechanism that enabled the barrels to be rotated without lifting the hammer, and also explored the idea of producing a tube-lock pepperbox. Neither of these two ideas is likely to be encountered by the average collector.

Another British self-cocking pepperbox with horizontal nipples was made for George and John Deane, of London Bridge, and was styled the "Improved Central Fire Pistol"; this weapon resembled a Cooper or Mariette pistol, but the vertical hammer struck upon the topmost nipple, instead of the lowest. Each barrel was normally fitted with a fore-sight, and a back-sight was formed in the lock-case of the pistol; as a rule, these arms are found as six-barrelled weapons, of about 70-bore calibre, and it is possible that they too, like the Blunt & Syms pistols in America, were imported or copied from a Continental design. History does not relate how manufacture and distribution were arranged, in the face of J. R. Cooper's patent for horizontal nipples, but production was so small that these arms were probably not serious enough as competitors to warrant action by Cooper.

A unique British design that also embodied such nipples is shown in Fig. 7; this self-cocking pepperbox is now extremely rare, and was protected by C. J. Smith's English Patent No. 10,667 of May 14th, 1845. The weapon used a small, square, metal detonating-box, in place of the conventional percussion cap, and this was fed from a tube-magazine in the lock-case.

Reference was earlier made to the existence of pepperboxes with modified double-action lock-mechanisms, and two such weapons have to be mentioned in the context of British arms, although not of English origin.

The first of these was unusual in having fixed barrels, a rotating striker, and in being rifled; it had a ring-trigger for cocking the revolving striker, and was fired by pulling a smaller trigger above the ring. The rifled barrels were tipped down on their hinged breeching for capping, and unscrewed *en masse* from that breeching, for loading. The weapon was patented by S. E. Robbins, of Vermont, U.S.A., as partly communicated to him by G. Leonard, of the same country; the British Patent (No. 1175 of May 24th, 1855) protected an American pepperbox pistol that is known to modern collectors as the "Robbins & Lawrence", and distilled the essence of what was best from U.S. Patents Nos. 6723 (of September 18th, 1849) and 7493 (of July 9th, 1850), which had been secured by G. Leonard, of Shrewsbury, Mass., U.S.A. Messrs. Robbins and Lawrence made pepperboxes under Leonard's patents, at Windsor, Vermont, and with such overall success as arms-manufacturers that a larger establishment was begun at Hartford, Conn., in 1853; unfortunately, the firm failed while this new plant was under construction, and the British patent was never developed.

The second of these double-action arms was Irish, but protected by a late British patent; W. N. Norman and J. Lubé (both of Dublin) were refused provisional protection as No. 694, of March 15th, 1860, for basic features of the design, but John Rigby and Norman together were eventually successful with a more carefully-drafted patent (No. 899, of April 10th, 1860).

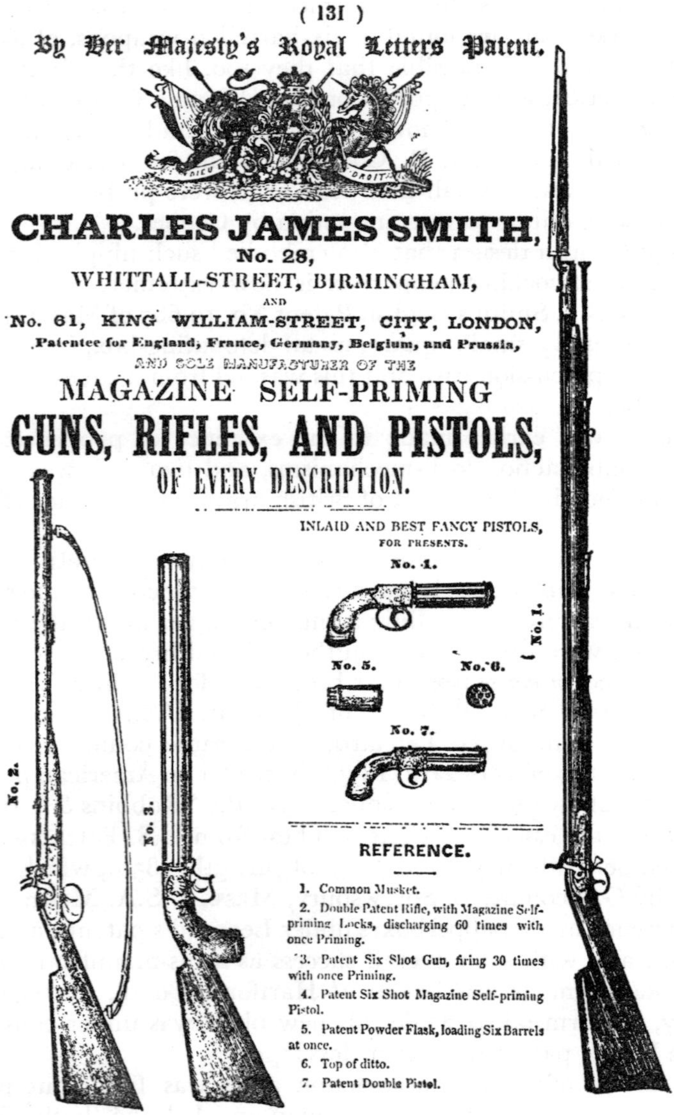

FIG. 7. *C. J. Smith's Pepperbox.*
Advertisement from Slater's *Commercial Directory for Birmingham,* 1848.

Pistols made under this patent are now only known to exist as pepperboxes, but the patented improvements were actually illustrated in the Complete Specification as applied to true revolvers; they related to the stocking of such a weapon, use of its own barrel as the rammer-lever, and to a thumb-operated safety-catch. Pepperboxes made under this patent have a vertical hammer, mounted below the line of sight, a saw-handle butt, and two triggers; pressure upon the front trigger cocks the hammer and rotates the barrels, the former is dropped by pulling the rear trigger. It is curious that John Rigby, who had exhibited superb conventional self-cocking bar-hammer pepperboxes (with a press-button quick-release for their barrels) at the Great Exhibition of 1851, should have been associated in this strange design, on which Norman's name usually appears alone, as "Maker & Inventor".

It may not be clear from what has gone before, that little interest was evinced by British gunmakers, or their customers, prior to 1840, in any pepperbox firearms. Clearly, such weapons did exist from about 1830 onwards, but the records of the London Proof House show that no "revolver" classification was even recognized before 1844. (H. L. Blackmore, *op. cit.* p. 255.)

Year	*Number of Revolvers submitted for proof*
1844	194
1845	145
1846	102
1847	163
1848	297
1849	386
1850	425
1851	954
1852	6,121
1853	13,916
1854/55	No separate account kept
1856	12,604
1857	9,211

Unfortunately, equivalent figures for the Birmingham Proof House no longer exist in any form that differentiates revolving from conventional pistols, prior to 1855.

Global figures for pistol barrels of all types have survived, but only for the years between 1855 and 1857. J. D. Goodman's *The Resources, etc., of Birmingham* . . . (1866) gave the totals as shown at p. 58, *post*.

Year	*Pistol barrels proved at the Birmingham Proof-house*
1855	28,776
1856	33,252
1857	35,593

Given that these figures must have included barrels for every type of pistol, whether revolver, pepperbox or conventional, and that any pepperbox presumably accounted for five or six proofs (dependant upon the number of its barrels), we submit that any figures for Birmingham-proved pepperboxes alone could only have repeated the London pattern revealed above. In other words, any revolving pistol was something of a rarity prior to 1852, and whether pepperbox or revolver.

The 1851 catalogue of the Great Exhibition of London, and its Supplement, however, showed nine gunmakers with specific entries for pepperbox firearms, giving some indication that commercial interest in the pepperbox design had begun to take shape by that year.

Stand 222	E. Trulock & Son 9 Dawson Street, Dublin.	Six-shot revolving pistols with cases complete.
Stand 226	J. Lang 7 Haymarket, London.	The Turnover four-shot and six-shot revolvers of all sizes.
Stand 234	C. Osborne 1 Lichfield St., Birmingham.	Large five-barrel revolving pistol, stock inlaid with silver and lion's head carved on butt.
Stand 236	W. & J. Rigby, 24 Suffolk St., Dublin.	Improved six-shot revolving pistols with detachable barrels, safety-bolt and other improvements, in case.
Stand 240	Westley Richards & Son Birmingham.	Revolving six-barrel pistols. Revolving rifle.
Stand 241	J. R. Cooper & Co. 24 Legge St., Birmingham.	Six-barrelled revolving pistol, central-fire, with safety-bolt. Twelve-barrel revolving pistol. One side of the stock is removed to show the working parts of the lock.
Stand 246	Brooks & Son 28 Russell St., Birmingham.	Four-barrel revolving gun (used in India). Six-barrel revolving pistols.

Stand 247	Tipping & Lawden Birmingham.	Six-barrelled revolving pistols, ivory stock, silver inlaid. Walnut stock, silver inlaid, and chequered stock.
Stand 307	Pecare & Smith New York, U.S.A.	Self-cocking and repeating pistols, with stocks of ivory and rosewood mounted with steel and gold.

The last pepperbox listed was an American pistol, with a folding trigger; the latter feature had been patented on December 4th, 1849 (as U.S. Patent No. 6925) by Jacob Pecare and Joseph Smith, of New York. Their pepperbox had a concealed hammer, and the mechanically-rotated barrels were also hidden by a metal sleeve; both ten- and four-shot arms were made in America, by the patentees, but there is no record of any attempt either to manufacture or to sell these weapons in the United Kingdom, despite their appearance at this Great Exhibition. Since the design required use of horizontal nipples, it is possible that its inventors were balked by the existence here of J. R. Cooper's Eng. Pat. No. 8347 (of January 21st, 1840) for that feature.

In concluding our remarks upon the pepperbox pistol, it should be made clear that such arms continued in production after the appearance, in 1851, of the Adams's self-cocking revolver dealt with in Chapters V, *et. seq.* We are confident, on this point, that the post-1851 proof-figures cited above included a mixture of conventional revolvers and pepperbox pistols, and that the demand for the latter remained appreciable well into the mid-1850's. Giving evidence before the Small Arms Committee in 1854 (see Chapter XI), a Mr. Millward cast some light upon this point:

Ques. Do you make pistols in your manufactory?
Ans. Yes, that is a leading branch of our manufactory.

Ques. What sort of pistols do you make, are they like Col. Colt's?
Ans. We made a great many revolving pistols during 1852 and 1853. They are six-barrelled pistols very different in pattern to Col. Colt's and of different construction.

Ques. They are revolvers?
Ans. Yes.

Ques. Are they efficient revolvers, do they revolve quickly?
Ans. Yes.

Ques. Do you make them by machinery?
Ans. Yes, throughout as far as machinery can be applied to such work.

Ques. At what price do you sell them?
Ans. The price is low enough now, but we can get a good profit and sell them at 12*s.* apiece.

Whilst conceding that Mr. Millward's evidence may have been loaded in favour of the Birmingham gun trade (as potential suppliers of revolvers to the Army), there seems no reason to question the truth of what he stated, and we suggest that these comments, taken with the proof-figures earlier cited, must dispose of any theory that the pepperbox pistol appeared in any numbers before 1852.

It may also be of interest to the reader to consider a listing (which is by no means exhaustive) of British retailers, whose names have been noted upon bar-hammer, self-cocking pepperbox pistols:

Abbey, Newport Pagnell.
Adkin, Bradford.
Allport, H., Cork.
Armstrong, Clonmell.

Baker, J. A., London.
Baker, J., Hereford.
Baker, T. K., London.
Bales, G. W., Ipswich.
Barnes, F., London.
Barrow, Preston.
Beckwith, W. A., London.
Bentley, J., Birmingham and Liverpool.
Berry, S. W., Worksop.
Blake, J. A., London.
Blanch, J. & Son, London.
Blanks, Rochford.
Blissett, J., London.
Bond, G. E., Thetford.
Bond, E. & W., London.
Boston, J., Wakefield.
Bourne, J., Birmingham.
Brooks, E., London.
Budge, Collington.

Calderwood & Son, Dublin.
Calvert, T. W., Leeds.
Clark, Leicester.
Clark, C., London.
Clough & Son, G., Bath.
Cogswell, B., London.
Cole, Boston.
Colgan, Limerick.
Collin, F., London.
Collins, J., London.
Colson, Stowmarket.
Conway, T., Manchester.
Cook, T., Bath.

Deane, G. & J., London.
Dooley, W., Liverpool.
Dowling, Dublin.

Edwards, W., Devonport.
Edge, J. W., Manchester.
Egg, D., London.
Egg, H., London.

Fenton, R., London.
Fletcher, T., Gloucester.
Fowler, T., Dublin.

George, London.
Gibbs, G., Bristol.
Green, W., London.
Grey, Shaftesbury.

Hanson, C., London.
Hanson, S., Doncaster.
Hart, Colchester.
Hewson, T., London.
Holland, H., Birmingham.
Hollis, Cheltenham.
Horton, Salop.

Jackson, T., London.
Jones, C., London.
Jones, R., Liverpool.

Kell, A., London.

Lacy & Co., London.
Ladmore, E., Hereford.
Lancaster, C., London.
Lang, J., London.
Lane, Leamington.
Lees, Perth.
Ling, W., London.
Long, J., Glasgow.

Manton, J. & Son Co., London.
Marrison, Norwich.
Maybury, C., Birmingham.
Mills, W., London.

Naylor, Sheffield.
Newark, Coventry.
Nixon, D., Newark.
Nixon, W. N., Birmingham.
Nock, S., London.

Osborne, C. & Co., Birmingham and London.

Parker, Field & Sons, London.
Parkhouse, J., Taunton.
Patrick, Liverpool.
Pegler & Son, Blandford.
Piper, C., Cambridge.
Powell, W. & Son, Birmingham.
Pridie, London.
Pritchard, W. & Son, Birmingham.

Reilly, E. M., London.
Richards, W., London and Birmingham.
Rigby, W. & J., Dublin.
Riley, London.
Riviere, H., London.
Rogerson & Andrews, Eton.
Roper, Halifax.

Scott, R., London.
Smith, C. J., Birmingham and London.
Smith, O., Derby.
Smith, R., Carlisle.
Somers & Sworder, Bishops Stortford.
Spencer, M. S., Lyme Regis.

Tatham, H., London.
Thompson, A., Edinburgh.
Tipping & Lawden, Birmingham.
Towl, Boston.
Towlson, T., Marlborough.
Turner, T., Reading.
Twigg, London.

Veisey & Son, Birmingham.

Welch, Banbury.
Weston, W., Brighton.
Wetherhead & Walters, Derby.
Wheeler, Devizes.
White & Bates, Birmingham.

Wilkinson, J. & Son, London.
Williams & Powell, Liverpool.
Witton, J., London.
Withers & Grant, Rugby.
Witton, J., London.
Wood, J., Lewes.
Wood, J., York.
Wood, J. W., Manchester.
Wright, C., Barnsley.
Wyatt, W., Romsey.

CHAPTER IV

TRANSITION-TYPE REVOLVERS

WE would have preferred to avoid the title "transition", even for the type of percussion revolver now commonly described by that name; all weapons are transitional, in the sense of being constantly improved, and it is doubtful if the sobriquet was used at all before J. N. George (*op. cit.*, p. 177) thus applied it. However, this designation is now so firmly established that it cannot well be ignored, and we therefore identify a "transition-type" revolver of the type isolated by George, thus restricting that classification to one family of British revolving pistols.

An example of that family is shown on Plate 6, and it consists exclusively of small-calibre, bar-hammer, self-cocking revolvers, predominantly of Birmingham manufacture, and embodying lock-cases and lock-mechanisms identical with those used in the pepperbox pistols described in Chapter III; in all cases, vertical nipples without partitions are fitted, and the barrel is not forged integrally with the frame. For convenience, two single-action revolvers are also described in this chapter, but they are included only as a means for emphasizing our view that George incorrectly dated his "transition-type" revolver classification, and that name is used throughout these pages to describe arms only of the type shown at the top of Plate 6.

As George saw the position, development of British percussion revolvers was inhibited, prior to 1851, by the existence of Samuel Colt's English Patents Nos. 6909 (of October 22nd, 1835) and 12,668 (of June 20th, 1849), which described in detail that inventor's methods for rotating the revolver cylinder by a pawl, and bolting it in alignment with the barrel, for firing. In America, Colt had enjoyed a virtual monopoly of mechanically-rotated revolver manufacture, through equivalent patent protection, until well after 1851, and George seems to have been convinced that a similar position obtained in Great Britain. As a result, in his view, the British manufacturers were driven to producing revolvers based on the pepperbox design, during the 1840's, because (he inferred) more advanced cylinder-rotation or cylinder-bolting mechanisms were barred to them until the expiry of Colt's English patents.

With the greatest respect for this particular author, we cannot accept his view. In our contention the transition-type revolver appeared during 1851, as the direct result of a sudden demand for cheap pistols resembling the successful (but more expensive) arms produced by Colt and Adams, and through the failure of British manufacturers, as a whole, to produce a good revolver design until driven to it. The expiry of Colt's two patents, in 1849 and 1864, respectively, had no relevance to this failure.

There has also been speculation about another possible origin for these transition-type revolvers, with some holding the view that because a few weapons appear to have been adapted from percussion pepperboxes, by cutting down the revolving-barrel cluster, and adding a single fixed barrel, then the dating of transition-type revolvers should be parallel with that of the pepperbox. By this reckoning, the first of these revolvers could have appeared in the years 1835–40.

Although we have handled many of these pistols, only one example alleged such an alteration, both specifically and at such a period, and this was traced in 1957. The case containing this particular transition-type revolver bore a label, which read "one pepperbox conversion made by Mr. Lang, Gunsmith, Haymarket to the order of Wm. Pepys, Esq., of the Poultry, May 1837. 30*s*.".

Faced with this exceptional, if doubtful case, we still prefer the view that the transition-type revolver really evolved at a later period than is commonly supposed, if for no other reason than that the cost of converting a pepperbox, re-proof, and possibly altering or providing a case for it on a "one-off" basis, could well have exceeded the cost of buying a new transition-type weapon. We must concede, as an exception, the case where a man owned a top-quality pepperbox, and was understandably reluctant to discard it as obsolete; we remain to be satisfied, however, that many such conversions were carried out, or that they were done before 1851.

In our view, no real demand for revolvers existed at all, in Great Britain, until the Great Exhibition of 1851, when the Colt and Adams revolvers, there exhibited, created such a demand. As a result of the interest aroused by various newspaper and magazine reports, many would thereafter have liked to own a Colt or Adams revolver, but could either not obtain one, or found the price too high; they therefore looked elsewhere for a pistol of similar appearance, and both gunsmiths and retailers were then faced with satisfying demands for a weapon not in quantity production.

To design, possibly patent, and then to produce an entirely new and cheap weapon takes, paradoxically, both time and money; the latter was available, but the continued existence of this Colt-inspired enthusiasm could not be guaranteed, in that peaceful year of 1851. A radically new weapon was thus largely out of the question, but an easy solution was to adapt the pepperbox design; this could readily be done, by substituting a short cylinder for the barrels, and by fixing a single barrel on to the end of the spindle or arbor. The addition of a small pin-stop (shown as fitted to the trigger in Fig. 1) was all that was necessary to accomplish a rough-and-ready indexing of the chambers, in line with the barrel, for firing. The resultant revolver possibly resembled a Colt to any uncritical eye, and all could be accomplished within the established manufacturing pattern of the Birmingham gun-trade suppliers.

The demand for this type of weapon is evidenced by those numbers still in existence, but a far greater number must have perished. Quality varied, the wealthier customer in that particular price-range demanding a better-made and finished pistol, and being prepared to pay for a brass or German silver frame, in place of the standard case-hardened iron. Such arms were supplied in mahogany cases, lined with baize or plush, and divided into compartments for the accommodation of cleaning and loading accessories. Better-quality metal was used in the lock components, and care expended to obtain a smooth-working action; cheaper arms, however, were prone to develop wear on the ratchet and thrust-lever (see Fig. 1), which defect ultimately made the pistol unsafe to fire, as the chambers then failed to index properly, in line with the bore.

Regardless of quality, the transition-type revolver suffered from those same basic defects as the self-cocking pepperbox, having a heavy trigger-pull, and a liability to occasional multiple discharge, due to the use of unpartitioned nipples. It is also probable that the escape of gas between chamber and barrel, in the cheaper weapons, was considerable, but this point may have been over-emphasized in recent years. The quite substantial quantities of such revolvers found in collections today, would not have sold if they were not capable of forcing a bullet through the barrel, with at least the same degree of effectiveness as a pepperbox.

Not all of the barrels were rifled, but nearly every example seen had this refinement, and the majority of muzzles were provided with a foresight. It has been said that this sight could be of little value, since the rising bar-hammer intervened in the line of sight, and it is a fact that some manufacturers off-set their hammers to one side of the lock-case, and so cleared the line of sight; however, the disadvantages of this intervention have been exaggerated, for any temporary obscuring of sight, by the rising hammer, is barely appreciable against a man-sized target. Some weapons even had a vertical slot, or rear-sight, cut through the body of the hammer, which enabled the foresight to be seen as the hammer rose, but this does not alter the fact that there was a purpose in fitting a foresight to the cheaper weapons, and that it had some value.

Many transition-type revolvers had the barrel simply screwed on to a thread at the end of the arbor, and only indexed by a vertical screw that passed up through the barrel-lug into the arbor; on those of better quality, however, and in order to give some rigidity to the barrel, a bottom-strap was usually fitted, connecting this to the lock-case. Two methods of uniting the barrel to such a bottom-strap were commonly used:

(*a*) the barrel-lug was indexed to the bottom-strap by an end-screw, and two screws were also passed through the barrel-lug and into the arbor, to give the whole a fair rigidity,

or (*b*) by securing the barrel-lug to the bottom-strap with a thumb-screw, after the lug had been screwed onto the threaded end of the arbor in a normal fashion.

Such weapons were seldom fitted with an attached rammer, the user being provided with a separate ramrod, in the pistol case, for arms of better quality. One pistol, however, which bore the name of Beattie, Regent Street, London, was exceptional in this respect, but not unique; the weapon had a bottom-strap connecting barrel to lock-case, and through the front of the barrel-lug was passed a plunger-type rammer. This plunger was spring-loaded, and returned to the "at rest" position, automatically, after being used to ram home the charges. The 6-in. rifled barrel was in 50-bore calibre, and slid on to the end of the arbor, where it was retained by a transverse key; a safety-bolt was also fitted, which prevented the hammer from reaching the nipples, when set, and a slot back-sight was cut through the hammer.

In turning again to our contention that revolvers of this transition-type should be dated after 1851, it must be conceded that John Deane's *Manual of the History and Science of Firearms* (London, 1858) suggested that conventional revolvers were in general use many years before that date; it is reasonable to mention, however, that none of the cited weapons were of British transition-type. Early weapons by Devisme, Herman, and Rissac were mentioned, but it is doubtful if many crossed the Channel, and it must be emphasized that Deane (whose book was largely plagiarised from T. Anquetil's earlier *Notice sur les Pistolets Tournants et Roulants*. Brussels, 1855) was really bent only to the task of attacking claims that Colt's was the first mechanically-rotated revolver. He gave scant regard to the means employed for that purpose, and certainly gave no details to suggest that arms of the transition-type were contemporary with these Continental designs.

In referring, however, to the English percussion revolvers exhibited against Colt at the Great Industrial Exhibition of London, in 1851, Deane stated of these "other Richmonds in the field", as he termed them, that ". . . the most remarkable were those of Joseph Lang, J. Barnett, and Robert Adams". He then proceeded to give some description of the three revolvers, but as Lang's and Adams's pistols are elsewhere dealt with, we will confine ourselves here to his statement about the Barnett weapon, which was (for reasons which we hope will become clear) the revolver in point for this chapter.

"The Barnett revolver (Deane wrote) is with uninterrupted rotation and does not require to be cocked by a distinct or special movement. Like the Lang it wholly repudiates the use of loading-rod and lever of the Colt system. The office of the cylinder pin is fulfilled in this pistol by a prolongation of the lower part of the barrel, which passing through the cylinder adapts itself to the body of the lock behind the former, below the barrel in front, this prolongation is armed with a spear-like point which serves to pierce the cartridges used with this revolver. The bore of the barrel is smooth, i.e. without rifling. . . ."

". . . as regards the vertical disposition of the nipples and the action of the small spring performing the office of the safety-bolt, it participates in the defects of Lang's revolver. Finally, the hammer from the manner in which it is established greatly interferes with the aim because the eye of the operator in order to find the front sight is obliged to direct its glance through an intersection made in the substance of the body of the cock, as through the pinule of the graphometer."

It seems to us apparent from these remarks, that Barnett's revolver was merely a transition-type revolver without any bottom frame-strap, but clearly, it was entirely novel to Deane and Anquetil when they described it. Since the weapon was ranked with Lang's and Adams's revolvers (see Chapters V and XIV) as worthy of remark when these descriptions were published (in 1855 and 1858, respectively), it seems highly improbable that the transition-type revolver was anything but novel in 1855.

If Deane was correct, however, in claiming that the Barnett weapon was "in the field" against Colonel Colt, before the 1851 furore over the merits of revolvers in general, then it is curious that the catalogue

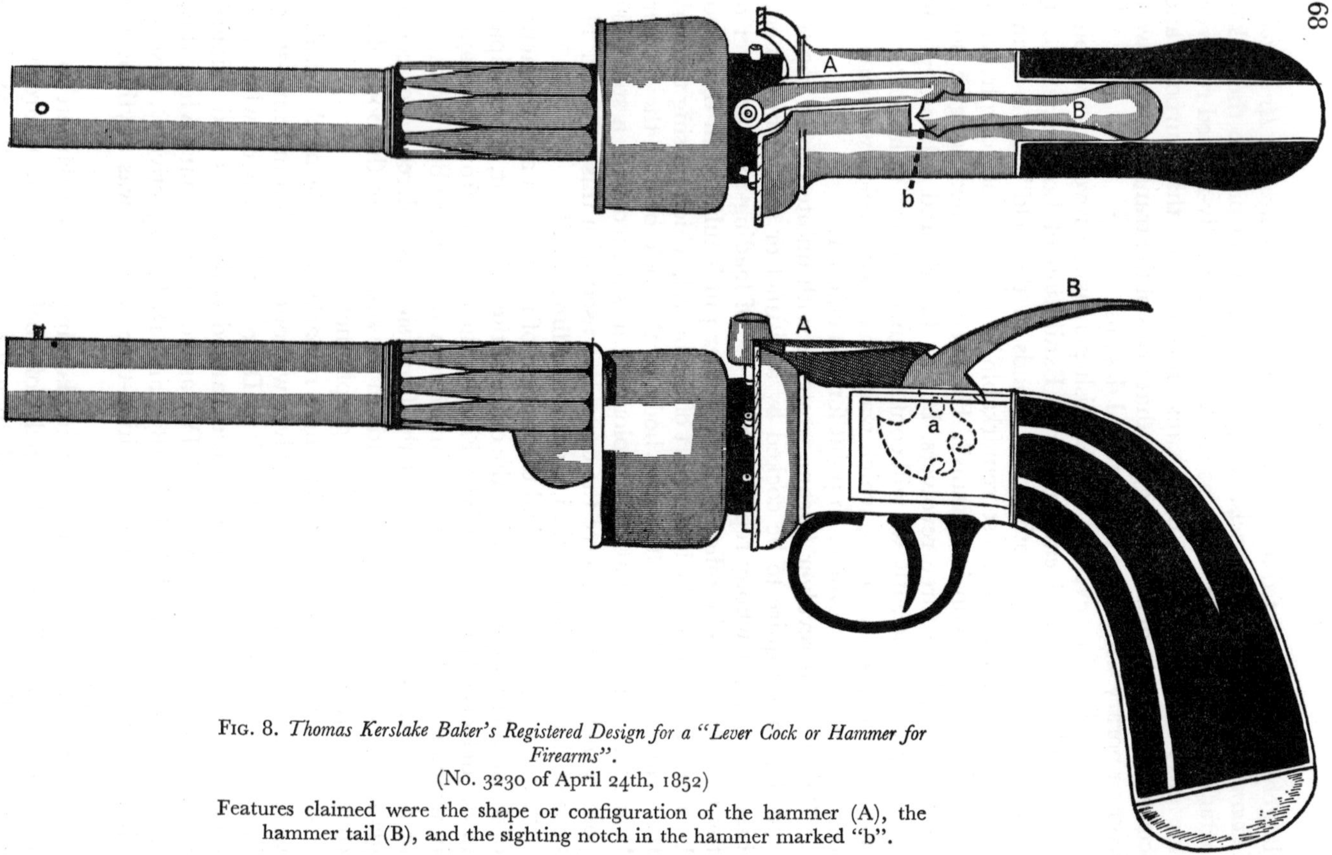

Fig. 8. *Thomas Kerslake Baker's Registered Design for a "Lever Cock or Hammer for Firearms".*
(No. 3230 of April 24th, 1852)

Features claimed were the shape or configuration of the hammer (A), the hammer tail (B), and the sighting notch in the hammer marked "b".

of the Great Exhibition gave Deane so little support. A close examination of the relevant Catalogue sections (with those in the Supplement to it) produced no firm evidence that the Barnett revolver was ever included amongst the exhibits, and a further glance at the proof-figures already shown in Chapter III strongly suggested that no striking demand for any revolving weapon (whether pepperbox or revolver) existed in Great Britain before 1851.

We regard this matter of the proof-figures as most damning to any claim that transition-type revolvers appeared prior to 1851. Specimens of these pistols survive in fair quantities, permitting their presence in the most modest of collections, and it is inconceivable that they could have been produced in such numbers before 1851, without their existence being reflected in the proof-figures.

In conclusion, attention is drawn to two revolvers similar to the transition-type, which were, however, single-action, and offered further proof that this type of revolver is unlikely to date before 1851.

The one likely to be encountered today is the revolver shown on Fig. 8, with its striking long-spur bar-hammer; this was off-set to the right of the line of sight (as to the bar), and also notched to provide a rear-sight; lock-cases of malleable cast-iron or German silver were used, and a weak sliding safety-bolt was fitted to some of these pistols, on the left of the lock-case.

A specimen arm (shown on Plate 6) is in 54-bore calibre, but 60-bore pistols are more commonly found; in either calibre, a barrel of a length between 5 and 6 in. was fitted, and this might be octagonal for its full length, or rounded as to the forward half. Rifling was multi-grooved, but the number of grooves was apparently left to the discretion of the craftsman rifling the arms.

Commonly, these revolvers are marked "Bakers Patent" on the hammer, and "Registered April 24, 1852" upon the top of their frame. In fact, no patent covered these arms, the reference being to T. K. Baker's British Registered Design No. 3230, of the date stated; Fig. 8 is taken from this design, which claimed only the particular configuration of hammer used. The lock-mechanism was a simple amalgam of features from a conventional single-shot pistol lock, and the rotation device used in bar-hammer pepperboxes; as to the latter, see Fig. 1.

Baker (a general gunmaker, at 88 Fleet Street, London) almost certainly had these arms made up for him by the Birmingham gun trade, and appears to have been the only retailer of them. Serially-numbered on frame or butt-strap, and in a range observed between 51 and 2,529, Baker's revolvers have survived in sufficient numbers to suggest a total output of some 3,000 arms.

A second single-action revolver is less to be expected in an average collection today, and is shown on Fig. 9; in this weapon also, the hammer was off-set to clear the line of sight, but it was unusual in having its barrel attached to the lock-case by a top-strap and hinge. On release, the barrel could be tipped up and the cylinder removed for reloading. Basically, the design was similar to a hand-rotated American revolver, earlier protected under D. Leavitt's U.S. Patent No. 182, of April 19th, 1839; the British arm was, however, the subject of B. Cogswell's Registered Design No. 3389, of November 16th, 1852. Specimens are rare.

In the context of British transition-type revolvers, it is submitted that these two weapons are further evidence for the contention that transition-type arms must have been made after 1851, and not at the dates suggested by Anquetil, Deane or George. It is impossible to believe that either Baker or Cogswell could have secured Registered Designs for their respective methods of securing an uninterrupted sight, if transition-type revolvers had been in general distribution in Great Britain during the decade before such protection was sought; their respective devices would have passed into the public domain long before the Design registration dates, in 1852, if such had been the case.

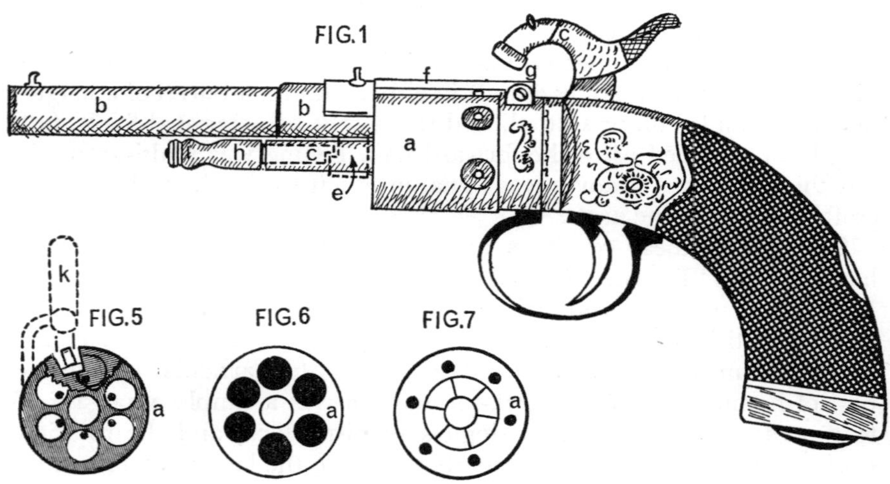

FIG. 9. *B. Cogswell's Revolver.*

Benjamin Cogswell's Registered Design No. 3389, of November 16th, 1852, for an "Improved six-shot Rifle Pistol". (Figs. 2, 3, and 4 are not shown). Features claimed in the Design were the shape or configuration of the barrel hinged at "g", and released for tipping up by pulling forward the catch "h"; this released the lump "e". The off-setting of hammer and nipples, to permit an uninterrupted line of sight, was also claimed.

CHAPTER V

ROBERT ADAMS'S REVOLVER

Model of 1851

AS the two preceding chapters have shown, 1851 was an important year for the development of British revolving arms. As a result of the activities of an American, Samuel Colt, great interest was aroused by some revolving-cylinder pistols that he showed at the Great Exhibition in that year, and the times also produced a native manufacturer, Robert Adams, whose pistol was pitchforked into prominence by the absence of any other British weapon in the same category.

Adams's revolver was a five-chambered, self-cocking pistol, in which barrel and frame were forged as one solid component; the cylinder turned in a rectangle carved out of the barrel/frame mass, and could be removed for cleaning, on releasing a spring-latch at the side of the lug connecting barrel to frame, by drawing forward the arbor. A spring safety-catch was mounted at the left of the hammer, and held this clear of the nipples when pressed in under the hammer-nose; the catch automatically sprang free when the trigger was pressed for firing. No rammer was fitted, the user loading his pistol by pressing home a charge into each chamber, with his thumb.

The lock-mechanism (see Fig. 10) was both simple and robust, consisting of hammer, trigger, pawl, sear, and three springs. Pulling the trigger (A), caused a linked sear (B) to push back the hammer (C) upon its pivot. The trigger itself rotated on another pivot, and had the pawl (D) hinged upon it; as the trigger turned, so that pawl rose to rotate the cylinder, by engaging a ratchet cut or mounted at the rear of the latter. The pawl was kept in engagement with the ratchet by a spring that also bore upon the lower end of the sear; the pressure of this spring also urged the sear against the notch or bent in the hammer-breast, by which the hammer was lifted for firing. At full-cock, a bulge upon the hammer-belly pushed the sear out of engagement

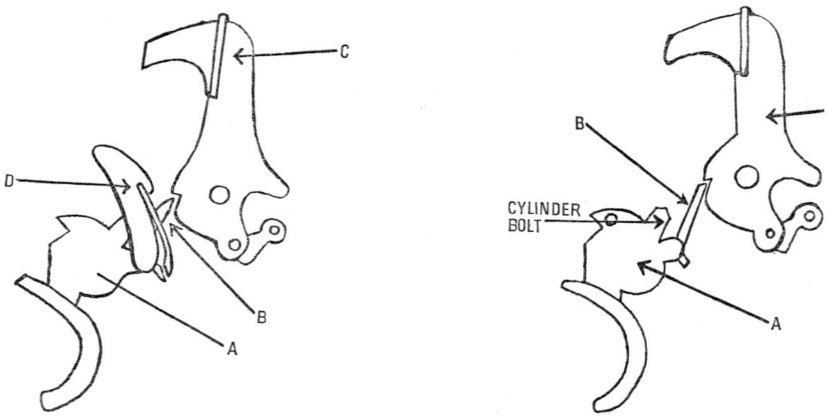

FIG. 10. *Robert Adams's self-cocking lock mechanism.*
The components are described below. At the left, the lock is shown with the the pawl removed, to show the sear.

with that bent, and a mainspring drove the hammer down upon a nipple. At this moment of discharge, the trigger had reached the rearmost point of its travel, and a lump upon it had been thrust up through an aperture in the frame, to engage a nipple-partition, and thus hold the chamber locked in alignment with the bore. Releasing the trigger, after firing, permitted a trigger-spring to return the former to the "at rest" position, dragging the pawl back down the ratchet face, to engage for a subsequent discharge.

It will be seen from the specimens shown on Plate 7, that Adams's revolver had horizontally-mounted nipples, which were buried in partitioned wells like those used in a Cooper or Mariette pepperbox. There is no record that Adams encountered any opposition, from Cooper, to the use of this refinement. Since, however, Samuel Colt's revolvers were constructed with exactly the same nipple seatings, it can be assumed that no real bar existed, in Britain, to their use for conventional revolvers.

Ingenious as the lock mechanism undoubtedly was, the important feature of Adams's revolver lay in its robust one-piece frame-and-barrel construction, and in view of the dominance which this patented feature gave the patentee over other British revolver manufacturers, between 1851 and 1865, we think it reasonable to devote some space to Robert Adams, and to those associated with him, in the launching of his pistol, namely, George and John Deane.

During the early 1840's George and John Deane were active in partnerships, which traded in London under the styles of "George and John Deane" and "Deane, Dray & Deane"; the latter partnership (formerly 'Jenner & Dray") is of smaller concern here, as it related to a prosperous ironmongery business, manufacturing and selling stoves, kitchen ranges, grates, iron fences, fenders and so forth, with gas-fitting and plumbing contract-work as a sideline. The premises for these activities were at 86 Chiswell Street and 119/120 Bunhill Row, initially, and later spread to Swan Lane and the London Bridge approach.

As to the pursuits carried on under the style of "George and John Deane" an advertisement taken from *Kelly's Post Office Directory* for London, 1846, (see Plate 38) sets the picture of the partner's activities as warehousemen and wholesale dealers in the products of Birmingham, at 46 King William Street, London Bridge.

Neither partner appears to have had any gun-making experience or background, although John Deane took a dilletante interest in the subject generally, and manufacture of firearms by the partnership probably stemmed, originally, from the Birmingham-made guns that formed part of their normal stock-in-trade as "birmingham & sheffield" warehousemen. Thus, apart from an obvious need to provide local "after-sales service" for such weapons, there would be a strong inducement for the partners to use those superb facilities available locally, through the London gun trade, for the stocking, screwing, and finishing of Birmingham-made weapons brought down to London "in the white". A natural extension of such contacts would be to have their own establishment for "setting-up" of Birmingham-made components, and, finally, to enter London manufacture on their own account.

When taking this ultimate step, the advantages of a vertical integration into their business of an already existing firm would be obvious, and the absorption (about 1845) of a firm called Groocock & Co. (see Plate 38) seems to have provided premises, equipment and experienced labour for a "Gun manufactory", at 30 King William Street. By late 1849, in fact, the Deane "gun-making" activities were on a scale sufficient to make their name anathema to the established gun trade in general.

The history of this trade-war is of no concern here, and it would be unnecessary to do more than mention its existence, were it not for a by-product of one squabble, which revealed the first mention of a then existing connection between the names of George & John Deane and Robert Adams.

The incident which established this connection between Deane and Adams, in 1849, occurred towards the close of the year, and can best be named as "the Scott Scandal". Very briefly, this matter involved the bursting of a Deane sporting gun, in the hands of Sir Claude E. Scott, and the gun trade generally took that opportunity to blacken the name of George & John Deane (with their firearms) by condemning their methods of manufacture and accusing them of tampering with barrels after proof. The bitterness of newspaper correspondence on the matter was quite extraordinary, and accusation or counter-accusation were widely debated in a truly envenomed manner.

Amongst those caught in this literary cross-fire (and heavily winged) were such giants as William Greener, Westley Richards and Colonel Hawker, but although G. & J. Deane won the day, they saw fit to clinch their victory (in January, 1850) by the publication of a small pamphlet, entitled "Deane's defence of themselves and their guns from charges brought against them by the Company of Gun Makers and Others".

Buried in this pamphlet were two references which, although tantalizing in their lack of any Christian name to establish the particular "Mr. Adams" concerned, did indicate that Robert Adams almost certainly had a gun-trade background and that he was probably the manager of the Deane gun manufactory in 1849.

The references occurred in letters published by *The Morning Chronicle* (of January 5th and January 14th, 1850) over the signature "W. A. Urquhart", who wrote in defence of G. & J. Deane's guns, and stated in his first letter ". . . I have recommended Deane's guns to several friends, believing his Sole Manager in the concern to be unequalled in his calling: that, with the exception of the barrels, *not locks*, he will challenge any of the elite of the trade, each with their hands and none other, to beat him in turning out a perfect gun . . ." In his subsequent letter, Urquhart stated that "I plead guilty to the charge of 'bias' for the best articles at the lowest price: *for Adams, Deane's manager* (our italics) because he supplies it. The trade are well aware that he knows his business, and would perhaps gladly see him anywhere rather than where he is: discharge him, and my 'bias' for Deane ceases. I have no 'bias' for him on the score of intimacy or self-interest, for I never enter the shop more than three or four times in a year, and then just before the shooting season commences. This equally applies to Deane, whom I have rarely seen there."

The point need not be laboured, for the next step in the Deane/Adams relationship was one of record (vide the *London Gazette*, July 8th, 1851, Vol. II, p. 1787):

"Notice is hereby given, that the Partnership heretofore subsisting between us the undersigned, George Deane and John Deane, as Gun Manufacturers at No. 30 King William Street, in the City of

London, has been dissolved by mutual consent, as and from the 25th day of March last. Dated the 7th day of July, 1851."

"George Deane"
"John Deane"

From this it would seem reasonable to assume that a new partnership between John Deane, Robert Adams, and John Deane (Jun.), trading as Gun Manufacturers under the style of "Deane, Adams & Deane", should be dated at, or shortly after, March 25th, 1851, that is to say, about a month after the sealing of Adams's English patent for his famous self-cocking revolver, and less than a month before the final closure of the original catalogue for the "Great Exhibition of the Industry of all Nations", which is now commonly called the "Great Exhibition of 1851".

We have no means of establishing exactly when the respective "G. & J. Deane" and "Deane, Adams & Deane" catalogue entries were sent in, but it is clear (from the first and second editions of the catalogue) that both partnerships participated in the Exhibition, having Stands 223A and 223 respectively for the purpose.

Curiously enough, the reason for this Deane, Adams & Deane partnership may not originally have lain in Adams's revolver at all, for none of the Exhibition catalogue entries lay any stress upon the presence of the weapon (at either of the stands), until the issue of the big, four-volume, souvenir catalogue in 1852. This massive, permanent record of the Exhibition was never intended for actual use within the Crystal Palace, and contained (in its fourth volume) an illustrated Supplement of revised catalogue entries, bound in with the report of the Commissioners, and the reports of the juries who had considered the various Classes, and made awards to some of the exhibitors.

However, this illustrated Supplement (at p. 1478) does make it clear that the Adams revolver had always been intended to be shown at the Exhibition, but that no special prominence was originally to be given to it. The Supplement entry identified a rather anonymous original cataloguing (for the "G. & J. Deane" stand) of an "electro-plated silver revolving pistol, ebony stock, silver studded" by reference to Plate 30 of the Supplement, and this showed Robert Adams's revolver, but placed it firmly on the "Deane, Adams & Deane" stand.

Generally speaking, the wide interest aroused by firearms shown at the Exhibition took both the Exhibition Commissioners and the gun trade exhibitors by surprise, for the original themes of the whole project had been envisaged as "Peace and Industry". Initially, the superb workmanship and decoration of the Spanish and French entries claimed the attention both of the public and of the journalists covering the

Exhibition, but the showmanship and enterprise of Colonel Colt (who imported and exhibited a considerable number of his own revolvers) soon brought the attention of all to a consideration of the merits of true revolvers, and despite the claims made by John Deane, seven years later (*op. cit.* p. 199), that three true revolvers were featured by English makers, in competition with Colt, it remains a fact that only one such exhibit (by Deane, Adams & Deane) was actually recorded in the Supplement to the catalogue. It is also significant that Adams's was the only revolver cited as competing with Colt's weapons, in the many newspaper and magazine articles commenting upon the firearm exhibits.

So far, in fact, was the true revolver from being a normal line offered by British gunmakers in 1851, that the *Illustrated London News* of September 20th remarked authoritatively that "the American revolving pistols differ from those commonly seen in England in the circumstances of their only having one barrel, a fixture, the compound breeching made to revolve, whereas in the English revolving pistol the barrels themselves turn on a pivot".

In the circumstances, therefore, we think that the original idea behind Deane, Adams & Deane's partnership could have lain in other directions than that of Robert Adams's revolver. His English Patent No. 13,527 of 1851 protected a breech-loading sporting gun, a rifle barrel, a modified Delvigne breeching for muzzle-loading rifles, a simplified rifle lock, and a safety catch, in addition to his self-cocking revolver. It may be, therefore, that the eyes of the partnership were originally fixed upon the rifle modifications as those items of the Patent proposed for serious development. It should be constantly borne in mind, moreover, that Robert Adams's patent was filed as "partly communicated" (we do not know which part, nor do we know by whom), and that the gun trade in general, at the time of the Exhibition, was in something of a ferment over the whole question of Service rifle contracts, and the type of weapon eventually to be officially adopted.

In such an atmosphere, the possibility cannot be overlooked that the Adams's revolver was originally written into the claims of English Patent No. 13,527, as an addition which the partnership might advantageously pursue as a side-line, and that the members of the partnership were as much surprised as anyone else (though considerably more gratified) by the furore aroused over revolvers at the 1851 Exhibition, and by the success of that revolver as the only serious competitor to the all-conquering Colonel Colt.

Due weight can be given to this view, when it is recalled that the "Deane, Adams & Deane" entry in those Exhibition catalogues sold and circulated whilst the show was actually in progress read simply "Patent spiral raised rib rifles. Patent safety stop lock guns. Patent gun locks. Fowling pieces. Indian and African Rifles, Pistols, etc.", and that the grounds for the award of a Prize Medal to the partnership were their "Double and single guns and pistols perfectly finished."

It can, of course, be argued against this view that manufacture of a weapon like the Adams revolver was not a process that could be lightly entered, and that considerable thought must have been given to the financing of such an operation, and to the machinery necessary for it. However, it seems clear from Supplement remarks upon the weapon itself, and from the methods still in use for its manufacture three years after the Great Exhibition (see Chapter XIII), that the original idea of the partnership may have been to supply the gun trade with main components of the pistol, and to leave those buyers to manufacture and finish the remainder of their weapon as best they could. On this sort of basis (as Chapter XIII will show) any tooling-up required by the partnership would, if necessary, have been comparatively modest so far as Adams's revolver was concerned.

In the event, however, "the tail wagged the dog" and patented rifling, gun-locks, breechings, sporting gun, and safety catches were rarely heard of again. Instead, the partnership made an all-out effort to develop Adams's revolver on an exclusive basis (both commercially and with an eye to its adoption by the Board of Ordnance), almost from the moment that serious public interest in revolvers began to be shown.

H. L. Blackmore (*British Military Firearms, 1650–1850*, p. 253; Herbert Jenkins, Ltd., 1961) recorded that a competitive firing-trial of a 32-bore Adams revolver, against a Colt "Dragoon" revolver (the single-action ·44-in. model of 1848) was arranged at Woolwich on September 10th, 1851, a month before the Great Exhibition itself was formally closed by Prince Albert. Witnesses included Lieut.-General Sir Thomas Downman; Major-General Fox; Colonel Dundas; Lieut.-Colonel Chalmers, Assistant Director General of Artillery; Lieut.-Colonel Burn; Captain Wingfield and Captain Anderson of the Royal Artillery; Brigadier-Major Walpole of the Royal Engineers; Captain Fox of the Grenadier Guards; Captain March; Mr. Lovell, "armourer-in-charge of stores at the Tower"; Mr. Laurence Jun. (son of the American Minister in London and probable coiner of the nickname "Peace maker" for Colt revolvers) and several others.

Regardless of whose pamphlet-record of the event is studied, both revolvers were of about equal merit at this trial (which lasted from 11 a.m. to 1 p.m.) as to accuracy and lack of misfires. However, Adams probably achieved a faster rate of fire, since he used his patented wadded ammunition, loaded by thumb-pressure, and had the good sense to bring a "spring feeder" or percussion-cap magazine with him. The match-point undoubtedly went to Colonel Colt, however, for a party of Sappers passed the range, on their way to an embarkation point for overseas service, and he presented a revolver to the officer-in-charge of the detachment. This example was not apparently followed by Adams, and, in view of this failure, it seems all too possible that the latter had to withdraw an Exhibition weapon to accomplish the trial at all.

According to a rather general reference in the *Illustrated London News*, and to contemporary broadsheets published by Deane, Adams & Deane, a further trial of the Adams was achieved at Enfield prior to the end of October 1851; possibly this was squeezed in amongst the rifle trials before a Select Committee on Small Arms, which fixed upon the Pattern 1851 Minie rifle adopted for the British Army. Alternatively, it is known that Colt "Dragoon" and "Navy" pistols went through a prolonged series of firing trials at Woolwich (before Lieut.-Colonel Chalmers) from October 13th–21st, 1851, and it is possible that the Adams revolver was joined in these tests.

On November 25th, 1851, after the Exhibition had closed, Colt gave his famous lecture to the Institution of Civil Engineers "On the application of Machinery to the manufacture of Rotating Chambered-Breech Firearms and the peculiarities of those Arms" and we have it on the authority of J. N. George (*op. cit.* p. 195) that Adams was present in the audience, and attempted to start an argument as to the relative merits and demerits of his own and the lecturer's revolver. It is a pity that a transcript of any such exchange has not survived, for although the published "Minutes of Proceedings of the Institution", for this occasion, actually make no reference to such an incident, at all, a contemporary pamphlet did indicate that Adams was present.

CHAPTER VI

DEANE, ADAMS & DEANE

WHATEVER the truth in stories of an attempted confrontation between Samuel Colt and Robert Adams, in that November of 1851, it is quite certain that more urgent problems faced the Deane Adams & Deane partnership, on a practical level, if they really did enter a race for markets, against the redoubtable Colonel, and were armed only with Adams's patent, and a few virtually hand-made specimens of his revolver.

Their first concern (sensibly enough) was to arrange that whenever Colt managed to achieve a trial or demonstration somewhere, the Adams revolver was also presented, and with appropriate appeals to national sentiment. The partnership's next step was to ensure that all possible benefit was squeezed from their weapon's successful debut at the Great Exhibition. To this end, they secured handsome prominence (at p. 1478) in the British section of a Supplement to the illustrated presentation souvenir catalogue, when it was published in 1852, but in company with Colonel Colt, who took an equal amount of space with the same publicity objective. In this entry Deane Adams & Deane boldly stated that their revolver was made in three sizes, namely "the target or holster size . . . ten inches long in the barrel, including the cylinder, the bore 38 balls to the pound and weight 2 lbs. 14 ozs. The next size is eight inches in length, 54 balls to the pound and weighs about 2 lbs. The smallest or pocket size is six inches long, 120 balls to the pound and weighs 18 ozs." Elsewhere in their entry they also mentioned that "rifles and carbines are made on the same principle and . . . are equally effective."

There is little point in setting out the entire Supplement entry, interesting as it undoubtedly was; some of the material is dealt with elsewhere in this book, and it only remains to quote one further claim from it, namely, that "all the parts of these revolvers with the exception of the hammer, trigger and springs are made by machinery."

This claim was definitely intended to mislead any reader, by suggesting that Adams's design was backed by the full weight of a mechanized factory, with specialized plant for its manufacture. In fact, as a cursory study of an advertisement featured in Chapter XIII will show, the partnership had barely achieved that level of operations even two years later. It was probably true that lathes and simple planing or slotting "machines" were used in manufacturing the revolver, as they could be used at the 30 King William Street manufactory in making some of the other weapons turned out there by the partnership. It is difficult, however, to believe that any machines specifically designed for Adams's revolver manufacture, and for no other duty, existed when that Supplement entry was written. In truth, Robert Adams filed no firearms-machinery patent for revolvers until September 1854, when the rifling and barrel-boring machines described in Chapter XIII were thus protected, and it is plain that manufacture of his pistol was commenced with only those facilities available at the London Bridge manufactory, and that output was very small.

Certainly, the London Proof House figures shown in Chapter III, for every type of revolver (including, presumably, pepperboxes), seem to bear out this opinion. It will be recalled that the figures totalled:

1850	425 revolvers
1851	954 revolvers
1852	6,121 revolvers
1853	13,916 revolvers

and it should be noted that Adams's self-cocking revolvers were nearly all proved in London, throughout their production life.

The 1851 figures would certainly have included those revolvers released from Colt's stand at the Crystal Palace, after the Exhibition had closed, and the 1852 figures would be a "crest of the wave" in diletante demand for such weapons, aroused by the Exhibition, when every gentleman's cabinet-of-arms would contain a revolver of some sort (be it pepperbox, transition-type, Colt or Adams), and the gun trade must have been extended to supply the market, whether at home or with officers serving abroad in the "Caffre War".

However gratified the Deanes, *père et fils*, may have been by the success of Adams's weapon, they do not appear to have sunk their own capital into the venture of producing it, but to have concentrated their efforts upon the tasks of marketing and publicity. As will be emphasized, in Chapter IX, Robert Adams himself seems to have assumed responsibility for finding the facilities and finance necessary to expand production of his revolver up to the level needed for both actual and anticipated demands. It appears, from the recital of facts in a Deed of Settlement for the later "London Armoury Company, Ltd.", that he arranged the necessary finance with "Hankey's" a family banking house at 7 Fenchurch Street, E.C., which comprised William Alers Hankey, Thomas Hankey, John Alers Hankey, Frederick Alers Hankey, Reginald Alers Hankey and Thomas Alers Hankey. Adams's financial arrangements were apparently made with the last-named, but presumably with T. A. Hankey acting as nominee for his bank.

The exact terms of the arrangements are unknown, but the same recital showed that Robert Adams (as patentee) assigned one-eighth part or share of his basic English Patent No. 13,527 to T. A. Hankey, on June 23rd, 1852, and that he retained the remaining seven-eighth parts or shares for himself. This split-up of ownership in the patent necessitated execution of a formal patent licence, of course, before Adams could then use what had previously been his own unencumbered property. Accordingly (on the same day), a licence was duly executed by Hankey and Adams, by which Robert Adams alone was exclusively licensed to manufacture and sell under the patent in England, Wales and Berwick-on-Tweed, subject to the payment of a royalty of undisclosed amount, to Hankey, on each weapon so made.

The royalty rates probably varied for each item covered by the patent, and as various permutations of use would be possible, i.e. rifles having the special locks only, or the locks and the rifling, or vice versa, and the revolver could possibly embody Adams's special rifling, the licence document probably contained quite a list of such rates. In recognition of the need to control payment of these royalties, it now became necessary to arrange that all items licensed for manufacture be appropriately numbered and marked by the manufacturer. The basis of such arrangements is examined in Chapter II, and there is no need to enlarge upon it here.

Pending the discovery of more definite information on the point, we believe that all Adams's revolvers were numbered serially, after this licence date of June 23rd, 1852, by reference to blocks of numbers allocated both to Adams and to other manufacturers licensed under his 1851 patent. These blocks were subsequently more particularly identified by adding a capital letter to the number to each of them; the letter-suffix "R" was most commonly used (although others will be noticed) and it seems certain that this particular letter was only applied to those revolvers made by, or specifically for Deane, Adams & Deane. In the case of other such serial numbers noticed in this book, it is usually possible to establish the use of a letter corresponding with the initial letter of a licensee's name; examples are the "L", "P" and "DHD" serial numbers described in relation to Harding's revolver, at Chapter XII. In the specific case of Robert Adams and the Deane, Adams & Deane partnership, however, we suggest that the letter "R" was used as denoting "Revolver", that is to say as being allocated to one specific item amongst all the items protected by his 1851 patent. Had he manufactured other devices so protected, Adams would (it is suggested) have used other letter suffixes upon them.

The two documents executed between Robert Adams and T. A. Hankey on June 23rd, 1852 appear also to mark the beginning of an end to Deane, Adams & Deane's partnership, for although it endured until 1856, the old manager/employer relationship must have faded as Robert Adams began to assume a still larger status in the gun trade.

Despite the Supplement entry earlier cited, true quantity-production of Adams's revolver actually commenced with a series of 34-bore (nominal 32-gauge) weapons of the type shown at the top of Plate 7, although it is not clear if such pistols antedated the Adams/Hankey licence above referred to. These arms bore the frame-legend "Adams's Patent Revolver No. . . .", had no serial number on their cylinders (in contrast to later practice), and we traced serial numbers between 56 and 204 upon them.

It is necessary to point out (although we anticipate details cited later in this chapter) that 38-bore revolvers appeared amongst the Deane, Adams & Deane "Holster" pistols during the 200–400 serial-range, and had virtually ousted 34-bore arms by the time serial numbers in excess of 1,300 had been reached. However, an exceptional 34-bore revolver was found in the 4,000(R) serial-range (see Chapter VII) and others may exist.

As to the original 34-bore revolvers, the retailer's name on the top strap of the frame, even in such early days, was not limited to that of Deane, Adams & Deane, although (for obvious reasons) they retailed the majority of Adams's pistols. Their barrels were octagonal, $7\frac{5}{8}$ in. long, and rifled with five broad grooves in a right twist; the case-hardened iron cylinder was $1\frac{7}{8}$ in. in overall length, and had a separate rotating-ratchet that was secured to it by two screws. On most specimens, barrel and cylinder were London-proved, the top frame-strap being engraved "Deane, Adams & Deane (Makers to H.R.H. Prince Albert) 30 King William St., London Bridge."

In certain details of construction, these weapons also differed from later versions of the Model 1851 revolver:

(i) In the first 300 pistols, the link between main spring and hammer was pinned permanently into position; it could not be removed from the hammer (in a fashion later adopted), without the aid of tools.

(ii) Trigger guards were lightly inletted into frame-metal at the rear, and also nicely faired to the frame-contour at their front.

(iii) The cylinder bolt (a lump on top of the trigger) was carefully made as a 90° segment of a circle, and a hole cut in the frame floor to permit its successive rise into engagement with the nipple-partitions, was shaped in the same manner; in later weapons, both lump and hole were rectangular.

(iv) A spring catch securing the arbor was not flat (as on later models), but curved outwards markedly at its tip, permitting use of the thumb to disengage this catch from its securing notch in the arbor.

(v) In the first 300 pistols, the trigger spring was a single *flat* spring, secured by a screw underneath the trigger guard, and not the "V"-spring used on all later models; a flat spring had been shown, and described, in English Patent 13,527 of 1851.

(vi) A larger button was formed at the front end of the arbor.

(vii) The nipple wells were bored as truncated cones, and not with parallel sides, as was the later practice.

The very first of these early arms had two small-diameter screws (with comparatively large heads) turned into each side of the frame, just forward of the standing breech, and close to the cylinder-aperture floor. These screws secured an "L"-shaped plate, which formed part of both that floor and the face of the standing breech; this plate was assembled sideways into a suitably dovetailed slot, and preserved from loss by the screws. Such methods of frame construction were abandoned in all pistols made after this first series, and Adams's reasons for using it initially are still uncertain.

The changeover from 34- to 38-bore calibre (during the 200–400 serial-range) caused no essential changes in appearance and, as may be seen in Plate 7, the butts of this whole series were more steeply pitched than was later fashionable, being almost at right angles to the frames, and with flared bases in which a cap-box was embedded. At an overall length of 13¼ in., and weighing 2½ lb. these were really a cavalryman's revolver, in whichever calibre they were produced; curiously enough, the triggers had a rounded, almost hemispherical contour, to which Deane himself objected (*op. cit.* p. 214), and which was not a feature of Adams's patent specification drawing at all.

In default of any evidence to negative the assumption, we believe that this weighty series was latterly interspersed with a number of

54-bore weapons, bearing the shorter, revised frame-legend "Adams's Patent No. . . .". This legend still had no letter engraved after it, and all principal parts were either stamped with the number in full, or in part if their size would not accommodate it; unusually, that number was also stamped upon the lower butt-strap. These revolvers mingled serially with the larger arms in the 600–1500 range, and after about No. 1300 a large capital "R" sometimes appears upon their lower butt-strap; as with the earlier 34-bore pistols, retailers' names other than that of "Deane, Adams & Deane" will occasionally be found in the whole series, although the weapons retailed by that partnership do outnumber all others. It is also possible that an 80-bore pocket weapon was briefly considered, at this point in time; revolver No. 821 was reported in this calibre, but was the only specimen noted.

Mechanically, these 54-bore pistols appeared to have most features of later revolvers (i.e. those with true R-suffixed serial numbers), rather than those of the early 34-bore arms mentioned above, although the butt-strap cap box was retained. However, they had the same five-groove rifling and top frame-strap legend as their predecessors, and were of smaller dimensions; the barrel length was $6\frac{1}{5}$ in., and the pistol measure $11\frac{1}{2}$ in. overall. In cases where a bullet-mould accompanied the arm, it was sometimes stamped "52" rather than "54" bore, but always carried an oval emblem upon one side of it; this recorded Robert Adams's Registered Design No. 3033, of November 28th, 1851, covering the shape of his bullet. In many cases, too, a screw used to secure the rear of the trigger guard was positioned from inside the frame, and this feature remained in frequent use until well into another series of Adams's Model 1851 revolvers, later described, which was serially-numbered in the 10000R range.

The partnership now received a rather unpleasant setback, by a letter published in the *United Services Gazette* (on December 25th, 1852) which revealed that they had at least three gravely-dissatisfied customers, and in the one quarter where they could least afford to have them, namely, as serving officers in the "Caffre" war theatre.

The letter (dated at Exmouth, December 22nd, 1852) was signed "D. Campbell" and probably came from a close relative of one of these dissatisfied customers (Captain Campbell, 73rd Regiment), who could also have been the donor of an offending pistol.

To the Editor of the *United Services Gazette*.

Sir,

Having read in your last issue some remarks on the superiority of Adams's over Colt's revolver, I am induced to send you a *Graham's*

Town Journal of the 2nd October . . . in which you will find the result of several trials to test their comparative merits, which took place lately, on the Cape frontier, under the eyes of very competent judges, and you will perhaps be surprised to find that Colt's weapon exhibited over its rival the same superiority that the latter did over the former at Woolwich.

How this difference is to be accounted for I know not; but it is clear that, on the occasion referred to, Colt's beat the other out of the field, and the agent for the latter was obliged to fall back, to maintain his ground, on the trials at Woolwich to which you refer in the *Gazette*.

I am, Sir, your obedient servant,
D. Campbell.

The letter was followed by details of three comparative shooting-trials between Colt and Adams revolvers, in which the latter weapons put up extremely bad performances (both as to range and penetration), which ended with a burst Adams revolver on the first occasion, in the Commissioner for British Kaffraria ordering a Colt revolver on the second, and in the exchange of his Adams's weapon, for a Colt, by the Cape Mounted Rifles contestant at the third trial.

All this, we should add, appearing in the *Gazette* a bare three weeks after a large Deane, Adams and Deane advertisement had appeared in the *Times Supplement* of December 9th, 1852, touting the severity of tests for a period extending over twelve months "of ADAMS'S PATENT REVOLVING or REPEATING PISTOLS, Carbines and Rifles", by a [*sic*] Select Committee of Her Majesty's Honourable Board of Ordnance, and claiming superiority for these arms over every other species of revolver as being established "beyond dispute".

John Deane and his son must have been rather sensitive to product failures of this type, after the earlier "Scott Scandal" sporting-gun incident, and any discussions about Graham's Town with Adams, are unlikely to have echoed the Christmas spirit that year.

In defence of Adams's revolvers, we would point out that the *Graham's Town Journal* report suggested that the ammunition used in the burst weapon (and also in the other two offending revolvers) was grossly incorrect as to the bullet diameter or the powder charge. We were already in possession of bullet moulds in stated calibres larger than that of the pistols for which they had been supplied, and the 1851 catalogue Supplement entry, referred to earlier, had specifically stated also that the "5 chambers . . . are bored two sizes larger than the gauge of the barrel, in consequence of which the ball having to pass through the diminished aperture of the barrel is forced into the grooves of the rifling, windage is thereby prevented and great accuracy and strength of shooting obtained". Yet a layman's report on the Woolwich firing-trial of September 10th, 1851, commented of Adams's pistol, that "the reports (were) very sharp as if the balls had been projected with great velocity", suggesting that this method of chambering gave no trouble when the pistol was correctly loaded.

It is probable, therefore, that any Adams model 1851 revolver thus chambered, and used with a slow-burning, damp, or too-coarse powder would exhibit the defects of the Graham's Town trials. Fine-grained, quick-burning powder and excellent percussion caps were an absolute *sine qua non* to successful practice with that revolver and its thumb-seated bullet: if, therefore, these early revolvers needed "hot" caps, FFg. powder, and careful loading with tight bullet-wads under well-seated bullets, the weapons used at Woolwich and similar trials probably received such aids in the hands of their inventor, or his brother, but may not (by a wide margin) have achieved them at Graham's Town, on the end of a long and tenuous supply line, or in hands unaccustomed to their use.

The Colt, of course, with a bullet literally sized into the chamber by a powerful compound-rammer achieves superb ignition even with rather bad ammunition. The initial resistance of the bullet to the idea of moving at all is considerable (as any reader who has tried to pull a Colt cylinder-charge will know), and we have never encountered cases of modern use, where firing one charge has jarred forward a correct bullet in either neighbouring chamber.

At some time in late 1852, or early in 1853, Adams leased additional manufacturing space from the South Eastern Railway Company. This was at No. 2 Weston Street, Bermondsey (see Chapter XIII), and it seems certain that the terms of the Adams/Hankey manufacturing licence were extended to revolvers made at this factory, and again by reference to their serial numbers. The partnership did not abandon their premises at 30 King William Street, which belonged to (or were leased by) the two Deanes, but the continued up-turn in British revolver production during 1853 (revealed in part by the earlier-mentioned London proof-figures) presumably reflected the beginning of production at the Bermondsey factory.

At about this time, too, Deane, Adams & Deane's revolvers began to appear in quantity with the letter "R" directly suffixed to their serial numbers. As earlier mentioned, this practice was not wholly new to the partners, for the letter "R" can sometimes be found engraved upon the bottom butt-strap of weapons antedating the commencement of Weston Street manufacture, but that letter was now moved to a more prominent position on the frame, immediately after those digits composing the serial number.

So far as we can establish, the direct "R"-suffix for Adams's revolver serial numbers began to appear on the frame shortly after a 4000 serial-range had been initiated (see Chapter VII); this is not to say that 4,000 revolvers had been made by that point in time, but to suggest only that the new series was arbitrarily begun at that number. Initially, too, this letter was enclosed by brackets, as in "4327(R)", but these parentheses were soon discarded.

In the 4000 series, too, we found the first 120-bore specimens of Adams's self-cocking revolver, despite his Exhibition catalogue supplement claim that the model was in production by 1852. These were handsome little weapons (see Plate 7), sometimes presentation-engraved, with $4\frac{1}{2}$-in. barrels and an overall length of $9\frac{1}{2}$ in. As a rule, the concealed rear trigger-guard screw was retained from the previous series, but their barrels were now rifled with only three broad grooves, in a right twist.

As it happened, Colonel Colt's London factory had also come into full production, in January of 1853, and, further, the Americans (smarting still under their treatment at the hands of the Crystal Palace jurors) also held their own equivalent to the Great Exhibition, in that year. Robert Adams seems to have crossed the Atlantic, as a result, with the three-fold intention of patenting his revolver in the U.S., of supervising a Deane, Adams & Deane exhibit at this New York "Crystal Palace", and of interesting the American Services in his pistol.

The first objective was attained as U.S. Patent No. 9694 dated May 3rd, 1853, and (from the manner in which the American specification was drawn) we are reasonably confident that Adams was in America when he signed it. The patent was a weak one, and despite the drawing and prolix wording of the Specification, all that was apparently patentable on Adams's revolver in the U.S. was his spring safety-catch, the joining of sear to trigger (with Adams's particular method for disengagement of sear from hammer-bent, by a cam-face on the hammer-belly), and the integral cylinder-bolt formed on top of the trigger. His solid, one-piece frame and barrel assembly, which assured for Robert Adams a fourteen-year dominance of the British revolver market, was omitted from this American patent as a specific claim, but whether omitted on the advice of his American patent agents, as being a direct infringement of an existing American patent, or whether because such a claim would have provoked widespread opposition to his whole patent, is not known.

The second objective of Adams's visit (that of exhibiting at the New York Crystal Palace) was apparently arranged through the New York house of Henry Tomes & Co., but it was a Deane, Adams and Deane fowling-piece which was exhibited and not Adams's revolver. The matter may, of course, have been earlier arranged through Lewis & Tomes of London, in circumstances where premature revelation of Adams's revolver, as a potential exhibit, might have prejudiced his pending American patent application. The fowling-piece achieved, for the partnership, an award of a Bronze Medal by "Jury E, in Class 8, Naval Architecture, Military Engineering, Ordnance, Armor and Accoutrements", but gratification may have been somewhat marred by the same Jury's award of a Silver Medal (the only one given for fire-arms) to Samuel Colt.

One other task, which Robert Adams may have undertaken, if he did make this journey to America in 1853, could have been to establish contact with the Massachusetts Arms Company of Chicopee Falls, Mass., U.S.A., who later made ·36-in. and ·31-in. calibre versions of a Beaumont-Adams double-action revolver; see Plate 15. At the date of Adams's visit, this Company (based in the old Ames Manufacturing Company factory) was being sued for infringing Colt's U.S. Patent for mechanical rotation of revolver cylinders; the case had opened in August 1851, and the Massachusetts Arms Co. was eventually to be barred not only from making the Wesson & Leavitt pistol that it had been primarily formed to promote, but also (with all other American arms-makers) from making any mechanically-rotated revolvers at all, until 1857. During 1853, however, Adams's self-cocking revolver may have been under consideration, by the Company, as an extra product to complement their single-action Wesson & Leavitt pistol.

There is also the possibility that Adams approached the Company as well-known suppliers of arms-making machinery, if his new London factory at No. 2 Weston Street was still in the process of being staffed and equipped. As a later newspaper report upon this factory shows (see Chapter XIII), a fair proportion of handwork went into Adams's revolvers, and its patentee may have considered ordering plant from the Massachusetts Arms Co. to reduce this.

Our suggestion that Adams also desired to interest the American Services in his revolver is purely conjectural. Colonel B. R. Lewis (*Small Arms and Ammunition in the United States Service*, Smithsonian Institution, 1956) stated that Adams "belt" revolvers were sold to the U.S. Army in 1856, but these were American-made (by the Massachusetts Arms Co.) and almost certainly American-sold; having regard to the date of sale, the arms were almost certainly double-action Beaumont-Adams pistols of the type described in Chapter X and shown, for example in Plate 13, and not self-cocking arms. However, in view of Adams's English experiences in demonstrating his revolver against the Colt (and of the American's enterprise in London) it is hard to believe that he would not carry war into the enemy's country and arrange U.S. Service trials during this visit, however long the odds against adoption.

In England, one further legal formality was probably also tidied up during that year, since the recital to the Deed of Settlement for the London Armoury Company Limited clearly stated that Adams had "also obtained and is now entitled to certain other Letters Patent for the use of his said Invention for Improvements in Firearms in England, Ireland and Scotland". Since Robert Adams's basic English Patent No. 13527 of 1851 was a pre-Patent Law Amendment Act protection, he had (in order to protect his invention throughout the United Kingdom) also to secure Letters Patent for Scotland and Ireland. His Scottish Patent was duly sealed on January 18th, 1853, and, although there are no details, it can be assumed that appropriate Irish protection was also secured. Whether Hankey shared in these two patents also, as to a share of one-eighth, is not recorded; on balance, and having due regard to the London Armoury Company document, it is doubtful.

CHAPTER VII

ADAMS'S SELF-COCKING REVOLVER IN FULL PRODUCTION

ALTHOUGH Deane, Adams & Deane claimed that Adams's self-cocking revolver was available in 38, 54 and 120-bore calibres during 1852, it is doubtful if that 4000(R) series introducing the smallest calibre actually appeared before the early months of 1853.

Since, however, it is difficult to make any definite claims as to the exact dating for various models of self-cocking arms made by Deane, Adams & Deane, it is proposed to trace their later history by reference to serial numbers engraved upon them. Lists of serial numbers, which follow, do indicate the general production-pattern followed by Robert Adams and his partners, and some notes are added upon specific arms which are either landmarks in the whole series, or of individual interest.

It will also become clear that these self-cocking weapons were numbered chronologically (without regard to calibre or type), but that blocks of serial numbers were occasionally set aside for use by licensed manufacturers (other than Deane, Adams & Deane), who also numbered their products chronologically, within such blocks. Indeed, these lists of numbers also bear out the view that it was Adams's patented one-piece frame (and not his lock-mechanism) which was the vital feature of his 1851 patent, and was, therefore, the weapon-component to be serially-numbered without regard to size or type, for the control of royalty payments in respect of it.

Principal dimensions varied little over the whole series of Deane, Adams and Deane's self-cocking revolvers, at least as to the three basic calibres. 38-bore pistols had barrels slightly over 7½ in. long, and were rifled with five grooves twisted to the right. Nominal 54-bore revolvers had barrels about 6¼ in. long, and were predominantly rifled with five grooves (also twisted to the right), although three-groove rifling can be encountered; until well into a 10000R serial-range, many pistols of this model were actually in 56-bore calibre, rather than 54 bore, and were so stamped (in minute figures) upon the breast of their frames.

120-bore pocket arms had $4\frac{1}{2}$-in barrels, and a three-groove, right-twisted rifling. Barrels shorter than these three standards will sometimes be met, and could apparently be obtained on special order.

In studying the lists which follow, those arms marked by an asterisk * were actually examined; the other weapons were reported by friendly collectors, or noted from catalogues, and full details were not always available.

Serial No.	*Bore*	*Retailer*
*4004 (R on lower butt tang)	120	Deane, Adams & Deane
4005 (R as above)	120	Deane, Adams & Deane
*4008 (R as above; see Plate 7)	120	Deane, Adams & Deane
4220(R) (Full number on frame)	38	—
4271(R) (Full number on frame)	38	Deane, Adams & Deane
4327(R) (Full number on frame)	34	Hollis & Sheath (Marked "Joseph Brazier" inside)
*4727(R) (See Plate 7)	54	Deane, Adams & Deane
5247R (Full number on frame)	54	Deane, Adams & Deane
5262R (Full number on frame)	54	Deane, Adams & Deane
5639X (Full number on frame)	38	Hollis & Sheath
6320X (Full number on frame)	38	Hollis & Sheath

The significance of these two reported "X"-suffixed serial numbers remains to be established. As later mentioned, Deane, Adams & Deane were forced by the demand for Adams's pistol, both to have some frame-forgings made specifically for them, in Birmingham and Wolverhampton, and to licence others to make the arm. However, such weapons had other letter-suffixes to their serial numbers, and if an "X"-series also existed, then it presumably arose through the granting of a licence to Hollis & Sheath (of Birmingham) to manufacture complete revolvers under Adams's 1851 patent, and more than two specimens should have been either encountered or reported to date.

Serial No.	*Bore*	*Retailer*
6487R	54	Deane, Adams & Deane
6826R	—	Deane, Adams & Deane
6830R	54	Deane, Adams & Deane
6835R	38	Moore & Woodward
6843R	38	Deane, Adams & Deane
*6871R	38	Deane, Adams & Deane
6904R	38	Deane, Adams & Deane

It will be noted that although Deane, Adams & Deane signed nearly all the pistols in this "R"-suffixed serial-range, there were exceptions (like 6835R, above), where another retailer's name was reported.

Pistol 6871R was of particular interest, in having a cylinder which rotated to the left, and not clock-wise in normal fashion. This feature was also encountered in pistols 10460R and 11107R (*q.v.*), and we have undocumented reports of other weapons with this feature which were also in the post-10000R serial-range. Why it should have been thought necessary to introduce such a modification, without extending it to all weapons, is not clear, for such an alteration required reversal to a number of manufacturing operations. During this series, too, the frame legend altered (for the last time) from "Adams's Patent No. . . ." to "Adams' Patent No. . . ."

Serial No.	*Bore*	*Retailer*
7107R	90	Deane, Adams & Deane
7287R	38	Deane, Adams & Deane
7402R	54	Deane, Adams & Deane
*7484R	54	—
7649R	38	E. M. Reilly, London
7852R	60	Deane, Adams & Deane

Pistol No. 7107R was the lowest "R"-suffixed serial number reported upon a pistol that was not in 38, 54, 56, or 120-bore calibre; the report might have been discounted, were it not for another 90-bore pistol described under the 9000R series, and for No. 7852R in this same group.

Serial No.	*Bore*	*Retailer*
8035R	54	Deane, Adams & Deane
8493R	38	Deane, Adams & Deane
8608R	54	Deane, Adams & Deane
8694R	54	Deane, Adams & Deane
*8831R	38	—
8923R	38	—
*8958R	54	Deane, Adams & Deane
9255R	38	—
9296R	54	Deane, Adams & Deane
9402R	54	Deane, Adams & Deane
9421R	54	Deane, Adams & Deane
*9449R	90	Deane, Adams & Deane
9748R	56	Deane, Adams & Deane
9980R	38	Deane, Adams & Deane

Revolver No. 9449R was the other 90-bore pistol earlier mentioned, and had been cased and retailed by J. D. Dougall, of London and Glasgow, although engraved with Deane, Adams & Deane's name and address. It had also, curiously enough, been adapted during its working-life, to use metallic rim-fire cartridges (in an extra cylinder), as well as the conventional cap-and-ball charges. This work had been expertly done, and the pistol re-proofed in Birmingham; rifled with a five-groove rifling, and with $5\frac{3}{4}$-in barrel, this weapon was quite clearly not made-up from 54-bore frame and cylinder components.

Serial No.	*Bore*		*Retailer*
10000R	38		Deane, Adams & Deane
10163R	56		Deane, Adams & Deane
10182R	56		Deane, Adams & Deane
10198R	38		Deane, Adams & Deane
10296R	54		Deane, Adams & Deane
*10398R	120		Deane, Adams & Deane
10460R	38		—
*10547R	36	Rifle	Deane, Adams & Deane
10844R	54		Deane, Adams & Deane
10895R	38		Deane, Adams & Deane
*10948R	24	Rifle	Deane, Adams & Deane

Apart from one report of a 24-bore rifle numbered 225, without an "R"-suffix, this 36-bore weapon, No. 10547R, was the first long-arm encountered. A cursory examination of the remainder of the lists will show no long-arms after serial number 12508R. This may, of course, be a fact peculiar to our rather small sampling, but the self-cocking rifles and carbines are undoubtedly rare. Barrel lengths were normally about 24 in., and simple leaf-sights were fitted; a detachable metal shroud covered the hammer, a rather crude spring-safety was present, and the trigger guards were flat-fronted. This latter feature enabled the firer to keep his supporting wrist clear of the flash (between chamber and barrel) on discharge, and may be seen in Plate 8.

Pistol No. 10460R was extremely unusual, in being fitted with a modified "hesitating" self-cocking lock mechanism, which had been patented by Robert Adams, as British Patent No. 2712 of November 22nd, 1853. In weapons so modified, trigger-pressure raised the hammer to full-cock, where it was automatically held; upon releasing the trigger, deliberate aim could be taken, and a further light pull upon the trigger would discharge the arm. This was a most ingenious improvement, disposing of one major defect in Adams's basic design, namely, the difficulty in accomplishing aimed fire with a self-cocking revolver. Adams did not file the Complete Specification to this Patent until May 22nd, 1854, and it is therefore quite possible that revolvers embodying this lock-mechanism were not made until the early months of that year.

The definitive article upon this patent is R. Bedford's "Percussion Revolvers. The Hesitation Action", *Journal of the Arms & Armour Society*. Vol. IV, No. 8, December 1963; these revolvers are extremely rare today, and No. 12041R (retailed by Wilkinson, of London) was the only other serial number reported by Bedford.

Serial No.	*Bore*		*Retailer*
*11107R	38		Deane, Adams & Deane
11410R	38		Deane, Adams & Deane
11457R	54		Deane, Adams & Deane
11526R (See Plate 8)	24	Rifle	Deane, Adams & Deane
11567R	38		Deane, Adams & Deane
11657R	54		Deane, Adams & Deane
11675R	38		Deane, Adams & Deane
11818R	24	Rifle	Deane, Adams & Deane
11844R	54		Deane, Adams & Deane
12033R	32	Rifle	—
12041R ("Hesitation" revolver)	54		Deane, Adams & Deane
12130R	54		Deane, Adams & Deane
*12148R	38		Deane, Adams & Deane
12152R	38		—
12193R	56		Deane, Adams & Deane
*12200R	24		Deane, Adams & Deane
12310R	38		—
12412R	54		Deane, Adams & Deane
12419R	54		Deane, Adams & Deane
12451R	54		Deane, Adams & Deane
12480R	38		Deane, Adams & Deane
12508R	80	Rifle	Deane, Adams & Deane
12777R	38		Deane, Adams & Deane
12868R	—		Deane, Adams & Deane

Revolver No. 12200R was another pistol of unusual calibre, and is in the J. B. Bell Collection. It is cased, with spare cylinder and all accessories. At an unloaded weight of 2 lb. 2 oz., and an overall length of 15¼ in., it seemed possible that this ·577-in. weapon was a cut-down rifle; the fact that trigger guard and butt-cap were made of bronze, rather than iron, also suggested such an adaptation. The rifling had unfortunately been too badly damaged for identification, and might have settled the point, but we are inclined to accept this weapon as a true revolver-pistol, made to special order from rifle forgings.

Serial No.	*Bore*	*Retailer*
13332R	54	Deane, Adams & Deane
13333R	54	Deane, Adams & Deane
13361R	90	Deane, Adams & Deane
13450R	54	Deane, Adams & Deane
13462R	54	Deane, Adams & Deane
13463R	54	Deane, Adams & Deane
13794R	54	—
13848R	54	Blanch & Son

It might be inferred, from the preceding details, that this 13000R series was one evidencing a large production-run in a single calibre; however, this effect arises only in our particular sampling, and any temptation to allocate this serial-range to one calibre of revolver must be resisted.

It is a fact, however, that these 13000R weapons exhibited points of detail which suggested that events of importance, in the activities of Deane, Adams & Deane, occurred during manufacture of that series.

Prior to serial numbers in the mid-13000R series, revolver-frames retailed by Deane, Adams & Deane were often found to bear internal "WT" or "HS" stampings, and this practice strongly suggested that frames were being made for the partnership, by William Tranter, or H. Hollis & Sons, of Birmingham. We have been able to establish that not all frames prior to (say) 13500R were so stamped, and would have been surprised had the contrary been the case, for it seemed to us most unlikely that Deane, Adams & Deane were incapable of making some revolver-frames on their own account, and were limited to "locking and stocking" a main component made for them, by others. Logically, the need to use Hollis or Tranter, as outside contractors for the main pistol-component, can only have arisen through the partnership's inability to meet all demands, from their own manufacturing resources in London.

It is, unfortunately, not clear when these Birmingham-made frames were serially-numbered during manufacture. One double-trigger Tranter revolver (a weapon which embodied Adams's patented frame, under licence, and is described in Chapter XVIII) was serially numbered 13702R, and retailed by Deane, Adams & Deane; this pistol had a $6\frac{5}{8}$-in. barrel, rifled with five grooves, and although the cylinder had London proof-marks, the barrel had escaped proof altogether. The existence of this revolver, and others like it, numbered with Adams's self-cocking weapons in a serial-range allocated to Deane, Adams & Deane, and not to William Tranter at all, certainly suggests a decentralized system of manufacture, in which errors of serial numbering (or in use of already-numbered frames) could occur. Such a system was perfectly viable by British standards, and consistent with the view that Adams's revolver was originally intended for just such a method of manufacture, that is to say, with "locking and stocking" to be carried out by anyone in the gun trade who had authorized access to the main component.

An advertisement set out in Chapter XIII, was dated July 18th, 1854, and certainly confirmed that Deane, Adams & Deane were making revolver-frames at Weston Street, London, by that date. There seems little point in such an advertisement, if Adams's revolver-frame was then still being made by Tranter and Hollis, and it is noteworthy that the "WT" and "HS" internal frame-stamps disappeared in the course of this 13000R serial-range. In addition, a new shape of frame began to appear during that series, which had a longer cylinder aperture, and butt less steeply angled to the frame, than its predecessors; in our small sampling, pistols Nos. 13463R and 13794R were definitely reported as having these two peculiarities, and were the earliest weapons to show transitional changes to a new frame-shape ushered in by the next serial-range.

In turning to the next range of serial numbers, therefore, it must be emphasized that a new frame-shape for Adams's self-cocking revolver had definitely emerged by about serial number 14000R. Mechanically identical with its predecessors, this new weapon had frame and trigger shapes shown in Plate 8 (*lower*), and latterly often embodied three other patented improvements described later in this chapter, although the weapon illustrated in fact lacks such improvements. A practice also developed, by which the frames were now stamped "D.A. & D." externally, in minute letters on the front frame. Bore sizes will also be found stamped there, but this latter mark had been introduced as early as 4000R serial-range, and was not therefore a new feature.

Weapons with the new frame-shape shown in Plate 8 are sometimes referred to, by collectors, as Adams's "Improved" revolvers, Model of 1854 or Model of 1855; having regard to the existence of pistols without the patented improvements later described, we are inclined to accept the earlier year as that for commencement of their manufacture. All pistols in the two lists which follow were marked as sold by "Deane, Adams

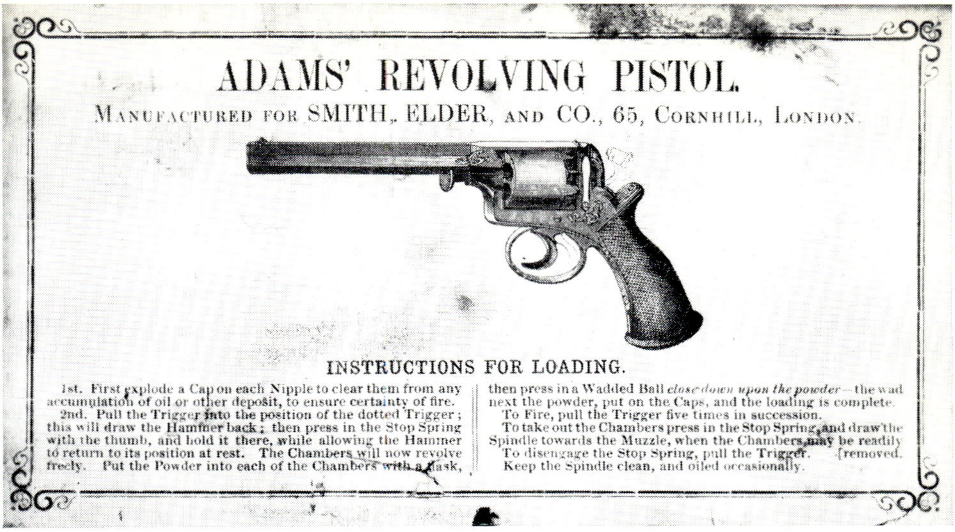

PLATE 9. ROBERT ADAMS' SELF-COCKING REVOLVER, MODEL OF 1851
(*Top*) Case-lid label for Robert Adams' self-cocking percussion revolver, Model of 1851.

(*Lower*) Continental version of Robert Adams' self-cocking percussion revolver, made by D. Herman (Liege). The pistol uses Adams' lock-mechanism, but not his patented one-piece frame and barrel. *Andrews Collection.*

Plate 10. Early Beaumont-Adams Revolver, with Adams' Rammer
(*Top*) Early Beaumont-Adams double-action percussion revolver, in 54 bore calibre, fitted with Robert Adams' patented lever-rammer (to Br. Pat. No. 2645/1854). Serially-numbered as B1629 14,950R. *A. N. Kennard Collection.*

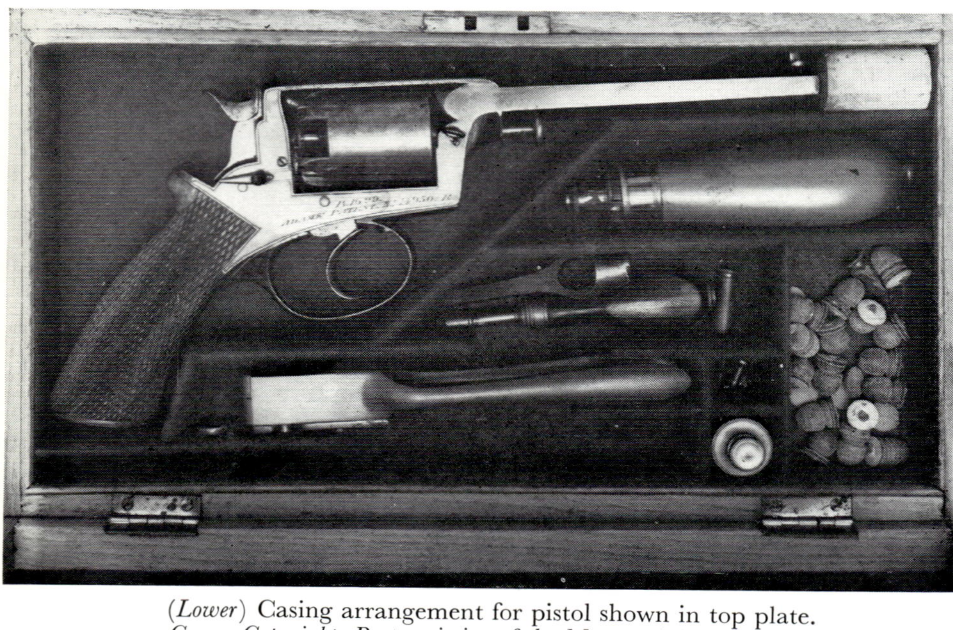

(*Lower*) Casing arrangement for pistol shown in top plate.
Crown Copyright. By permission of the Master of the Armouries.

Plate 11. Variant Beaumont-Adams Revolvers

(*Top*) 38 bore double-action Beaumont-Adams percussion revolver, built upon the original 1851 frame, and fitted with J. Brazier's rammer. Serially-numbered 9580R, and retailed by "Deane, Adams & Deane". Probably a conversion from a self-cocking Adams revolver. *E. J. Turner Collection.*

(*Lower*) 54 bore Beaumont-Adams percussion revolver, built upon the "Improved" frame, and fitted with J. Brazier's rammer. Retailed by Wilkinson & Son, London, and probably gold-damascened by that firm. The decoration has obscured the serial-number. *Andrews Collection.*

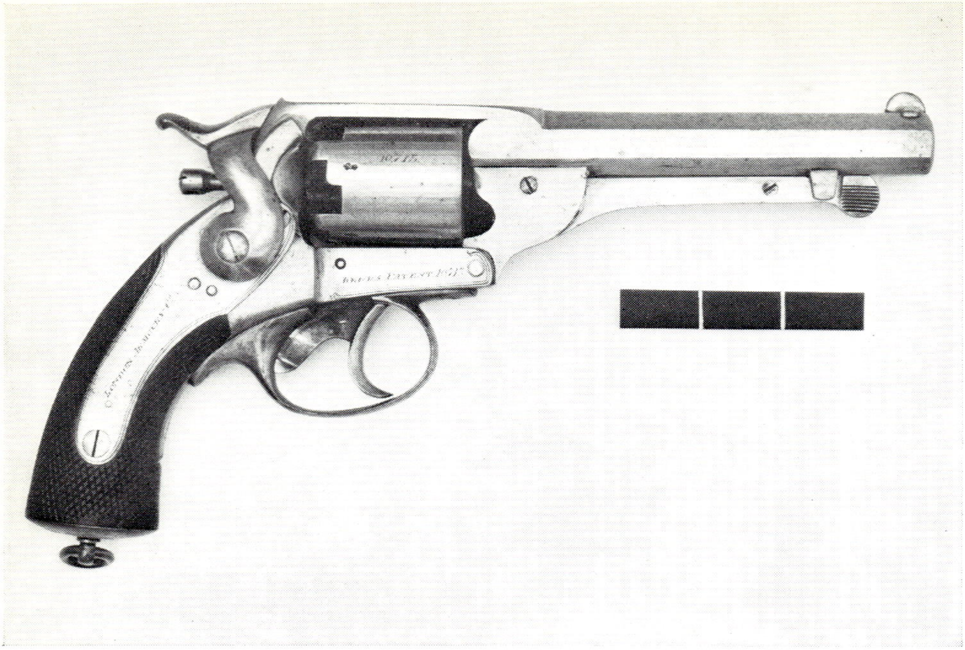

Plate 12. James Kerr's Revolver

(*Top*) A double-action percussion revolver, in 54 bore calibre, made by the London Armoury Company (Limited) under James Kerr's Br. Pats. Nos. 2896/1858 and 242/1859. Serially-numbered 10,715.

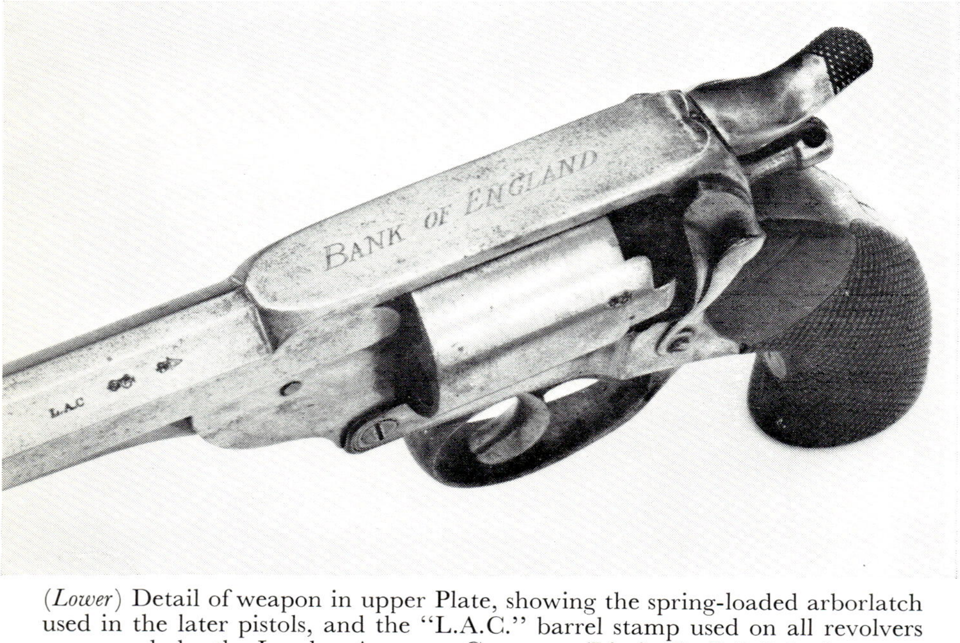

(*Lower*) Detail of weapon in upper Plate, showing the spring-loaded arborlatch used in the later pistols, and the "L.A.C." barrel stamp used on all revolvers made by the London Armoury Company (Limited). *Bell Collection.*

Plate 13. Beaumont-Adams Percussion Revolvers

(*Top*) Five-shot double-action Beaumont-Adams percussion revolver, in 38 bore calibre. Serially-numbered B1272 14576R, and retailed by "Adams & Co., London". An early specimen, probably made in Birmingham immediately prior to commencement of manufacture by the "London Armoury Co. Ltd." *Darwent Collection.*

(*Lower*) Standard 54 bore Beaumont-Adams double-action percussion revolver, serially numbered B16550 32223R. Has "L.A.C." barrel-stamp, and was retailed by "Witton Bros., Gt. St. Helens, London". *Darwent Collection*

(*Bottom*) Standard 120 bore Beaumont-Adams double-action percussion revolver, serially-numbered B17596 33343R. Has "L.A.C." barrel-stamp, and was retailed by "Deane & Son, 30 King William Street, London Bridge". *Darwent Collection.*

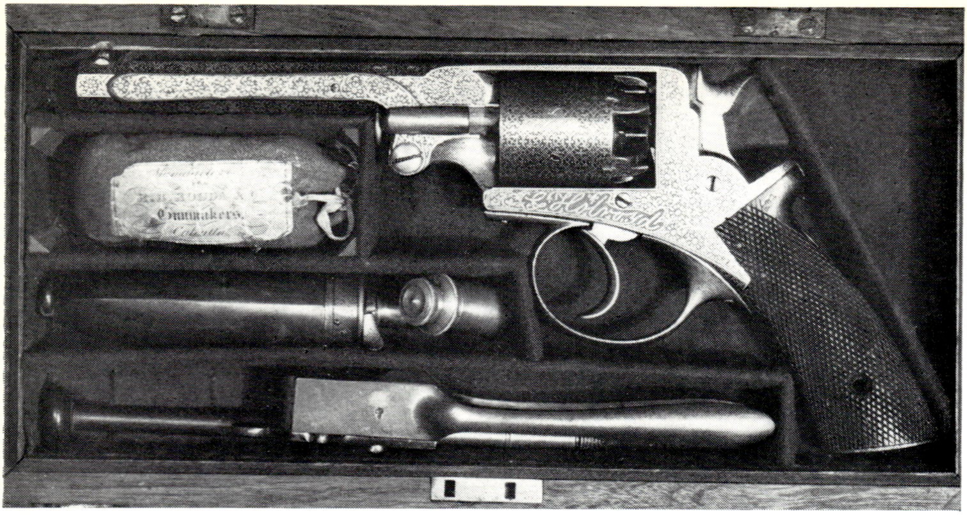

PLATE 14. DECORATIVE BEAUMONT-ADAMS REVOLVERS

(*Top*) Gold-damascened double-action Beaumont-Adams percussion revolver, in 54 bore calibre. Originally a British "War Department" issue service revolver (No. B.7497. 23,143R), made by the London Armoury Company Ltd., and probably decorated in India. *Private Collection.*

(*Lower*) Fully gold-plated and presentation engraved double-action Beaumont-Adams percussion revolver, in 54 bore calibre, with carved ivory butt. Retailed by E. M. Reilly, London, but the serial-number (41036) suggests that it was made by one of the Birmingham licensees under Robert Adams' patent of 1851. *Andrews Collection.*

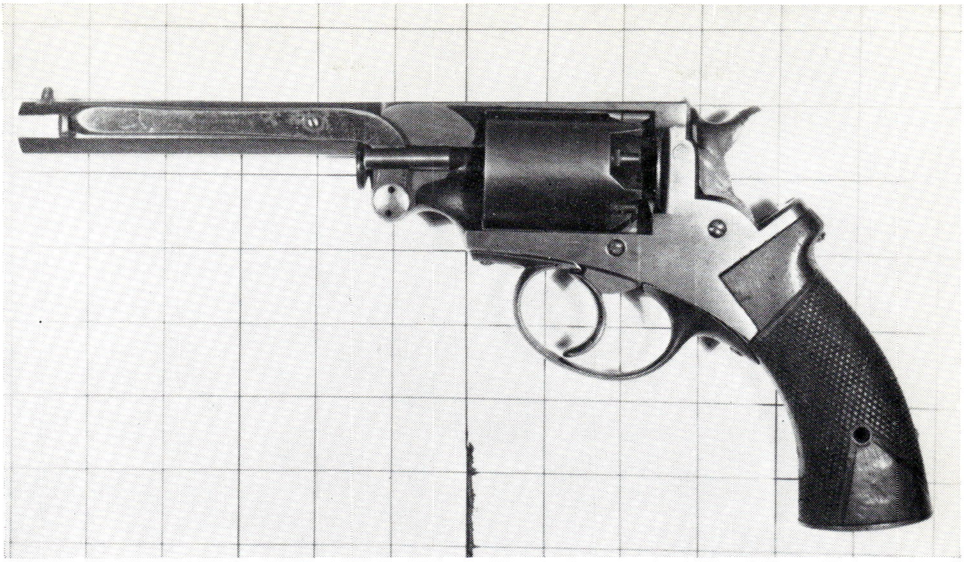

Plate 15. American-made Beaumont-Adams Revolvers
(*Top*) ·36 calibre "Belt" revolver, made by the Massachusetts Arms Company, under the U.S. patents of Adams, Beaumont and Kerr. Serial No. 12. *Roger C. Peterson Collection.*

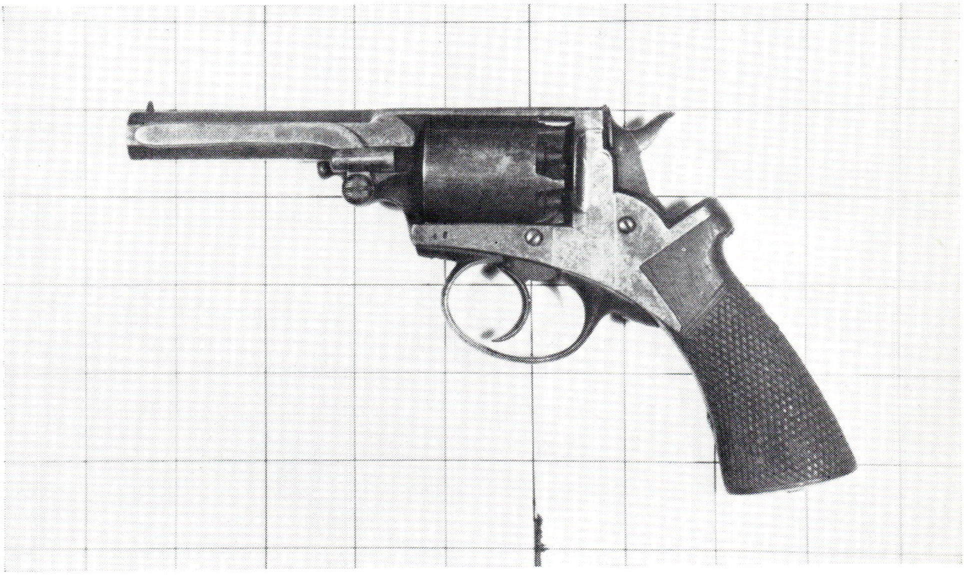

(*Lower*) ·31 calibre "Pocket" revolver, made for the "Adams Revolving Arms Company", by the Massachusetts Arms Company, under the U.S. patents of Adams, Beaumont and Kerr. Serial No. 40. *Roger C. Peterson Collection.*

PLATE 16. COLT PERCUSSION REVOLVERS
(*Top*) Five-chambered, single-action Colt percussion revolver, Model of 1849, in ·31 calibre. Made in the London factory, and serially-numbered 7937.
Tower Collection.
Crown Copyright. By permission of the Master of the Armouries.

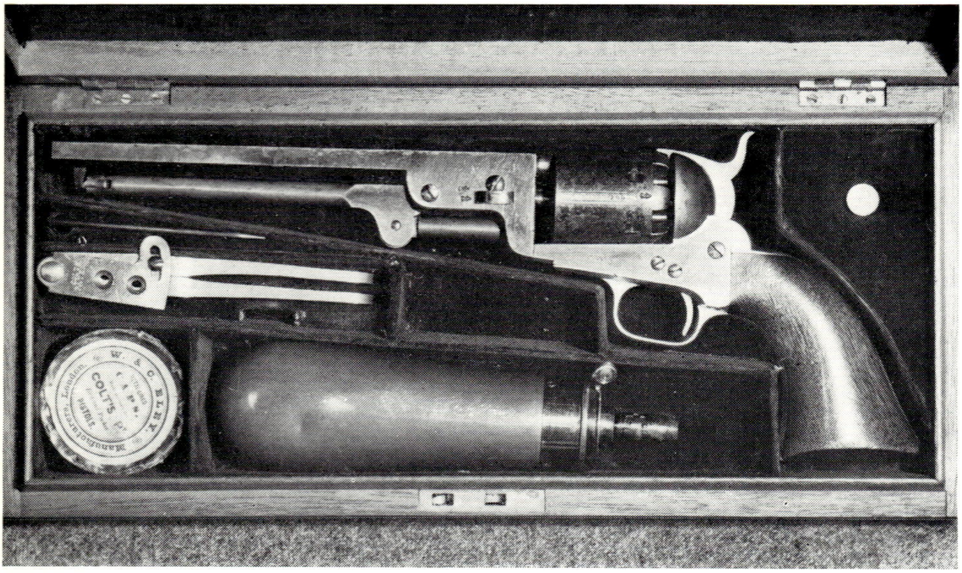

(*Lower*) Cased six-chambered, single-action Colt percussion revolver, Model of 1851, in ·36 calibre. Made in the London factory, and serially-numbered 199. Known also as the "Navy Model". *Private Collection.*

& Deane", at their 30 King William Street address. 120-bore arms are also known to exist, but no specimen could be traced for detailed examination.

14031R	54 bore
14118R	54 bore
14158R	54 bore

It is noteworthy that very few self-cocking revolvers were traced in this 14000R serial-range, and that the same serial-range was used for some of the first Beaumont-Adams double-action revolvers. These latter arms are dealt with in Chapter VIII, and there is no need to describe them here; it is sufficient to point out that early Beaumont-Adams revolvers with serial numbers in the 14000R, 15000R and 16000R ranges were made, and that Adams's new solid-frame can thus be seen as used in both self-cocking and double-action revolvers during and after 1855.

15274R (see Plate 8; *lower*)	54 bore
15438R	38 bore
15627R	54 bore
15763R	54 bore

Pistol No. 15627R was the first reported as having a sliding safety-bolt, in addition to Adams's old spring-safety; one weapon in a previous series (No. 13450R) had been reported as having a "screw-in" safety-catch, but no specimen was traced for examination.

Adams's original thumb-seated bullet was never satisfactory for use in a weapon likely to be jarred about in its holster, and his spring-safety (though adequate in careful hands) was similarly unsuitable for active service. Both defects were now remedied.

A simple one-piece lever-rammer was hinged upon the left side of the pistol-frame. Commonly, such a rammer is now called "Rigby's rammer", and it is a fact that John Rigby (of London and Dublin) secured British Patent No. 1976, of September 11th, 1854, which covered a similar rammer amongst various inventions; the device was simple, and appears as Fig. 11. Rigby filed no drawings with the Provisional Specification to his patent, but when (on March 10th, 1855) he filed an illustrated Complete Specification, the drawing clearly showed that Adams's self-cocking revolver was intended to be the vehicle for this rammer; it is plain, however, that a revolver without the new 1855 frame was used in preparing that drawing.

In truth, John Rigby's rammer was rarely used for Adams's revolvers (although later copied in Belgium), and a device patented by Robert Adams himself was normally fitted; see Fig. 12.

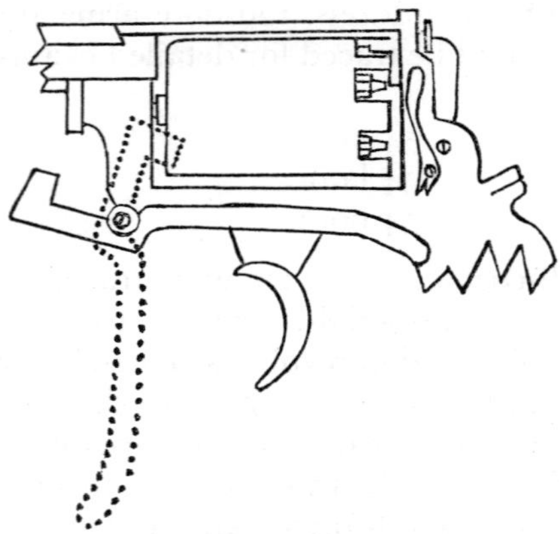

FIG. 11. *John Rigby's rammer.*
(Br. Pat. No. 1976/1854)
Note that the travel of the lever, for commencement of ramming, is only through 90°; compare with Adams's rammer, in Fig. 12.

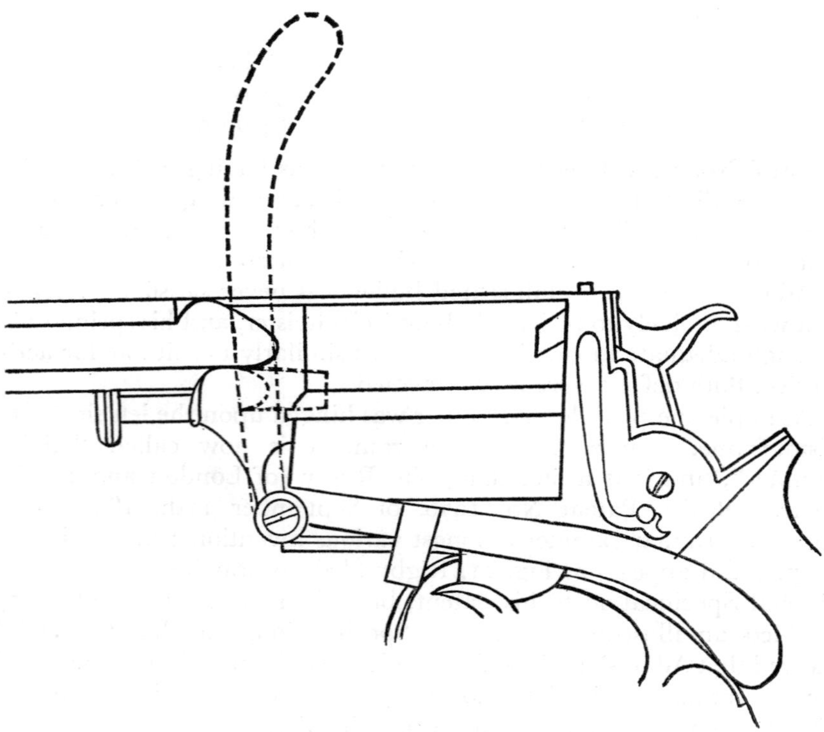

FIG. 12. *Robert Adams's rammer.*
(Br. Pat. No. 2645/1854)
Note that the travel of Adams's lever, for commencement of ramming, is through nearly 270°; compare with Rigby's rammer, in Fig. 11.

The rammer shown in Fig. 12 is almost invariably described as "Rigby's rammer", by modern collectors, despite its parentage. In defence of the practice, however, it must be noted that Adams's rammers will be encountered with the words "Rigby's Patent" actually stamped upon them, in a small oval.

Adams's rammer was protected by Br. Pat. No. 2645, of December 15th, 1854, and is generally found upon improved 1854 frames, and in a serial-range 15000R–19000R; this was a range that embraced both self-cocking Adams's and double-action Beaumont-Adams pistols. Due to the ease with which it could be mounted, however, the rammer was also added to weapons not originally fitted with it; self-cocking revolvers Nos. 8608R and 12152R were in this category, and other such weapons exist.

In addition, the patent also protected the sliding safety-bolt already mentioned, and an improved arbor-latch. The bolt (sliding in a groove on the right rear of the pistol-frame) could be engaged with notches in the rear of the cylinder, and locked that component so that capped nipples did not rest beneath the hammer; the latch was a captive screw, that could only be turned out far enough to free the arbor for withdrawal. Adams's old spring safety-catch was retained in these new self-cocking arms, and was also used (with rammer, bolt and latch from the 1854 patent) on those first Beaumont-Adams double-action revolvers described in Chapter VIII; see Plate 10.

In concluding our notes upon the serial numbers used for Adams's self-cocking revolvers, little further assistance can be offered to the reader, as to the course of events. Only one pistol was reported (as No. 16915R) for the 16000R serial-range, against six early Beaumont-Adams double-action revolvers in the same group.

No self-cocking pistol was reported or traced in either the 17000R or 18000R serial-ranges, although a 120-bore pistol (marked as retailed by Deane, Adams & Deane) was reported to have serial number 18730T. "T"-suffixed serial-ranges were used by William Tranter, upon two types of revolver patented by him which also used Adams's patented frame; this particular pistol may therefore have been either misreported, or so corroded that the serial-suffix was misread.

Serial numbers in the 19000R range were only found upon self-cocking revolvers made by the "London Armoury Company (Ltd.)" which succeeded Deane, Adams & Deane's partnership; see Chapters IX and X.

In view of the scarcity of Adams's self-cocking revolvers with serial numbers above 15000R, it seems certain that Deane, Adams & Deane ceased production of this weapon, altogether, either late in the 16000R series, or early in a 17000R range; thereafter (it appears from names engraved on the revolvers) manufacture of Adams's patented frame was in the hands of successor firms.

In addition to the "R"-suffixed serial-ranges, however, three other blocks of serial numbers were also used, and these had "Y"-, "B"-, and "T"-suffixes respectively. Their dating is still uncertain, but it is believed that the "Y" series were begun during 1854, that "B"-suffixed serial numbers were in use before May, 1854, and that "T"-suffixed numbers were briefly used by William Tranter, of Birmingham, around 1857 or 1858.

It appears, therefore, that Deane, Adams & Deane not only licensed other firms to make Adams's patented frame for locking and stocking in their own London factory, but also licensed two manufacturers to make complete self-cocking revolvers, and allocated special blocks of serial numbers to them. Presumably, differing scales of royalty payment were applicable, as between mere frames and complete pistols.

The "Y"-suffixed numbers were only encountered in a 20000Y range, and that letter-suffix and serial-range were also used by William Tranter in the first series of solid-framed double-trigger revolvers made under his own patent of 1853, and described in Chapter XVIII. Since Adams's patented frame was also used for the double-trigger weapons, this "Y"-suffixed serial-range was allocated to Tranter, for the control of royalties payable by him, and he also used it upon a few self-cocking Adams's revolvers made by him. Five of these weapons were traced:

1. 20229Y was a 54-bore pistol, retailed by W. & J. Rigby, of Dublin, and embodying a W. Tranter rammer (which the latter had patented in December, 1853) stamped with the bore size of the weapon.
2. 20254Y, another 54-bore pistol, was retailed by B. Cogswell, London, and marked "Adams' Patent, London"; it also had the Tranter rammer.
3. 20426Y was a 56-bore revolver, retailed by Allport, of Cork.
4. 20588Y, again in 54 bore, also had a Tranter rammer, and had been retailed by B. Cogswell.
5. 20750Y was a 54-bore pistol, retailed by Blanch & Son, of London; this frame bore the words "Adams 1851 Patent", one above the other, in an oval stamp.

It is not known why Tranter was licensed to make complete Adams's self-cocking revolvers (as opposed to the barrel/frame component only), or why he chose to supply these retailers with it, in competition with double-trigger revolvers protected by his own Br. Pat. No. 212/1853.

The "B"-suffixed serial-range for Adams's self-cocking revolver began at about 30000B, and was allocated to a family firm, Joseph Brazier & Son, of Wolverhampton. "J. & R. Brazier" (of the Ashes, Wolverhampton) had exhibited at the 1851 Exhibition, and were makers of the finest gun-locks; they employed some of the best lock-part forgers and die- or swage-makers in the country, and had been prize-medallists at the 1851 Exhibition. The Brazier name was not generally associated with revolver manufacture during the early 1850's, however, and Joseph Brazier appears to have been the first member of this family to engage in it.

How, or why, the firm became licensees under Adams's 1851 patent is not known, but Brazier apparently made both the patented frame and complete self-cocking revolvers of the 1851 pattern.

30245B was the lowest serial-number traced for this series, and was used upon a 54-bore self-cocking revolver, with a 6½-in. barrel which bore the name "Joseph Brazier & Son, Wolverhampton". Almost certainly, this 54-bore model was the version of Adams's self-cocking revolver most commonly made by Brazier, and weapons Nos. 30245B, 30327B, 30552B, 30839B, and 30849B were all recorded in that calibre. Two 120-bore pistols were also reported, as Nos. 30422B and 30441B, but this variant is undoubtedly scarce. These Brazier weapons were sometimes used as a vehicle for the patented rammer shown in Fig. 15; this was not patented by Joseph Brazier until April 5th, 1855 (as Br. Pat. No. 760) and is dealt with in Chapter VIII, as an adjunct to the early Beaumont-Adams revolvers upon which it more usually appeared. However, rammers numbered "291", "314" and "319" were reported, respectively, upon the first 120-bore pistol and the last two 54-bore weapons mentioned above.

Amongst the 30000B serial-range, two "R"-suffixed numbers were found, upon 54-bore revolvers that were not fitted with Brazier's rammer, but had certainly been made by Joseph Brazier. Revolver No. 30669R had been retailed by Deane, Adams & Deane, and 30703R had been sold by George Gibbs, of Bristol; both barrels were a scant 6½ in. long, and rifled with three grooves in a right twist. A top-strap legend upon the first weapon read "Deane, Adams & Deane, London", instead of the familiar "Deane, Adams & Deane, Makers to H.R.H. Prince Albert, No. 30 King William St., London Bridge".

Careful examination of both pistols was necessary to reveal their association with Joseph Brazier, but despite the use of an "R"-suffixed serial number upon them, both were stamped (in small letters, on the underside of the frames) with the legend "I. Brazier". Another 54-bore pistol was reported in this high "R"-suffixed serial-range, but was not examined; this 30604R weapon was catalogued as retailed by Dickson, of Edinburgh, but as having "Rigby's rammer". It would be interesting to learn whether, or not, the minute "I. Brazier" legend appeared upon it.

As it happens, a small contemporary record of Brazier's activities in manufacturing Adams's self-cocking revolver does still exist. Richard Prosser, C.E., was employed by the British Patent Office, and died on May 1st, 1854, whilst preparing an introduction to a new edition of the *Abridgement of Specifications relating to Firearms*. His papers are still preserved in the Patent Office Library (Ref. PU20*/15217), and they include a manuscript note, which reads:

"Deane's Pistols Adams' patent	Large & Middle ✓ size *5 shot 5 shot*	Small size *5 shot*	
	£9–14/– each	£10–10/– each	25%– { fitted in cases complete with implements"
	£7–5–6 net	£ 7–17–6 net	

"Adams' patent made by Brazier } –	plain pistols –	no cases 90/–	in cases 105/– each net
	best ,,	110/–	126/– each"

"Colt's patent	1	2 ✓	3	4	5
	6 shot	6 shot	5 shot	5 shot	5 shot
	£9	£8–10	£6	£5–10	£5 in cases complete
	£6	£5–10	£4–10	£4–5	£4 without cases but with bullet moulds"

"The two pistols marked ✓ are those you had – Deane's case has much more in it – as wads, caps etc."

"20% for £100 worth
15% ,, a less sum
– since has allowed
25% for any quantity"

This note is unfortunately not dated. However, apart from the interesting price details that it records, the document clearly establishes Joseph Brazier as a manufacturer of Adams's self-cocking revolver, and provides the strongest inference that such a relationship existed before May, 1854. It is clear, also that Brazier's prices were well below those charged for "Deane's Pistols", and that he did not normally offer more than one size of revolver.

To conclude these notes upon licensed manufacture of Adams's self-cocking design, something can be added upon the use of his ideas overseas.

No Empire manufacturing capacity has ever been alleged to have existed, and although it would have been theoretically possible for overseas buyers to import only the basic frame (for locking and stocking by their own workshops), there is no evidence to suggest that this course was adopted either. A few of the larger merchant-houses in London, Liverpool and Birmingham did export Adams's revolvers, and sometimes claimed that the arms were specially made up for them; pistols are also occasionally found with local-agency claims engraved on the barrel, together with addresses in the larger Colonial cities. An example of the former category had a case-lid label shown in Plate 9; the firm in question had one of its principal outlets in Australia.

It is certain, however, that manufacture was arranged at Liège, and by licence under a Belgian patent; the date of commencement for this arrangement is uncertain, and the serial-numbering of Belgian-made revolvers does not suggest block-allocations of numbers to various licensees, although at least three manufacturers were involved.

Only 54- and 120-bore pistols appear to have been made, and the dimensions varied slightly between weapons; the 54-bore revolvers measured 11½ in to 12 in. overall (with barrels between 6 in. and 6½ in. long), whilst the smaller calibres measured 8 in. to 8½ in. overall, and had barrels 3¾ in. to 4¼ in. long. Even similar pistols made by one manufacturer varied in dimension.

Serial numbers had no letter-suffixes in the British style, and were principally below the 10000 level; one pistol by A. Francotte was reported at serial number 72436, but was so far above the last number traced below it (No. 10163), that it may have been misread.

Normally, each top frame-strap carried the maker's name, together with an acknowledgement to the inventor and his associates; a typical legend read "A. Francotte & Cie, Liège. Licensed by Deane, Adams & Deane, London". Less commonly, the frame was also marked "Adams Patent 1851", and on one occasion "W. Tranters Patent" was also present; Liège proof-marks were always in evidence, and the pistols were sometimes sold by Deane, Adams & Deane, with a case-lid label acknowledging their Belgian manufacturer. The licensees so far noted were Auguste Francotte, C. Dandoy, and Pirlot Frères, all of Liège; exceptional weapons, marked "Zaoue, Marseilles" and "LePage Frères a Paris", were probably made in Belgium for their French retailers. Adams held a French Patent (No. 12247 of November 20th, 1851), but there is no evidence that he actually worked it in France.

In a number of cases, amongst our small sample, lever-rammers were fitted to the revolvers, and a surprising proportion of these arms had gold-inlaid decorations, and ivory butts. At a casual glance, where no rammer was fitted, the weapons could readily be mistaken for a British-made specimen, so closely were they modelled upon Deane,

Adams & Deane's product; in fact, however, the trigger guards were more rounded, and the triggers less concave than those fitted to the British arms.

One Continental manufacturer produced an interesting variant of Adams's self-cocking design, discarding the patented one-piece frame, but embodying a useful and ingenious lock-modification.

It will be appreciated that Adams's cylinder bolting-lump on the trigger, which engaged a nipple-partition to index the cylinder for discharge, restricted his weapons to the use of an odd number of chambers; considerations of weight, against the need for a reasonably large calibre, meant that a five-chambered revolver was normally made under Adams's 1851 patent. In a Belgian weapon, now to be described, the bolting-lump was eliminated, and an extra chamber provided as a result of that elimination.

The pistol was made by Herman of Liège, and is shown in Plate 9; this weapon is contained in a mahogany case, together with one single-shot pistol, and a plate inlet into the case-lid was marked "Confidential". Bearing both Liège and London proof-marks, the revolver had Herman's name on the frame, but was retailed by T. E. Mortimer, of London, whose name appeared on the barrel. The lock mechanism is sketched below, as Fig. 13.

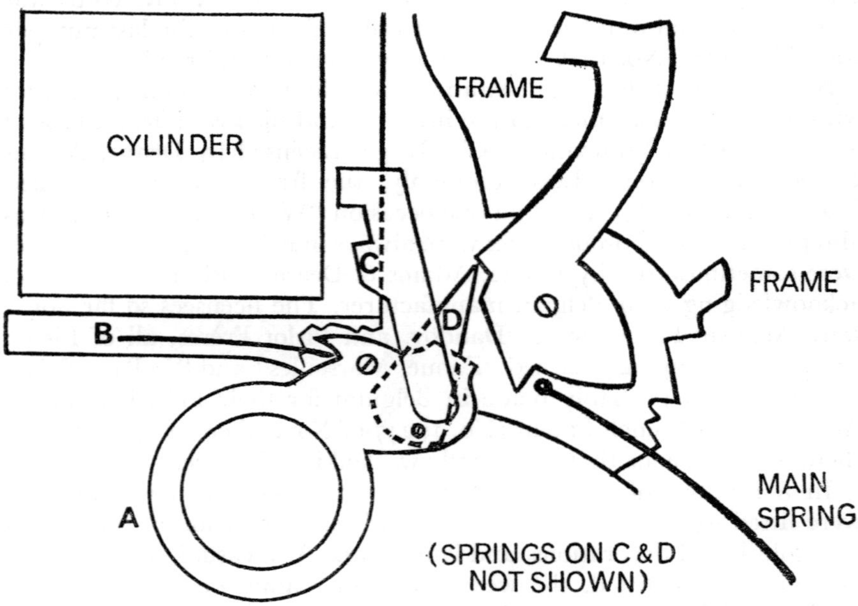

FIG. 13. *Herman's lock mechanism.*

The ring trigger A is situated in a slot in the base of the revolver-frame, and retained in position by a screw passing through both. As the index finger pulls back the trigger against spring (B), the pawl (C), which is pivoted to the outside of the ring trigger, rotates the cylinder by means of its top projection; this engages with a ratchet on the rear of the cylinder and, as the trigger moves backwards, it pushes the cylinder round until a nipple is almost opposite the hammer nose. Further slight backward movement of the trigger brings a nipple into line with the hammer nose, and at that point the lower projection on pawl (C) also engages with another spur on the cylinder-rotation ratchet. Thus the need for Adams's bolting-lump (on the trigger) is eliminated, and an effective cylinder-lock provided. In the meantime the hammer has been raised by this backward movement of the trigger, due to the lifter (D) which is pivoted internally in the trigger; this has a small spring on its back, which presses against the internal surface of the ring-trigger and causes the lifter to engage in a bent on the hammer. As the trigger moves backwards, the lifter pushes the hammer backwards, but there is a projection on the hammer which (when the hammer has reached its point of maximum backward travel) engages against the lifter and pushes it forward out of engagement with the hammer bent. The hammer then falls, under pressure of the mainspring, to strike a nipple.

The 6½-in. barrel of Herman's revolver was rifled with eight grooves, and screwed on to the forward end of the arbor; a retaining screw was turned through the front of the barrel-lug, and into the front of the frame, to index the barrel like that on a transition-type revolver. The rammer was of a two-piece type, similar to the Kerr rammer shown on Plate 14, and was retained by a screw threaded sideways through the barrel-lug; on the specimen illustrated in Plate 9, this retaining screw has been sheared off, and the rammer lost.

CHAPTER VIII

EXIT "DEANE, ADAMS & DEANE"

ROBERT ADAMS'S partnership with John Deane and his son was to be dissolved in August, 1856, shortly after the introduction of a double-action revolver that was based upon his own patented frame and self-cocking lock mechanism. This new weapon is now called a "Beaumont-Adams" revolver, in recognition of the two men who conceived it, and was to be seriously developed only after the partnership had terminated. Clearly, however, the introduction of a new revolver had to embrace a number of interesting events in the partnership history, and this chapter is therefore devoted to the first arms; later versions of this Beaumont-Adams pistol are dealt with in Chapter X.

Chapter VII traced a history of Adams's self-cocking revolver, between the introduction of a 120-bore series (in 1853) and a likely cesser of production at the 16000R serial-range; introduction of a new frame-shape was also noted, and 1854 was suggested as the year in which that refinement (later with three improvements of detail) was probably applied to Adams's pistol.

The preceding year had obviously been materially successful for Deane, Adams & Deane; in recollection, however, it must also have been recognized as a year of great disappointment in one important aspect, for it was then that the sharpest division had existed between an undoubted commercial success, and a real failure to "sell" their weapon as a Service side-arm.

In 1853, Europe had seen the seizure of Moldavia and Wallachia by Russia from Turkey, British naval operations in the Baltic, and destruction of a Turkish fleet at Sinnope. 1854 saw the drift ended, Great Britain and France at war with Russia, and in September the landing of Allied forces to the north of the Crimean naval base at Sevastopol, which were intended to invest the port, and destroy a Russian fleet bottled up in the harbour by an Allied naval blockade. Many serving officers chose to buy Adams's self-cocking revolver for their personal side-arm, and the various U.K.-based Volunteer Associations and units were also active in 1854 as a result of the war; when it is further realized that more than 83,000 Britons emigrated to Australia and New Zealand alone, in that year, and accepted that many must have taken an Adams's revolver with them, a substantial demand for the pistol seems established.

Yet, the sittings of a Parliamentary Committee considering "the cheapest, most expeditious and most Efficient Mode of providing Small Arms for Her Majesty's Service" ended early in 1854, and brought no promise of Service purchases to the partnership; the Committee's report appeared in March, and contained a good deal of welcome publicity for Adams's revolver, which is detailed in Chapter XI. However, the end of 1854 saw no formal Service adoption of any revolver, to replace the various single-shot pistols then issued. In every case during that year, where the British armed forces had seen fit to make special purchases of such arms, they were needed in limited quantities only, and it was the Colt Company which filled the order (*vide* Blackmore, *op. cit.* p. 254).

Admittedly, any gap between the two competitors, in their race for a commanding position should Service adoption of the revolver be decreed, had closed considerably by the end of 1854. Colonel Colt had a factory in production during 1853, with which the Deane-Adams partnership could not have hoped to compete seriously, if formal adoption had occurred in that year. However, by July, 1854, the partnership had a viable revolver-manufactory by British standards, with trade suppliers to assist them. They could have faced the Colt Company on something approaching equal terms, if the Services had reached any decision on adoption, but still no ruling came.

In the light of subsequent events, it seems probable that the partners began to consider production of a double-action revolver, as soon as their Weston Street factory was in satisfactory operation; it is not clear whether the new 1854 frame was introduced as part of their experiments in this direction, or whether they would have launched it for self-cocking arms, using Adams's patented improvements of 1854, in any event.

Indeed, it must have been clear to the partners, at the filing date of Adams's patent, that only a revolver capable of being cocked by the thumb would ever be accepted by the British Army. Self-cocking arms were useful in a mêlèe, but Colt's single-action revolver obviously enjoyed greater military confidence than Adams's design; moreover, the setting of a spring-safety, by using trigger-pressure to raise the hammer for it, was a dangerous exercise for the clumsy, and made Adams's revolver particularly undesirable for use on horse-back.

It is curious that Deane, Adams & Deane made no attempt to overcome this particular defect in their self-cocking revolver, during the years when they really were committed to that design. The Complete Specification to Adams's English Patent No. 13527 of 1851, had actually shown a picayune hammer-spur, suitable for easing back the hammer when engaging his spring-safety, but that lead was never followed in production versions of his revolver. At least one self-cocking pistol survives, which has had a spur grafted onto the hammer for that very purpose, but it seems likely that this was added at the instance of a former owner, and not by the designer.

However, on February 20th, 1855, Lieutenant Frederick Blackett Edward Beaumont, R.E. (of Upper Woodball, Barnsley, Yorks) filed patent-application for a double-action lock-mechanism that was actually based upon Adams's self-cocking design. The application eventually became British Patent No. 374 of that date, to protect a weapon free of those defects which had barred Adams's revolver from official issue to the armed forces, and which also embodied all three of the detailed improvements from Adams's own patent of 1854.

Since neither Adams nor Beaumont filed drawings with the Provisional Specifications to their patents, when these were respectively filed on December 15th, 1854, and February 20th, 1855, no evidence exists as to the revolvers then envisaged for using their inventions. In view of the speed with which "Beaumont-Adams" revolvers became available for test, it is assumed that Deane, Adams & Deane provided facilities and revolver-frames for the development work, but (as later shown) it is conceivable that some of the preliminary work was carried out by another manufacturer.

When, however, Adams and Beaumont filed Complete Specifications to their patents, the drawings made it abundantly clear that both men were developing one weapon. The Complete Specification to Adams's British Patent No. 2645/1854 was filed on June 15th, 1855, whilst Beaumont achieved the same step, for his British Patent No. 374/1855, on July 19th; both sets of drawings showed the same pistol, and establish a common interest in it as existing by June, 1855, at the latest.

It is necessary to labour this point, because a small group of revolvers (later discussed) were apparently made, with Adams's frame and Beaumont's lock-mechanism, by Joseph Brazier of Wolverhampton. Since the Provisional Specifications to Adams's and Beaumont's patents were unillustrated, there remains the possibility that Lieutenant Beaumont flirted, briefly, with another possible developer of his pistol.

In the terms of Beaumont's Complete Specification, however, no doubt can exist of a firm association between the inventor and Deane,

Adams & Deane, by July, 1855. Revolver-locks "now manufactured by Messrs. Deane, Adams & Co.", with the cylinder-rotation system "heretofore practised" by that firm, were expressly recognized, and Beaumont really claimed only the combining of hammer and trigger by adapting Adams's long sear, a connecting hook on the hammer-breast, and the famous "short sear" through which thumb-cocking was achieved. He did not claim the self-cocking system used in Beaumont-Adams revolvers, nor any general arrangement of revolver lock-mechanisms to permit double-action operation.

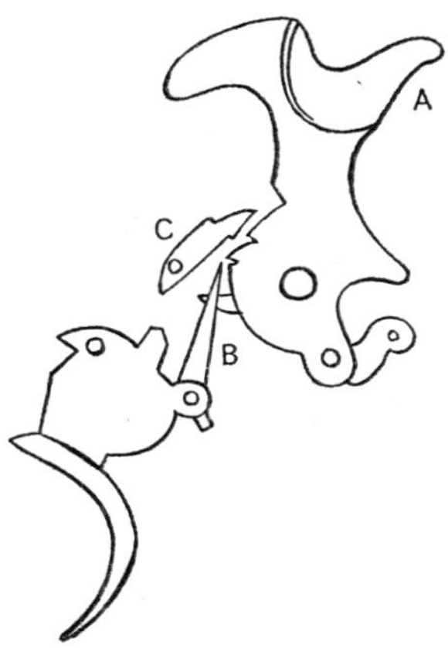

FIG. 14. *Beaumont's lock mechanism.*
(Br. Pat. No. 374/1855)
The Figure shows Beaumont's lock-mechanism as patented: compare with the later production-version in Fig. 18. (Pawl omitted, for clarity.)

The lock mechanism actually patented by Lieutenant Beaumont is shown in Fig. 14, and should be compared with Adams's self-cocking design (Fig. 10), and with the version more generally produced (Fig. 18); it can be seen that all basic components of Adams's lock were retained, but ingeniously modified to permit thumb-cocking.

By reference to Fig. 14, it may be seen that when the hammer (A) is thumb-cocked, the short sear (C) takes into a bent in the hammer to hold it at full-cock; Adams's old long sear (B) now has a vertical slot machined through it (which engages a hook formed on the hammer-breast), and has been pulled up close to (C) by the act of cocking. On pressing the trigger, (B) pushes (C) out of engagement with its bent, and the hammer falls. Self-cocking operation remains as described in relation to Fig. 10, the long sear (B) ensuring that (C) does not foul or catch the hammer, in self-cocking travel.

Lieutenant Beaumont's design was so well-developed by February 1855 (the date of filing his patent-application), that H.M. Board of Ordnance ordered 100 of his pistols for test, in the following March (Blackmore, *op. cit.* p. 255). This was a remarkable act, for the patent was not even sealed until May 11th, and no Complete Specification was filed until July 19th; on that date, however, Lieutenant Beaumont assigned his patent to Robert Adams personally, subject to the payment of a royalty for every pistol made under it.

H. L. Blackmore (*op. cit.* p. 255) recorded a further Board of Ordnance order, for 2,000 revolvers, later in 1855. This must have given great satisfaction to the partners (who had laboured since 1851 to secure such a substantial recognition for Adams's design), but that event almost certainly sowed the seeds of dissolution within the Deane, Adams & Deane partnership. As Chapter IX will show, men of greater substance became interested in Adams's methods for manufacturing arms, and his association with his partners ended in the following year.

To supply Beaumont's design for the Board of Ordnance, and for any private buyers, a pistol shown in Plate 10 had been introduced; it can be seen from the Plate, that Adams's new 1854 frame was used, together with the rammer, safety-bolt and arbor-latch from his 1854 patent, but that the old spring-safety was still retained. Presumably, this latter fitting was regarded as less objectionable when used with a thumb-spurred hammer.

As a result of Beaumont's act in assigning his patent to Adams, however, a change in serial-markings emerged with this new weapon. Adams's "R"-suffixed serial-range continued to control royalties payable under the Adams/Hankey licence of 1852, but a "B"-prefixed number was now also engraved upon the pistol frame, immediately above the "Adams's Patent No. . . . R" legend. This new number, of course, controlled payment of royalties to Beaumont, under the assignment of his patent to Adams.

A short list of early serial numbers is set out below, and shows that some confusion attended introduction of the new "B"-prefixed serial numbers, upon the first pistols. These weapons are sufficiently rare for difficulty to be experienced in compiling a list even as short as this, but it became clear, by actual examination, that a "B"-prefixed number was not always engraved upon the earliest of them.

Number		*Bore*	*Remarks*
B.241	15917(R)	54	Adams's rammer
	16082R	54	Adams's rammer

Number		*Bore*	*Remarks*
B.651	16457R	38	Marked "WD", with broad arrow
	16636R	54	Adams's rammer
B.727	16691R	38	Adams's rammer
B.1145	14447R	54	Adams's rammer
B.1272	14576R	38 (see Plate 13, *top*)	Kerr's rammer
B.1629	14950R	54 (see Plate 10)	Adams's rammer
	15426R	54	—

(N.B. All of these weapons bore the name of "Deane, Adams & Deane", save B.1272 14576R. This latter had been retailed by "Adams & Co."; see p. 120 *post.*)

It will also be noticed, from this list, that the "B"-prefixed numbers ran consecutively from a 200 series, and (so far as can be established) they continued to do this, for as long as a Beaumont serial number was used at all. These numbers are further considered in Chapter X, and it is enough to note here that the "London Armoury Company" later discarded their use at the close of a 22000 numbering.

The list also reveals, however, that Adams's "R"-suffixed serial numbers, for these early Beaumont-Adams revolvers, fell back to a 14000R series, after beginning at the 16000R level; after this set-back, however, they continued to mount consecutively until (as shown in Chapter X) their ultimate disappearance at a 41000 range.

No reason can yet be offered for this curiosity of Adams's serial-numbering amongst the first of these Beaumont-Adams revolvers; self-cocking revolvers in the 16000R serial-range are almost unknown, and it can be assumed that nearly all the frames in that range were set aside for the new pistol. However, frames in the 14000R range were certainly used for self-cocking revolvers (as shown in the previous Chapter), and this was also the series at which Adams's new improved frame appeared. There seems no logical reason why the "B"-prefixed Beaumont numbers should have been started with the 16000R Adams's range, if 14000R frames were still available for use with Beaumont's lock-mechanism.

It is probable, however, that an impending break-up of Deane, Adams & Deane's partnership had some bearing on the matter. Robert Adams began to form his new "London Armoury Company (Ltd.)" during February 1856 (see Chapter IX), and formal dissolution of the partnership occurred on August 30th, of that year; in the months immediately prior to this latter event, it may be that production of the revolvers was split between two or more premises, and that confusion occurred, as a result, in numbering the various frames.

Formal announcement of partnership dissolution appeared in *The London Gazette* (p. 3019), on September 5th, 1856:

> "Notice is hereby given that the Partnership heretofore subsisting between us the undersigned, Robert Adams, John Deane the elder and John Deane the younger, heretofore carrying on business at No. 30 King William-street, in the City of London and Weston-street, in the Borough of Southwark, as Gun Manufacturers and Dealers in Fire Arms has been mutually dissolved—Dated the 30th day of August, 1856."
>
> "John Deane"
> "Robert Adams"
> "John Deane, jun."

At that date, only two models of the Beaumont-Adams revolver were in production by the partnership. 54-bore pistols (as shown in Plate 10) had a 6-in. barrel and an overall length of about 12 in.; the larger 38-bore model had a $7\frac{1}{2}$-in. barrel, and an overall length of 14 in. Either version may be encountered with the War Department mark shown in Plate 17 (*top*), and Adams's rammer was usually fitted; both calibres were London-proved, and had three-groove rifling, uniformly twisted to the right. In the larger pistol, this was officially stated to achieve one turn in 18 in.

Between disappearance of "Deane, Adams & Deane", and emergence of the "London Armoury Company (Ltd.)", a few Beaumont-Adams revolvers were marked as retailed by "Adams & Co., London"; two pistols of the type shown in Plate 10 were examined, namely B.2360, 17706R (now converted from 54-bore percussion, to use ·442-in. centre-fire metallic cartridges), and B.2671, 18031R (a 54-bore "WD" marked percussion revolver), whilst a more fully-developed arm (No. B.1272 14576R), of the type shown at the top of Plate 13, was also located.

In later years, a firm trading as "Adams & Co." was run by Henry Adams in Finsbury; there is no record of such a firm prior to 1870, so that it is unlikely that early Beaumont-Adams pistols were retailed so late. It must be assumed, therefore, that Robert Adams himself used that trading-style, briefly, before the "London Armoury Company (Ltd.)" began to trade on its own account.

Although we are virtually committed to the view that Robert Adams alone assisted Lieutenant Beaumont in the successful development of this new lock-mechanism, a contrary view can be argued. There are, for example, early Beaumont-Adams revolvers quite definitely not produced by Deane, Adams & Deane, and also a literary reference that clearly suggests a different course of events.

In 1858, John Deane (*op. cit.* p. 215) wrote that Adams's self-cocking revolver had been "modified with double-action by Tranter, Birmingham; Mangeot, Brussels; Beaumont, London; Brazier, Wolverhampton", but did not enlarge upon this statement. We have no information upon Mangeot's activities, but Deane's remarks were demonstrably true, in 1858, as to Tranter and Beaumont; there seems no reason, therefore, to doubt their accuracy in regard to Brazier, who was an early licensee under Adams's patent of 1851.

It is difficult to understand why Beaumont allowed manufacture of his lock-mechanism by any manufacturer other than Robert Adams, to whom he assigned his patent, and with whom he became sufficiently friendly to act as trustee for "London Armoury Company" shares transferred to Mrs. Adams. In the case of William Tranter, Deane's reference may have been directed to British Patent No. 1913/1856 (see Chapter XVIII), but the Beaumont-Adams revolvers associated with Joseph Brazier include a type that can logically only have been made whilst Deane, Adams & Deane were still in business.

This version of the Beaumont-Adams revolver is shown in Plate 11 (*top*); it is quite rare, but so distinctive in appearance that it cannot be overlooked when encountered. Adams's old-fashioned, pre-1855, frame was used (without the new rammer, safety-bolt or arbor-latch from his 1854 patent), and the original spring-safety was discarded.

In view of the self-cocking arms made by Brazier (and discussed in Chapter VII) a "B"-suffixed 30000 serial-range might reasonably have been expected for these weapons, but only one Beaumont-Adams Revolver (No. 30479B) was actually reported in that range. Most of the pistol frames had "R"-suffixed serial numbers, but in a range well below these 30000R serials earlier noticed on two self-cocking revolvers made by Brazier. In fact, the serial numbers that we noted as used on Beaumont-Adams revolvers made by Brazier fell into a 9000R serial-range, apart from one weapon reported to be numbered 14219R.

Such a chaotic pattern of serial numbers strongly suggested that some, at least, of these revolvers had actually been converted to use Beaumont's lock-mechanism, after initial production as self-cocking arms, and the revolver shown at the top of Plate 11 seemed (in view of its serial number) to have been so converted. Adams's self-cocking frame of pre-1855 pattern had barely sufficient room within it for the single-action short sear as shown in either Figs. 14 or 18, but could be made to accommodate such a lock-component by filing a curve in the hammer-breast to accommodate the sear when the hammer was down.

Upon every such weapon encountered, a rammer sketched as Fig. 15 (and visible in Plate 11) was fitted; this had been patented by Joseph Brazier, as British Patent No. 760, of April 5th, 1855, and was also fitted, upon occasion, to those Adams's self-cocking revolvers that he made under licence.

The Provisional Specification to Brazier's patent was filed on April 5th, 1855, and was unusual in containing a drawing of his rammer; only part of the accompanying revolver was included, but when his

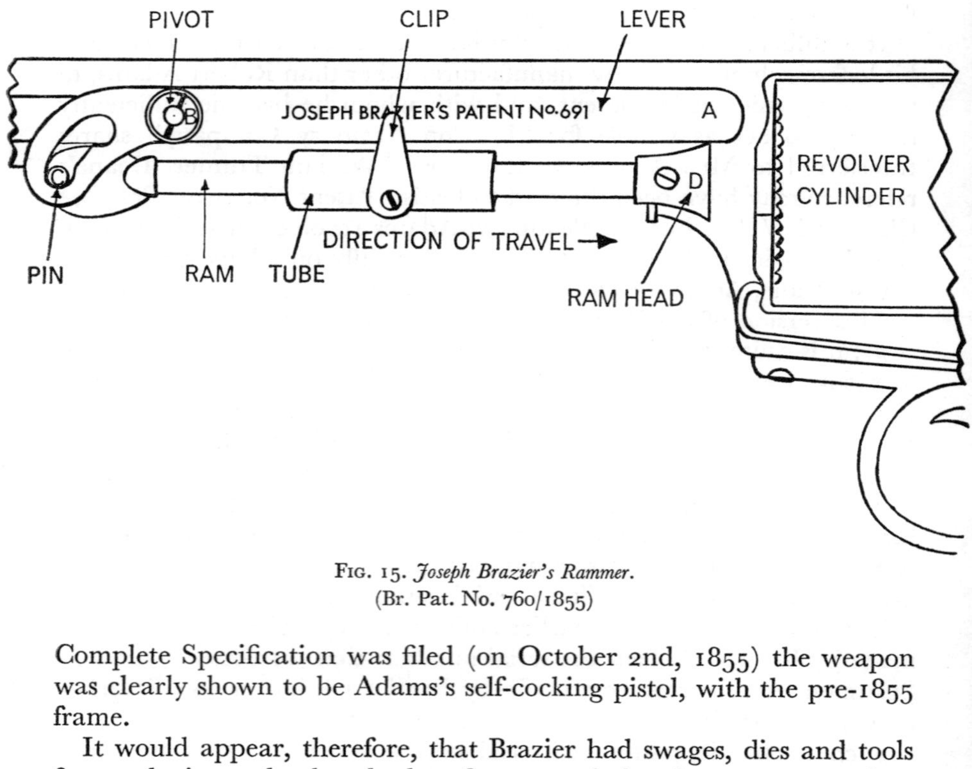

FIG. 15. *Joseph Brazier's Rammer.*
(Br. Pat. No. 760/1855)

Complete Specification was filed (on October 2nd, 1855) the weapon was clearly shown to be Adams's self-cocking pistol, with the pre-1855 frame.

It would appear, therefore, that Brazier had swages, dies and tools for producing only the obsolete frame, and that he saw no need to alter that configuration, when (by whatever means) he became entitled to manufacture Beaumont's new lock-mechanism. It is also clear, that Brazier saw no reason to ask for a licence to use the new safety-bolt and arbor-latch protected by Adams's patent of 1854, and that (like later manufacturers) he regarded Adams's old spring-safety as unnecessary with Beaumont's lock mechanism.

The appearance and operation of Brazier's rammer are sketched in

Fig. 15, and the rammer-lever invariably bore that patentee's name, together with a serial number quite separate and distinct from those familiar numbers engraved upon the frames and cylinders of revolvers to which it was fitted. Rammers numbered "291", "314", and "319" were noticed in Chapter VII, as mounted upon Brazier-made Adams's self-cocking pistols; amongst Brazier-made Beaumont-Adams revolvers were noted rammers numbered "241", "384", "441", "457", "483", "550", "602", "681" and "902". It is not clear that these numbers recorded actual production-figures for the rammer, since it was not uncommon for patentees to simulate demand for their invention, by adding an initial figure to their serial numbers.

As to the operation of this rammer, it may be seen from Fig. 15 that raising the tip (A) frees the rammer-lever from its retaining spring-clip and depresses the opposite end of that lever about the pivot (B); the limit of travel for (A) is forward (through 90°) to a roughly vertical position, and the pin (C) (trapped by a curved slot, cut in the lever) is urged towards the cylinder of the pistol, as (A) rises. The rearward movement of (C) in turn thrusts a ram into the chamber, along a horizontal line dictated by the tube brazed onto one side of the barrel, and seats the ball or bullet previously placed there, when (A), (B) and (C) are in vertical alignment. Depressing (A) withdraws the ram from the chamber (again by virtue of (C) and the curved slot), and permits the lever to be restored to its retaining spring-clip, as shown in the sketch.

Both 38- and 54-bore pistols (the former with barrels over 7 in. long, and rifled with five grooves) were encountered amongst these distinctive Brazier-made or converted Beaumont-Adams revolvers, and 80- and 120-bore arms have also been reported to us. A surprising proportion of them were retailed by Wilkinson & Son, the London swordmakers, and this firm decorated (or had decorated) some of the arms with a coarse gold "Damascene" decoration of the type shown on the lower weapon on Plate 11; however, our specimen Brazier pistol at the top of Plate 11 actually bears the address of Deane, Adams & Deane themselves, upon its barrel.

Wilkinson & Son appear to have favoured Brazier's rammer, and, at about 1858, they also used it (as shown below in Plate 11) upon later Beaumont-Adams revolvers embodying Adams's improved 1854 frame. This refinement was not difficult to accomplish, since the points of anchorage for Joseph Brazier's rammer were solely upon the pistol barrel, and it was as readily fitted to weapons originally made without rammers as to arms with rammers less powerful than the Brazier design. Adams's self-cocking revolvers have already been noticed in the first category, and these Beaumont-Adams pistols with the improved 1854 frame are found in the second; early Tranter double-trigger arms, originally fitted with detachable rammers, were also thus improved.

Apart from the presence of Brazier's rammer, these Wilkinson-retailed Beaumont-Adams revolvers with improved frames exhibited no differences from the same model as sold by Deane, Adams & Deane; all encountered were in 54 bore, and were presumably bought from the partnership by their retailers, and passed over to Brazier for the addition of his rammer. In most cases, they show no signs that they were originally fitted with Adams's rammer, so it is assumed that they left the Weston Street factory "in the white", and were finished by Brazier.

CHAPTER IX

THE LONDON ARMOURY COMPANY, LIMITED

ALTHOUGH the announcement of formal dissolution for the "Deane, Adams & Deane" partnership came in August, 1856, events leading up to it were in train well before that date, and *The Times* actually reported issue of a prospectus for the Company which replaced that partnership, on February 21st. It would serve no useful purpose to explore all the legal reasons for setting up this Company, in place of the old relationship, and it is enough to emphasize that definite practical advantages did exist in the step.

Obviously, an injection of additional capital into the partnership's arms-making activities must have seemed necessary, as H.M. Board of Ordnance moved towards the placing of orders for substantial quantities of Beaumont-Adams revolvers, and (see Chapter X) other irons may also have been warming in the fire. However, a partnership structure became increasingly difficult to operate, both safely and satisfactorily, as the number of partners increased, for "sleeping" partners (i.e. pure investors) were equally liable with those partners managing the joint business, on all debts incurred by the partnership, and to the last penny of their personal fortune if need be; in such circumstances, another source of capital had to be sought by Adams and his associates.

To obtain this additional capital, therefore, a limited liability Company was substituted for the partnership; management of day-to-day matters was thus placed in the hands of properly-qualified officers and employees of the Company, and the purely investing shareholders were protected from having to meet any losses, or liability for Company debts and contracts, in excess of the value of their paid-up shareholding. Fortunately, certain requirements of the Joint Stock Companies' Registration Act, 1845 (7 & 8 Vict. c. 110) involved registration of information as to the objects and shareholdings of this Company, and this material has survived at the Public Record Office, in a Register covering the activities of Company Number 532—the "London Armoury Company, Ltd.".

This first London Armoury Company was provisionally registered as a Joint Stock Company on February 9th, 1856, the promoter being Robert Adams, who gave his business address as 2 Weston Street, Southwark, and 30 King William Street, London (where the firm of Deane, Adams & Deane was still situated); initially, the Weston Street address was used as a registered place of business for the new Company, and provisional Directors, Officers, and Trustees, were duly appointed on April 8th, 1856.

Nominal share capital of the Company was £60,000 (in 3,000 twenty-pound shares), and by June 13th, 1856, the necessary Declaration was filed, stating that three-quarters of that nominal capital was in the hands of more than twenty-five shareholders, who had paid up at least 20 per cent of their holdings in cash. The majority of shareholdings were small, but amongst them may be noted the holdings of Robert Adams (250 shares), Frederick Edward Blackett Beaumont (25 shares), John Deane (125 shares), William Harding (5 shares), and James Kerr (25 shares), as representing the gun-trade interest, whilst men from other trades and professions, ranging between "Advertisement Contractor" and "Wheelwright", also held shares in the venture. Significantly, the occupation of "Merchant" figured prominently in that list, and it is also interesting to note that John Adams (Robert's brother) was not, and never became, a shareholder in the Company. The fact that John Deane, Senior, took shares in it two months before formal dissolution of the old Deane, Adams & Deane partnership, suggests that the latter did not originally terminate on bad terms, despite the opinion of J. N. George (*op. cit.* p. 220) to the contrary; it is as well to dispose here, also, of the legend that a "John Kerr" (as J. N. George mistakenly stated) was ever co-director of the Company with Robert Adams.

Two days before the filing of the Declaration mentioned above, and in compliance with the then required procedure for registration of a joint-stock company, Trustees for the London Armoury Company (Archibald Hamilton and John Gibson) had filed an Abstract of a Deed of Settlement, setting out the terms of formation of the Company; it is this Deed of Settlement, corresponding to a modern Company's Memorandum and Articles of Association, which shows us something of the terms upon which Robert Adams was then intended to be connected with it.

The Deed recited that Robert Adams was an established gun-manufacturer at Weston Street, Tooley Street (specializing in the manufacture of revolving pistols patented by him), that he had invented improved machinery for such manufacture and was solely and absolutely entitled to the stock-in-trade and the machinery used in that factory, and that he was engaged in completing certain "valuable" Government and other contracts for the manufacture and supply of firearms. The document further recited that the South Eastern Railway Company "now required" the Weston Street premises and we assume from this that Adams was in occupation there as lessee only, and that developments in the approach lines to what is now London Bridge Station required the eventual demolition of his factory. As a result of this, Adams had built a new factory on freehold land (which he had bought from the South Eastern Railway Company in Henry Street, Bermondsey), and was erecting "a Steam Engine and other machinery" thereon; from this wording, it is clear that the engine and machinery at Henry Street were not operational by June 11th, 1856 (when the Deed was executed), and that revolver manufacture either continued at Weston Street, *pro. tem.*, or was carried out by other gun-trade manufacturers, on Adams's behalf.

As a result of efforts by the staff of the Central Library, Metropolitan Borough of Bermondsey, we had rather greater success in establishing the position and extent of the new Henry Street factory, than we had in relation to Adams's Weston Street premises; see Chapter XIII. Henry Street still exists (although it has been renamed Holyrood Street), and is a short cul-de-sac off Bermondsey Street, which can be readily placed by any rail-passenger who cares to glance out of his carriage window, as it approaches the canopy of London Bridge Station. The buildings now labelled with the names of Messrs. Kasenit, Ltd., Messrs. Ash & Co., Ltd. and Thomas Nelson & Sons, Ltd., which can be seen from the railway, either constituted the factory erected by Robert Adams, or occupy the site of it.

In addition to his factory-building, on the rather awkward wedge-shaped site shown in Fig. 16, Adams also occupied (according to the rate records) all thirteen of the small houses shown as "Griffith's Rents" on Fig. 16, and, since the London Armoury Company was later listed for rating as occupying them, he probably also held twelve railway arches, between Arch No. 881 and No. 895, on the opposite side of Henry Street and facing his factory. Whether Adams used "Griffith's Rents" to house his workers, or used them (with the railway arches) as temporary working-accommodation pending completion of the new factory, we do not know; such explanations seem reasonable in the circumstances, but problems of access (or easements of air and light to the "Rents", threatened by his new building) may have forced him to purchase the houses in order to proceed with the main structure.

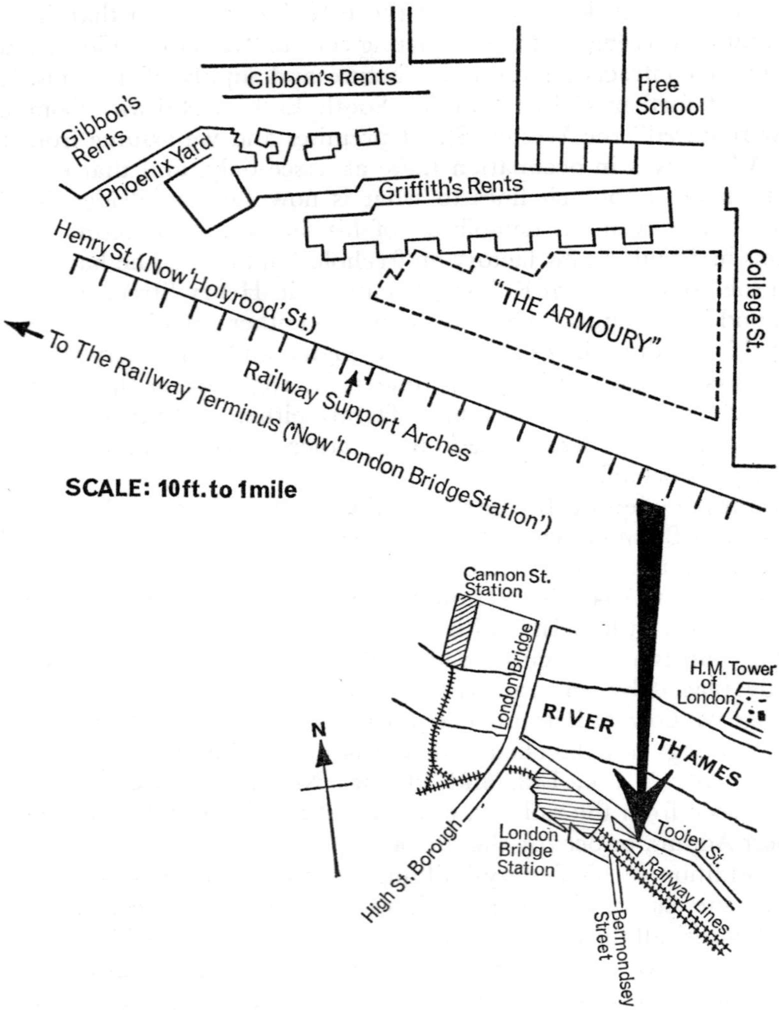

FIG. 16. "The London Armoury"

Site of the Armoury traced from London Bridge—St. Saviour's Dock—Snowsfields—Tanner St., section of the "East Division of Southwark Pavements & Improvements Plan, 1853" (courtesy of the Metropolitan Borough of Bermondsey, Public Libraries Department and reproduced by their permission. Copyright reserved), with the suggested outline of the factory building superimposed thereon in broken lines.

The heavy outline in Fig. 16 was traced directly from an 1853 "Pavements and Improvements" plan mentioned in our caption, but a map drawn in 1853 obviously could not feature the premises erected by Adams; we have therefore superimposed on that tracing the rough outline of a "Dogs Food Manufactory", which was later shown (on a smaller-scaled Ordnance Survey Map of 1872) as occupying the area labelled "Vacant Ground, the property of the South Eastern Railway Co." on the 1853 plan. This building was, to the best of our knowledge and belief, "the Armoury" built by Robert Adams and referred to in the Deed of Trust for the London Armoury Co, Ltd.

The Deed then proceeded to set out facts (to which we have earlier referred) relating to the pistol patents, namely, that Adams was not solely entitled to the famous patents for the weapons known to modern collectors as the "Adams" and the "Beaumont-Adams" percussion revolvers. By a Deed of Assignment dated June 23rd, 1852, he held $\frac{7}{8}$ of his basic patent (No. 13527 of February 24th, 1851), which covered the Adams self-cocking revolver mechanism and the vital solid barrel-frame assembly, and he was licensed to manufacture and sell under that patent by a Licence of the same date as the Assignment, whereby he paid a royalty of an undisclosed amount to Thomas Alers Hankey, who was entitled to the remaining eighth of the patent. The famous Beaumont improvement to Adams's revolver lock mechanism (No. 374 of February 20th, 1855), by which the revolver could be thumb-cocked at will, had been assigned (subject to a royalty payment) to Adams, by Beaumont, on July 19th, 1855.

Upon the complete registration of the London Armoury Company, Adams undertook to transfer most of the assets of his business to the Company, upon terms casting some light upon the value of his services and business at that time.

Firstly, the two principal patents referred to above (together, in effect, with any other of his British patents relating to firearms or their manufacture) were to be assigned to the Company, together with the Licence from T. A. Hankey, the benefit of Adams's Government and other existing contracts and the goodwill of his business, for the sum of £5,000. In return, Adams was to be paid a royalty, "not exceeding" 5*s*. each, for all firearms made under the two principal patents; this royalty was to include the royalty payable to T. A. Hankey under his Licence, and also that payable to Beaumont under the assignment of his patent, but it was to be subject to any reduction which might be agreed by the three parties (i.e. Adams, Hankey and the Company), and Adams expressly agreed to accept for his share of the royalty under the Hankey licence, the same amount that Hankey and the Company might thereafter agree upon as the sum payable to Hankey for his eighth share of the royalty. Foreign patents were not assigned, but no licences were to be granted by Adams, under them, without the agreement of the London Armoury Company as to the amount of any royalty reserved in such licences. It should be noted here, that the inventor abandoned his British Patent No. 2712 of 1853 in November, 1856, so that the Company must have expressed no interest in developing the idea covered by that patent.

Secondly, Adams was to devote his whole time and attention to "the Manufacturing Department" of the Company, for a period of seven years, and was to act as Manager, receiving as his remuneration for these services a sum equal to 7½ per cent on the net profits of the Company made in that time.

Thirdly, the stock and machinery in his Weston Street factory were to be purchased from Adams, by the Company, at a "fair and agreed" valuation (later recorded as £20,020 6*s*. 5*d*.), while the new steam engine and machinery in course of erection at Henry Street were to be taken over by the Company at their cost to Adams. The freehold factory at Henry Street was to be leased to the Company, at a yearly rent of £600, for a period of 28 years from March 25th, 1856, with an option for the Company to break the lease at 7, 14 or 21 years, or to purchase the freehold at any time within the first 14 years.

Finally, Adams was to subscribe for, and hold, 150 shares in the London Armoury Company during his services as Manager, and he agreed not to transfer any of these shares during his term of service, except to settle them upon his wife or family.

Part 2 of the Schedule of the Deed of Settlement revealed that (with an authorized capital of £60,000), the Company actually commenced operation with £45,700 subscribed or guaranteed, and although the Company had powers to raise an additional £150,000, if desired, this power was only exercised once throughout its operations, when (in September, 1858) the nominal capital was increased by issuing one thousand £10 Preference Shares.

The Directors of the Company were:

Archibald Hamilton	(Merchant)
Robert Adams	(Gunmaker)
Richard Ashton	(Merchant)
William Dray	(Manufacturer and Engineer)
Archibald Cockburn	(Merchant)
George Fry	(Solicitor)
John Shorter	(Merchant)

all of whom gave business addresses in or near the City. The Auditor was another City merchant, George May, and the Secretary of the Company was an Auditor by profession, one Thomas Finlayson.

According to *Kelly's Directory* for London, 1857, the Company was at least settled into "The Railway Arches" at Henry Street when that edition was proofed, and curiously enough a listing for "Deane, Adams & Deane" continued, but now for 30 King William Street only. The *Index of Assignees, etc.* also recorded the securing of an interest by Adams in James Kerr's revolver-rammer Patent No. 1722 of 1855, in 1857; this protected the famous compound lever-rammer shown in Plate 13, which was mounted on all Beaumont-Adams revolvers, made thereafter by the Company.

In 1858, *Kelly's Directory* for London still recorded the Company as located in the "Railway Arches" at Henry Street (with Robert Adams as "Manager"), but also contained a personal listing for him at No. 2 Henry Street, and had no listing for the two Deanes, at any address, in relation to the manufacture of firearms. Further, no evidence can be traced that Adams, in 1858, carried out the obligations that he had undertaken in the Deed of Trust as to vesting his patents in the Company.

In fact, there appears to have been some difficulty over Adams's position almost from the start, for the copy of a Resolution creating the above-mentioned Preference shares was filed with the Registrar of Joint Stock Companies (on September 20th, 1858) by Frederick William Bond signing as "Manager", despite the managerial status accorded to Adams in the advertisement featured as Plate 36 (*Top*). This does not necessarily mean that Robert Adams had severed all connection with the Company by that date, but *Kelly's London Directory* for 1859 (which must have been proofed in 1858) showed him back in business on his own account (as a wholesale gunsmith and rifle-manufacturer), at 76 King William Street, but the London Armoury Company, Ltd. at "Henry Street, Bermondsey Street", with F. W. Bond as "Manager".

Indeed, 1859 seems to have been a year for a general realignment of interest and power within the Company, and against the interest of Robert Adams. Thus, James Kerr filed applications for two British revolver patents in December, 1858, and January, 1859, which duly became British Patents 2896/1858 and 242/1859, respectively, and the 54-bore Kerr weapon (see Chapter X) "manufactured by the London Armoury Company, Ltd."—to quote one of Kerr's advertising broadsheets—was demonstrated at the Royal Small Arms Factory, Enfield, on April 25th, 1859. Conversely, the firm of "Deane & Son" appeared (in *Kelly's Directory* for 1859) at the old Deane, Adams & Deane premises of 30 King William Street, as "Gun Manufacturers", and John Deane, Senior, had completed the disposal of his London Armoury shareholding by April of that year.

Whatever the terms on which Adams parted from the London Armoury Company, it seems clear that he did eventually complete those formal undertakings in respect of his own and Beaumont's patents; it will be recalled that he had undertaken to assign these to the Company, with other assets, for £5,000. The Patent Office have now destroyed their *Index of Assignees*, but it did contain seven entries, in 1859, which related to those two patents and must have referred to this important matter. If the parting was not amicable, the Company had a useful lever, to force Adams's compliance, in their agreed terms of payment; £2,000 was payable to Adams on complete registration of the Company, but the balance was payable by instalments "to be agreed".

Robert Adams's name and retail address at 76 King William Street appear upon a surprising number of Beaumont-Adams revolvers, some of which were made by the London Armoury Company, Ltd., itself, and also upon conventional shot-guns or rifles (see Fig. 17). It also seems probable that he remained a substantial supplier to the general gun trade over a brief period, for when, in March, 1863, the London Armoury Company broke their lease to his Henry Street factory (as they were entitled to do), Adams duly reoccupied the building. As it happened, a slump developed throughout the British gun trade after 1863, and Adams disposed of the Henry Street factory in 1865, to a firm of printers, Thomas McCorquodale & Co. Thereafter, the premises passed (in 1868) to those "manufacturers of dog foods", in whose hands they appeared upon the 1872 Ordnance Survey map already mentioned.

On quitting the Henry Street factory, Adams also left his warehouse and retail shop at 76 King William Street, and removed to No. 40 Pall Mall; this was a retail shop only, and despite an earlier success with a breech-loading sporting-gun design, at the London Exhibition of 1862, and apparently approved status as "Maker to the Royal Family" (see Fig. 17), he ceased traceable operations as a London-based gunmaker very shortly after the conclusion of the period under study in this book. It is as well to record, also, that he held no shares in the "Adams Patent Small Arms Co., Ltd.", which was formed in 1864 to promote *John* Adams's revolvers, and somewhat ironical that the same year should have seen a refusal of even provisional patent-protection for his last recorded revolver-design.

As to the London Armoury Company, upon Adams's departure, neither the remainder of 1859 nor the year 1860 yielded much of record. It is possible that this latter year saw the Company receive an order for 1,000 "Deane and Adams" revolvers, since these were delivered to the Virginia State Ordnance Department, on January 21st, 1861 (C. E. Fuller and R. D. Steuart. *Firearms of the Confederacy*. 1944. p. 246). In that year also, James Kerr (the Company's Superintendent) moved from Bedford Terrace to the Henry Street factory, and

FIG. 17.

Advertisement from "The Volunteer's Book of Facts."
W. H. Blanch. (Ed.). London, 1863.

worked directly with F. W. Bond, Adams's successor as Manager; presumably this move was prompted by the need to supervise preparations for production of Kerr's revolvers (see Chapter X), but he was a skilled gunmaker, and his presence may have been needed for more conventional manufactures.

In the following year of 1861, external events began to shape the Company's future, for (on April 12th) the Confederate attack on Fort

Sumter, at Charleston, S.C., made civil war in America inevitable, and the first resultant orders for small-arms reached Birmingham (Goodman, *op. cit.* p. 417) on May 9th. The London gun trade was as promptly favoured, for Captain Caleb Huse, C.S.A., was reporting to his superiors on preliminary discussions with the Company for Enfield rifle manufacture by May 21st, 1861 (Fuller & Steuart, *op. cit.* p. 215); however, his call was always primarily for rifles and little survives, in detail, on the demand and supply of London Armoury Company revolvers during the entire Civil War.

About April, 1863, the Company moved from Henry Street to new factory premises at Victoria Park Mills (between Grove Road and Wick Lane), in Old Ford Road, London, and surrendered their old factory to Robert Adams, as earlier described. This was no sudden or precipitate decision, however, for the new factory appears to have been elaborately tooled and organized. H. W. S. Cleveland, a contemporary American author, stated in a preface (dated November, 1863) to the revised 1864 edition of his *Hints to Riflemen* that, in addition to the American technicians who installed certain machinery made for the Royal Manufactory at Enfield, "another American has since started the works of 'The London Armoury Company' in that city, and now is superintending the construction of an immense armoury at Birmingham, and nearly all the machinery and tools for both these establishments were made at the Ames Manufacturing Company's Works at Chicopee"; this was presumably Chicopee Falls, Mass., U.S.A., and the Birmingham "armoury" would almost certainly have been that of the "Birmingham Small Arms Co., Ltd." In view of the comparative predominance of muzzle-loading "L. A. Co." Enfield rifles dated 1863 (or after), and the known reputation of the Ames Mfg. Co. as manufacturers of rifle-making plant, it seems probable that the new factory was primarily intended and designed for the production of such arms, with fully interchangeable parts.

About a year later, in April, 1864, the Registered Offices of the Company were moved from the factory to retail premises at number 36 King William Street, and James Kerr moved from his position as "Superintendent" at the factory, to this new address. The reasons for this move are not known; as will later appear, it is unlikely that Kerr was sent to these retail premises in an attempt to increase business, for the Company was still fully occupied with Confederate Army orders, at that time.

However, on April 12th, 1866, a special General Meeting of the Company was held, and the resolutions of a similar meeting held on March 22nd in that year, were confirmed, namely, that the London

Armoury Company, Ltd. be wound up voluntarily. In view of the rather stormy start to the London Armoury Company's history, *vis-à-vis* the fate of Robert Adams (its "Promoter"), it is of interest to note that the Resolution to wind up "voluntarily" was not actually filed with the Registrar of Joint Stock Companies until October 26th, 1866, and that it specifically recorded the confirming of an Agreement dated February 28th in that year with "certain dissentient shareholders". No details have survived about this latter Agreement. Six liquidators were subsequently appointed, but their powers were exercisable by any two of them. Since Frederick William Bond was one of these liquidators, it is interesting to learn that a new "London Small Arms Company, Ltd." (formed by the famous Bond and Barnett gunmaking families) took over the factory premises at Old Ford Road later in the year, for their own operations under the Schneider and Snider breech-loading rifle patents.

The final winding-up meeting of the Company was held in August, 1867, when the Winding-Up Account was placed before the meeting, and the first London Armoury Company, Ltd. was dissolved. The retail address of the Company, at 36 King William Street, was taken over briefly (until premises at 54 King William Street were ready) by "James Kerr and Company", who were careful to describe themselves as "successors" to the London Armoury Company, Ltd. So faithful did they remain to this name that, in March, 1894, when a *second* London Armoury Company, Ltd. was formed by the Kerr family, the stated object of their new company was to take over the business *then* carried on "under the firm or style of James Kerr and Company and the London Armoury Company".

CHAPTER X

THE LONDON ARMOURY COMPANY (LIMITED) AND ITS REVOLVERS

IT seems probable that Robert Adams's departure from the London Armoury Company, in 1858, was caused by the Company's decision to develop manufacture of infantry-rifles, at the expense of his revolvers. In its own day, the Company was best known as a rifle-manufacturer, rather than for the revolvers that it made, and something has to be said here about these activities. It is difficult satisfactorily to account for certain Birmingham-made Beaumont-Adams revolvers, unless it has been grasped that pistols were really a side-line, and often made by sub-contractors because the Company was too busy to produce them.

Ironically enough, Robert Adams may himself have contributed to the initial switch in emphasis from his revolvers, for it appears that he brought to the Company, on joining it as Manager, an interest in a breech-loading rifle. The "*Index of Names of Assignees, etc. 1853–66*" (since destroyed by the British Patent Office) contained an entry to support this assertion, for it had recorded, in relation to the year 1856, an interest as held by "Adams, Peabody and the Sharps Rifle Company", in English Patents Nos. 14052 and 14085 (of March 31st and April 22nd, 1852, respectively), together with British Patent No. 712, of November 11th, 1852.

These three patents protected the principal features of a lever-operated, falling-block, breech-loading rifle, invented by Christian Sharps, an American. This weapon enjoyed great success in the United States, and a carbine version had been submitted for British Army tests, as early as April, 1854; C. H. Roads, *The British Soldier's Firearm, 1850–1864*, (Herbert Jenkins, Limited, 1964) traced a number of attempts to secure British adoption of it, culminating in a Board of Ordnance Order for 6,000 carbines. These arms were delivered between May 31st, 1856, and March 31st, 1858, but their source is still obscure, for C. H. Roads (*op. cit.* p. 280) merely remarked that they appeared to have been manufactured by Robbins & Lawrence. This firm were indeed substantial arms manufacturers, (at Windsor, Vermont and Hartford, Conn.) into the early 1850's, but as mentioned in Chapter III, they were in financial difficulties after 1853, and it seems unlikely that they could have provided the arms.

In the light of "*Index*" entries mentioned above, and uncertainty as to where these carbines actually came from, it seems at least arguable that Robert Adams was involved in some part of their manufacture when he joined the London Armoury Company, and that the Company contributed to their completion in its turn. In 1857, another "Index" entry appeared in relation to these patents, for "Adams & others"; there seems little purpose in any renewal of rights, if the carbines were entirely manufactured in America.

Generally, however, the London Armoury Company is best remembered for an excellent "Enfield, Pattern 1853, Rifled Musket", which it made for H.M. Board of Ordnance, and also for private buyers; this was the British Army's issue rifle (a single-shot, muzzle-loading weapon, in ·577-in. calibre), and the Company apparently specialized in a "Long" version, with 39-in. barrel. When, exactly, Board of Ordnance confidence was thus established is not known, but C. H. Roads (*op. cit.* p. 88) recorded one Board order for 30,000 rifled muskets as dated February 21st, 1857, and this was almost certainly filled by the Company. The latter was also sole British commercial supplier of these weapons with fully-interchangeable parts; a Government factory at Enfield was also designed for such supply, but was not fully equipped in the first years of the Company's life.

This Royal Small Arms Factory was provided with specially-imported American machine-tools, but even after 1859 (when it became fully operational) Service demands greatly outstripped production. As a result, the London Armoury Company seems to have received a series of official contracts, and also to have enjoyed orders for the same Pattern 1853 arms from both civilian and Volunteer buyers. Interest in target-shooting was considerable, during the 1860's, and the Company therefore provided for it, with a specially-rifled version of the P.1853 weapon. This was called the "Kerr" rifle, and was in a smaller ·45-in calibre, rifled with a gain-twist after British Registered Design No. 4368, of May 10th, 1861; that Design had been secured by James Kerr, the Company's Superintendent.

It appears that this civilian market was not without its perils, and *The Times* reported one unpleasant incident as occurring at the Wimbledon ranges, in July, 1862. Mr. Ross (a contestant for the Queen's Prize) submitted his London Armoury rifle for the necessary tests, and it was rejected by the range authorities as failing to meet trigger-pull and bore-size requirements for his contest. It was necessary for entrants in this Queen's Prize shoot to use rifles identical with "as issued" P.1853 Enfield rifled muskets, and Mr. Ross was not the only marksman to have his privately-made rifle rejected that day.

Mr. Hamilton (the Company's Chairman) was unwise enough to write a letter to *The Times* protesting against these rejections, and prompted a letter of rebuttal from the Wimbledon authorities, alleging that some of the forty-odd rifles rejected had had "reject" stamps, from the Enfield factory, on their barrels.

The London Armoury Company's letter of reply was published on July 8th, 1862, and claimed that these particular "reject" barrel-stamps meant only that the rifles had been withdrawn from the Royal Small Arms Factory, Enfield, before complete acceptance. It was denied that the offending barrels had been rejected as unsafe. Such an episode was clearly damaging, however, and correspondence was wisely dropped at that point.

As mentioned in Chapter IX, another factor contributing to the Company's prosperity had been the outbreak of war in America.

Captain Caleb Huse, C.S.A. (a Confederate purchasing-agent) had arrived in London, and contacted the Company, by May, 1861. He wished to buy Enfield rifles in substantial quantities, and his first order upon the Company was for 10,000 rifled muskets, at £3 16*s.* 6*d.* apiece. This order could not be processed for nearly a year, however, and was not put in hand until March, 1862. W. B. Edwards (*Civil War Guns*, the Stackpole Company, 1962) revealed that Company output, when Huse arrived, was about 1,300 rifles per month, but that a Board of Ordnance order was absorbing most of them. The Board would not release the Company from this contract, and the Confederacy had to wait until it had been filled.

In approaching the London Armoury Company, Captain Huse enjoyed either good judgement or extraordinary luck. Archibald Hamilton (Chairman of the Company) was a Confederate sympathizer, and also a principal in the London merchant-house of Sinclair, Hamilton & Co. This house acted as purchasing-agents for Captain Huse (at a 2½ per cent commission) and bought considerable quantities of arms and equipment, for the Confederacy, from manufacturers all over Britain. It seems likely, therefore, that Mr. Hamilton was largely responsible for the London Armoury Company's virtual commitment to the Southern cause, after mid-1862.

Indeed, this dependence was no bad thing for the Company. A slump gripped the British gun trade after September, 1863, when

Northern agents suddenly stopped buying arms in England; American arms-manufacturers could thereafter meet all Federal demands, and, as a result, only those firms supplying the Confederacy continued to work at full-stretch here. This activity dropped away after the close of 1864, when Federal blockade measures against Southern ports finally became effective, but the London Armoury undoubtedly had a most useful period of activity until that date.

It might be thought that the Company's eventual liquidation stemmed from a Confederate failure, as it disintegrated, to pay for what was ordered. In fact, this does not appear to have been the case, for although it is true that payment sometimes occurred considerably in arrears, and that Southern obligations to British manufacturers were once as much as £500,000 in excess of funds immediately available, payment was always made in due course.

Responsibility for this financing, in the teeth of enormous difficulties, undoubtedly lay with the Confederacy's European bankers, Messrs. Fraser, Trenholm & Co., who were London-based, and of great reputation and integrity. Left with worthless Confederate paper when the South finally capitulated, these bankers grimly honoured every Confederate draft upon them, and to their own ultimate ruin.

Faced with such facts, a student must inevitably wonder why the London Armoury Company went into voluntary liquidation in April, 1866. Clearly, the Company was not insolvent, for the liquidation could not then have been officially permitted as a voluntary one, whilst subsequent success by the "London Small Arms Company, Ltd.", in operating the Victoria Park Mills factory profitably, must argue that there was business available for the London Armoury, if desired.

In all the circumstances, it seems most likely that the Company was somehow involved with a growing financial crisis in the City of London, which was shortly to be climaxed (in May, 1866) by the "Black Friday" collapse of Messrs. Overend & Gurney; the latter was a large firm of discount bankers whose speculative activities had ended in their direct involvement with far too many of the projects for which they had originally arranged finance. Prior to failure, Messrs. Overend & Gurney had tried for some time to reduce their enormous obligations, and many clients were under pressure to redeem the paper held by these bankers. Having regard to various city houses with interests in the London Armoury Company, Ltd., it is suggested that the Company was put into liquidation by the principal shareholders, who thus sold a profitable concern to raise cash for their own commitments to Overend & Gurney.

There is no certain evidence to support this suggestion, but it is noteworthy that the Hankey family (who originally financed Robert Adams's self-cocking revolver) were involved in the Consolidated Bank Limited, and that this was one of the houses which closed its doors, temporarily, during that "Black Friday" crash. It is not known if the Company's bankers (The Union Bank of London) was similarly affected.

Enough has now been said to establish the Company's position as leading rifle-manufacturers for nearly a decade, and also to show the existence of an ever-present temptation to ignore revolvers, in favour of another weapon. Formed primarily to develop and market the Beaumont-Adams revolver, it would seem that latterly the Company lacked either means or inclination to make it in the quantities sometimes needed.

Between 1856 and 1866, three types of revolver were intermittently produced; two had been inherited from Robert Adams, and a third (the first to be considered here) was patented by the Company's Superintendent, James Kerr; see his British Patents Nos. 2896 (of December 17th, 1858) and 242 (of January 26th, 1859). Mr. Kerr was a good revolver shot, and had earlier demonstrated "Adams Revolvers" before both British and American committees; see P. E. Dove, *The Revolver: Its Description, Management and Use*, Edinburgh, 1857.

Kerr's revolver is shown in Plate 12, and his first patent was aimed at producing a single-action weapon, so simple that any gunsmith or armourer could keep it serviceable. To that end, the lock-mechanism was built as a detachable component (similar to those back-action locks used on cheap shot-guns), which Kerr's second patent modified ingeniously to a double-action design. To circumvent Adams's patent for the solid-frame, Kerr formed his barrel and top frame-strap as one component, and mated this to a one-piece butt and lock-case with suitable screws. By British Patent No. 1722, of July 28th, 1855, he had already designed a most effective lever-rammer, but had assigned this patent to Robert Adams (in 1857) for use upon the Beaumont-Adams revolver. Kerr's weapons therefore had another rammer of his design, mounted beneath their barrels; this was an adequate device, but more easily broken than that used by his rival.

The London Armoury Company's reasons for taking up Kerr's revolver remain obscure, in view of the success enjoyed by their Beaumont-Adams pistols, and of heavy commitments to manufacture rifled muskets. However, a 54-bore version was tested at Enfield (on April 25th, 1859), and performed creditably at ranges up to 300 yards. It is possible, of course, that Robert Adams was being difficult about assigning his revolver patents to the Company (as he was bound to do), and that the existence of Kerr's revolver, and its testing at Enfield, were used to make him more amenable. Certainly, there is no record of further attempts at pressing the Services to consider this design, after Adams had assigned his patent rights.

The Company produced Kerr's pistol in 54- and 80-bore versions; single- and double-action revolvers are found, but the latter are more common. Both versions were five-chambered, and had 5¼-in. barrels, rifled with five grooves twisted to the left. Initially, a screw-in arbor-latch was used, but later revolvers had a spring-loaded latch of the type visible in Plate 12; this design was also unusual in having the arbor inserted from the back of the frame.

A very early revolver, in the J. B. Bell Collection, has an extremely shallow rake to the butt, but a rather more pronounced curve was almost immediately adopted; after the first 200 revolvers had been made, the butt design shown in Plate 12 emerged, and was maintained in use thereafter. The legend "London Armoury" was engraved upon the frame (and sometimes on the lock-plate), together with the words "Kerr's Patent" and a serial number. In addition the letters "L.A.C." were impressed upon the top left-hand flat of the octagonal barrel (see Plate 12); as with Beaumont-Adams pistols, Kerr's revolvers were also sold to various retailers, and their names were sometimes engraved upon the top frame-strap. London proof-marks were always present.

It appears that a single serial-range was used, without any letter-suffix or prefix. Exceptions to this rule have been reported as having "P" or "S" prefixes, but no specimen could be located for examination. As to the serial-range itself, this apparently began at No. 1, and continued to about the middle of an 11000 series.

In the serial-range below 200 there was some evidence that one batch of single-action 80-bore revolvers was supplied to a Volunteer unit, such pistols engraved "1st Sussex Artillery Volunteers" being encountered, with an additional unit-number upon them. In general, however, the major purchaser of this pistol was the Confederate Army.

Due to values now attaching to all Confederate arms, Kerr revolvers with an alleged Confederate Army association are claimed in many collections. Such claims must always be viewed with reserve, and are sometimes demonstrably suspect. As an example, revolver No. 11319 was alleged to be a Confederate weapon, but no details were given as to its markings. However, revolver No. 11293 was found to be marked "London Armoury Co., James Kerr & Co., successors", and this was a trading-style adopted in 1867, after the Civil War had ended. It would seem unlikely that any revolver with a higher serial number could have a true Confederate background.

C. E. Fuller and R. D. Steuart (*Firearms of the Confederacy*, Standard Publications Inc., 1944) published an official Confederate record that 900 Kerr revolvers (with spare parts and flasks) were landed at Wilmington, N.C., on October 31st, 1864. Neither Messrs. Fuller & Steuart, nor the equally able Albaugh, Benet & Simmons (*Confederate Handguns*, Riling & Lentz, 1963) could produce documented evidence of other shipments. General references to cargoes that included "revolvers" have been traced in various Confederate Reports, but these gave no identities for the weapons involved.

On this sort of basis, it would be rash to attempt any dogmatic statement about serial numbers for Kerr revolvers actually used by the Confederacy. Surviving army rosters give the most accurate information, and de Witt Bailey III (whose assistance we most gratefully acknowledge) has established from them the issue of revolvers in 1700, 3800 and 9000 serial-ranges, with a majority of weapons in the latter group. Other researches, by Messrs. Albaugh, Benet & Simmons, showed weapons in these groups, but allied with them arms in almost every 100-series between 300 and 11000. Clearly, a great deal of work remains to be done upon this matter.

Kerr also appears to have hoped that the Federal government could be interested in his revolver, and secured U.S. Patent No. 39409 (of August 4th, 1863) to cover all principal features of the double-action design. In fact, however, only sixteen Kerr revolvers are known to have been purchased by the North; *U.S. House Executive Documents*, v. 12. 40th Cong. 2 Sess. 1338, 954–956 showed these weapons as bought from Messrs. Schuyler, Hartley & Graham, in the desperate days of December, 1861. Presumably, the pistols were ordinary civilian stock.

The late J. N. George (*op. cit.* p. 219) also stated that Kerr's revolver was adopted by the Portuguese Army. Only two weapons were noted as having a possible association with that country, and both were single-action pistols in 54-bore calibre. One, serially numbered 3475, was marked as made by the Orbea Brothers, of Eibar, Spain; the second, serially numbered 5142, was simply marked "Fabrica de Duranco". Both were inferior, in workmanship, to weapons bearing the London Armoury Company name, although Spanish versions of Kerr's revolver have been encountered that were no wit inferior, in this respect. If the weapon was adopted by Portugal, it is probable that supplies were made in Spain, and not by the London Armoury Company.

Over the years of its existence, however, this Company's revolver-manufacturing activities were based upon the patents formerly held by Robert Adams. It was not until 1859 that the "*Index of Names, etc.*" contained entries relevant to his assignments of these patents to the

Company, but in due course it secured title to Adams's English Patent No. 13527/1851, Beaumont's Br. Pat. No. 374/1855, and Kerr's Br. Pat. No. 1722/1855. These patents, of course, respectively protected Adams's one-piece frame and self-cocking lock mechanism, Beaumont's double-action refinement to the lock, and Kerr's rammer; no interest was recorded in Adams's patent of 1853, for the "hesitating" lock mechanism, or that of 1854 covering his sliding safety-bolt, rammer, and screw-in arbor-latch. As can be noted in the Appendix, all British patents were allowed to become void during the Company's life; presumably, they were regarded as expendable, because their use by others was improbable when divorced from Adams's frame. The basic patent of 1851, however, being secured before the Patent Law Amendment Act, 1852, was for a fixed fourteen-year term, and required no renewal.

In working these patents, the Company appears to have ignored self-cocking revolvers, and to have supplied such weapons only on special order. Four London Armoury self-cocking revolvers were traced, two in 120-bore calibre and two in 54-bore calibre, all of which were reported to have "Rigby's rammer" fitted to them. To base such an assumption, about the Company's activities, upon only four specimens of the revolver must appear risky, but even this minuscule sample hardly suggested that the self-cocking pistol was produced in quantity, as compared with Beaumont-Adams weapons. The 120-bore revolvers were in a 19000R serial-range, whilst the 54-bore arms were in a curious 33000R serial-range that is later remarked upon; in our view, these weapons could be ordered from the Company, but were not carried in stock by it.

As to its double-action Beaumont-Adams revolvers, the Company offered these in 38-, 54- and 120-bore calibres; in every case known to us, Kerr's rammer (after his Br. Pat. No. 1722/1855) was used, except for pistol No. B.4970 20455R, retailed by Deane & Son, which was reported to have "Rigby's rammer".

Before dealing with this weapon in its three calibres, something must be said about the lock-mechanism and method of construction used for it.

P. E. Dove (*op. cit.*) called Beaumont's lock mechanism "the Action", and listed its components as shown in Fig. 18, namely:

"A. The main spring
B. The swivel
C. The hammer
D. The short scear
E. The short scear spring
F. The long scear

G. The lifter, with the spring which works both lifter and long scear
H. The trigger
I. The trigger spring

The four screws or pins—hammer pin, trigger pin, short scear pin, and short scear spring pin."

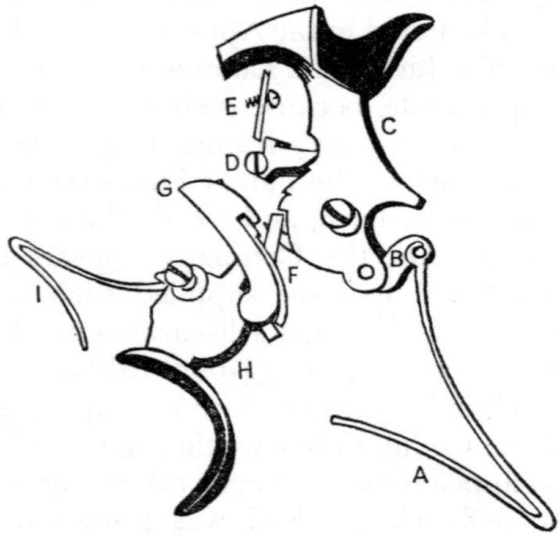

FIG. 18. *London Armoury Company's version of Lieutenant Beaumont's lock mechanism.*

This developed version of Beaumont's lock-mechanism may be compared with the one actually patented (see Fig. 14), and it will be seen that the only real alteration made by London Armoury Company designers was to substitute a shorter "short sear" than that used by their predecessors.

It would be difficult to improve upon P. E. Dove's succinct statement of the sequence of operation.

"The mode of action" (he wrote) "may be seen by looking at the engraving. Thus—

"When the trigger is pulled the long scear pushes up the hammer, and the lifter pushes round the ratchet of the cylinder. As the hammer goes up, its circular portion, called the belly, pushes the long scear forward, so that it goes out of the notch, and the hammer falls on the nipple or cone. When the pistol is cocked by pulling back the hammer, the short scear goes into the full-bend notch, and the long scear is pulled up close to the short scear by the hook on the hammer, which by this action draws back the trigger. When the trigger is then pulled, the long scear pushes the short scear out of the notch, and the hammer falls. After one shot has been fired the finger must be slackened so as to allow the trigger to go forward to its original place, which it does by the action of I, the trigger spring."

Very little information has survived, in respect of manufacturing methods used by the London Armoury Company to produce their Beaumont-Adams revolvers. However, Dove's little book did contain some paragraphs upon the subject, and these can be usefully compared with those methods to be described, in Chapter XIII, as used three years earlier.

"The Frame," (Dove wrote) "which includes the barrel of the pistol, is made of a single piece of the best quality of iron that can be procured, known as Marshall's iron. It is not welded, but cut out of the solid. It has no joint in it of any kind, but is one piece of iron. The angles are therefore as secure as the strength of the metal can make them; and the frame being centred once for all, is worked true in all the after processes of boring, rifling, and fitting.

"To give an idea of the processes through which the parts of a pistol are carried in the course of manufacture, we may take the frame as an instance. The frame is first forged to very nearly the outward shape which it is to retain. It is then a solid piece of iron, with a thick block at one end and the barrel at the other. It is then centred, that is, a point at each end is marked and struck as the centre of future operations. The centre keeps everything true. It is then planed on the outside. It is then slotted; that is, the place for the cylinder is cut out of the solid block. It is then bored and rifled, then filed up and fitted. It then receives the action (lock, &c.); is then stocked and proved; has the bolt and lever-rod fitted; is polished, engraved, blued, screwed together, and then shot."

Nothing in these paragraphs suggests any radical departure from those earlier manufacturing processes discussed in Chapter XIII, except that Dove's words indicated various frame apertures as cut at a later stage of manufacture in 1857, than during 1854. In both years, the frame was stated to be planed on its outside, and it is possible that the London Armoury used a slightly more advanced machine than their predecessors had done. Messrs. Nasmyth & Gaskell had introduced their "Steam Arm" shaping-machine in 1836, and this simple planer (in which the tool travelled across a traversing table) may have been the type used in 1854; by 1857, heavier and more versatile machines, with fixed tools and travelling tables, could be had from Nasmyth, and would have been admirably suited to the London Armoury Company's needs.

P. E. Dove was less expansive about the manufacture of revolver-cylinders, but a little information can be gathered from his remarks.

"The Cylinder" (he wrote) "is also made of Marshall's iron, or of steel. If of the former, it is case-hardened. . . . The central hole, through which the pin passes, is bored so as to make the cylinder touch the pin only at the two extremities, thereby lessening the friction. At the back of the cylinder is the ratchet, to give the revolving motion to the cylinder when acted on by the projecting lever of the trigger action, called, technically, the lifter. The cylinder, when finished, is case-hardened. The ratchet is made of steel."

Formation of this "central hole" in the fashion described by Dove was of definite advantage to users of Beaumont-Adams pistols. Black powder rapidly fouls some muzzle-loading revolvers to a point where the cylinder has to be manually assisted in rotation; we have never experienced this handicap, when using a London Armoury Company weapon.

In turning to the three sizes of revolver available during the London Armoury Company's life, it must be emphasized that some weapons in each calibre were made by Birmingham manufacturers and not by the Company. This point will be enlarged upon, in due course.

The largest pistol, that in 38 bore, disappeared at quite an early point in that overall serial numbering allocated to the Company's own products. A revolver numbered 40570 had the highest serial number recorded but (as later explained) was not of Company manufacture; regular Company production in this calibre could not be traced beyond serial-number B.20700 36470R, and the weapon so numbered bore the name of Gasquoine & Dyson, in Manchester. It was accompanied by a bill of sale dated July 28th, 1863, but the bill-head was that of Thomas Newton, and it can be assumed to have recorded the sale of a second-hand pistol. On the basis of a dated weapon, admittedly in 54 bore, which had a serial number quite close to that cited, it is suggested that this Gasquoine & Dyson 38-bore revolver was made quite late in 1859, and that the London Armoury Company (whose "L.A.C." barrel-stamp was impressed upon the barrel) made few (if any) 38-bore revolvers after that specific point in overall serial-numbering of Beaumont-Adams pistols, whatever their Birmingham licensees may have done.

Measuring 14 in. overall, and having 7½-in. barrels (rifled with three broad grooves twisted to the right), these enormous weapons were a legacy from the past. It would appear that Robert Adams had all necessary tools and die for their manufacture, because the Services were originally interested in them, and that the Company (having inherited this equipment) would for a time supply the arm on request. C. H. Roads (*op. cit.* p. 302) published a list of arms still regarded as "serviceable" at about 1862, by the Military Store Office at H.M. Tower of London, and the 38-bore service revolver was then included; in fact, it is doubtful if the weapon had ever seen serious use, even by that late date.

However, a possible revival of interest was recorded by the Minute Book of the Ordnance Select Committee, and dated January 27th, 1864. Tests were then in progress with 38-bore versions of "Deane & Adams revolving pistol", arising from a decision to send eighteen revolvers, with detachable carbine butts, for Australian police service. The Committee secured specimen ammunition from the Superintendent, Royal Laboratory, for sealing to govern manufacture, and this (with waterproof bag, drawings and specification) the Committee duly approved. In view of the test-data given, it seems probable that the Committee accepted a solid-based, single-cannelured, cylindro-conoidal bullet of 244 grains weight, which measured ·66 in. in length, ·492 in. in diameter, and was composed of a 19 to 1 lead/tin alloy. Propelled by 20 grains of medium rifle-powder, the bullet was wax-lubricated and embodied in a "skin" cartridge.

Best known of the three models, was a 54-bore pistol with 6-in. barrel (shown in the midst of Plate 13), for that version outnumbered the combined totals of other calibres, and is still extremely popular with muzzle-loading enthusiasts.

Although this 54-bore revolver was bought for the British armed forces (and possibly also for the Honourable East India Company, prior to 1858), it is difficult to establish either the identity of those to whom it was issued, or the quantities involved. During the period of concern, an army officer normally bought for himself any revolver that he fancied, provided that it took the Service ammunition; logically, therefore, Service-issue should have been to cavalrymen and senior N.C.O.s, but it is a curious fact that C. H. Roads's list of arms accepted as "serviceable" by the Military State Office, around 1862, described 54-bore pistols as for "Sea Service", in distinction from the still-listed 38-bore "pistol, revolver, Deane's, with Beaumont's improvement". The same author (*op. cit.* p. 300) also recorded overall stocks of serviceable revolvers on January 1st, 1862, at home and abroad, as 14,576 weapons; Colt revolvers would have been represented in this total, since they too were regarded as "serviceable" by the Tower, but the bulk of it must have related to Beaumont-Adams weapons, of 54-bore calibre.

In view of the Board of Ordnance orders for Beaumont-Adams revolvers, mentioned in Chapter VIII, it could be anticipated that issue pistols would fall into definite serial-ranges, and this did prove to be the case.

A pistol marked "Adams & Co." (No. B.2671 18031R) was the lowest serial number encountered for this 54-bore Service version, and thereafter the following serial numbers were established for weapons signed by the London Armoury Company:

B.3663	19077R*	B.7726	23379R	B.14265	29929R
B.3700	19114R	B.7939	23594R	B.14903	30567R
B.5042	20337R	B.8401	24136R	B.15506	31170R
B.7497	23143R	B.888?	24509R	B.18257	34004R

In view of a civilian 54-bore revolver that is marked "J.R. 1858", and serially-numbered B.15451 31115R, it seems quite possible that, with one exception, all of the weapons listed above were supplied before 1859. The last pistol on this list had a badly defaced top-strap legend that read "London Armoury Company . . . Victoria", and if this did in fact refer to the Company's Victoria Park Mills address, then that weapon could not have been made before 1863.

In appearance, these Service revolvers differed very little from their civilian counterparts, and had the same 6-in. barrel, rifled with three grooves in a right twist. Early revolvers were marked "London Armoury Company (Ltd.)" on the top-strap, but this was later shortened to "London Armoury", and the broad-arrow marking. In some cases, a lanyard hole was provided through the wooden butt, and it was apparently standard practice to heat-blue the hammer, trigger and rammer, as well as bluing cylinder and frame in the normal manner; civilian versions of this revolver had a bright case-hardened finish to the lock and rammer components. On one highly-ornamented "W.D." Beaumont-Adams revolver (shown in Plate 14) the gold damascening was, of course, applied privately.

Although there were Service complaints about difficulties in keeping Beaumont-Adams revolvers serviceable, and of a tendency for their barrels to lead-up, the British Army does appear to have persevered with this design for some years. A 54-bore bullet-mould (still preserved in the R.S.A.F. Pattern Room, at Enfield) is engraved "Mould for last approved Bullet for D. & Adam's Revolver Pistol", together with the "W.D." mark, and is dated April 3rd, 1867; the ammunition for this weapon did not become formally obsolete until February 2nd, 1882 (W. C. Dowell, *The Webley Story*, The Skyrac Press, 1962). However, stocks of serviceable revolvers appear to have run down markedly, from their pinnacle on 1st January, 1862 (earlier noticed) for in 1869 existing stocks were examined with a view to their conversion for use with a metallic centre-fire cartridge; Minute Books of the Director General of Ordnance then noted only 5,500 Colts and 7,000 Adams's revolvers as available. Colt weapons were rejected, since they would not use the ·450-in. cartridge accepted for a new John Adams's revolver, then recently adopted, but a number of Beaumont-Adams pistols were converted; in our previous list of "W.D." revolvers, the weapon marked with an asterisk had undergone this process.

Apart from service weapons, it was unusual for the Company name to appear on any Beaumont-Adams revolver (whether made by the Company, or for it), and so the names and addresses of retailers all over the United Kingdom can be found upon 54-bore pistols. In appearance, these arms were identical to the Service version (except as already noted), but pistol No. B.10631 26294R had a 4-in. barrel, and suggested that special barrel-lengths were available to order. In normal course, extra engraving was the only decoration applied, and (unlike certain Birmingham manufacturers later mentioned) it was uncommon for the Company to provide anything but a blued finish; one revolver, numbered 36090R, was silver-plated, and exceptional in that respect. Where a pistol-case was ordered, the Company seems to have provided a standard oak-case, which it secured from the "trade"; casing arrangements shown in Plate 10 and 14 were normal, but one or two owners had special cases made for their weapons, which were larger and more elaborately equipped.

As to the smallest Beaumont-Adams revolver, that in 120 bore, the series does not appear to have begun until about serial-number B.8641 24254R. These were charming little replicas of the larger weapons (see Plate 13, *bottom*) having 4½-in. barrels, and were almost all signed by retailers. In listing serial numbers, one concentration of 120-bore pistols was observed at a B.1700 33000R serial-range, and another in the 41000 series, but generally this version was distributed through the whole spectrum of serial numbers, and no other evident production-runs could be established for it.

An overall picture of the practice adopted for serially numbering Beaumont-Adams revolvers, during the London Armoury Company's life, can only be offered in terms of our own survey. Those serial numbers now to be quoted were taken from individual pistols, and it must be appreciated that the serial-ranges given can be only approximate; collectors will inevitably find revolvers with other serial numbers, whether smaller or greater than those given below for commencement and termination of each serial-range.

Although it seems certain that the pistols were numbered consecutively, within their respective serial-ranges, it cannot be assumed that the serial numbers were actually related to the quantities produced; thus the Birmingham manufacturers later cited as using serial-ranges of

60000 and 100000 were probably allocated these blocks of numbers to avoid confusion with serial-ranges then used by others, and not because 60000 or 100000 pistols had actually been produced. It follows, necessarily, that the dates for using these serial-ranges must have overlapped, and that revolvers in each of these two cited ranges may have been produced at approximately the same time.

For a general proposition, we recognize two types of Beaumont-Adams revolvers as made in Britain during the life of the London Armoury Company (Ltd.). Regardless of calibre, the first had London proof-marks, three-groove rifling twisted to the right, and an "L.A.C." barrel-stamp of the type visible in Plate 12; these we term, collectively, "London Armoury" pistols. The second type, again without regard to calibre, had a preponderance of Birmingham proof-marks, five-groove rifling twisted to the left, and if an "L.A.C." stamp was present at all, it was not of the type shown in Plate 12; these arms we termed "Birmingham" pistols, for reasons later stated.

In all, the following five serial-ranges were established:

(i) Twin serial numbers, already noticed, with a "B"-prefix to the first, and an "R"-suffix to the second, in a range from B.3663 19077R up to B.20700 36470R.
Buried within this serial-range were two others:
(*a*) A single serial number, without prefix or suffix, in a range from 30337 up to 32575
and
(*b*) A single serial number with an "R"-suffix, in a range from 33101R up to 36090R.

(ii) Twin serial numbers, without prefix or suffix, in a range from 21002 36801 up to 22680 38477.
Married with this serial-range was another, with a single serial number, in a range between 37237 and 39257.

(iii) A single serial number, without prefix or suffix, in a serial-range from 40080 to 41281.

(iv) A single serial number, with a "P"-suffix, in a serial-range from 60131P up to 60474P.

(v) A single serial number, with a "C"-suffix, in a serial-range between 100076C up to 101014C.

It can be assumed that where twin serial numbers were used, the first recognized Beaumont's patent and the second had regard to Robert Adams's patent for the one-piece frame; the eventual disappearance of Beaumont numbers, altogether, must have occurred when that patent was allowed to become void, on February 20th, 1862. However, the significance of those pistols with single serial numbers in sequences that march with twin serial numbers has yet to be explained, *vis-à-vis* Beaumont's patent.

Revolvers in serial-ranges (i) and (ii) above were generally of the "London Armoury" classification earlier defined, but those weapons in serial-range (i)(*a*) had the "Birmingham" characteristics; Robert Adams's name and address were well represented upon the "London Armoury" pistols, and one (No. 34523R) had a presentation date, in 1860, upon it. In serial-range (ii), pistol No. 39257 was, however, the highest serial number traced upon any pistol with our stated "London Armoury" features. There is no direct evidence on the point, but it is suggested that the London Armoury Company itself made those arms for which we use its name, and that the "Birmingham" weapons were (for reasons made clear in our remarks upon the remaining serial-ranges) produced either by Pryse & Redman or Calisher & Terry, of that town.

Revolvers in serial-range (iii) were all of this "Birmingham" type, although a few actually had London proof-marks, and one specimen (a 54-bore pistol, No. 40080, retailed by Holland, London) had a "P. & R." stamp upon the frame. It is suggested that these were the initials of Messrs. Pryse & Redman, of 84 Aston Street, Birmingham, and that this firm manufactured all the arms in serial-range (iii). One magnificent pistol from this group is shown in Plate 14, and is gold-plated, with an ivory butt; serially numbered 41036, it was retailed by E. M. Reilly & Co., of London.

It is possible that one of our serial-ranges should either have been allocated to, or extended to include pistols with serial numbers in a 50000 range. However, only one doubtful report was received in support of the existence of this group, which was of a revolver retailed by Purdey, of London, and numbered 55540.

Serial-range (iv) was therefore allocated to revolvers with a "P"-suffixed serial number in the 60000P range; all specimens reported were of the "Birmingham" type, and it is suggested that Messrs. Pryse & Redman were again the manufacturers. "P" was a serial number suffix used by that firm upon a series of Harding percussion revolvers discussed in Chapter XII, and it appears reasonable to suggest that they also used that suffix for Beaumont-Adams weapons. The serial-range was also interesting in containing a reported 80-bore pistol (No. 60360P) retailed by C. Ingram, of Glasgow; this was a calibre which the London Armoury Company never used.

Serial-range (v) again consisted of "Birmingham" type revolvers, although all specimens examined were London proved, and Robert Adams retailed six out of the seven specimens traced; a pistol numbered

100406C had his last address (in Pall Mall) upon it, and is evidence for dating manufacture of the group in 1865 or 1866. This "C"-suffix serial numbering probably used by Messrs. Calisher & Terry, since that firm actually advertised that it manufactured Beaumont-Adams revolvers, we therefore feel it safe to suggest (with a confidence that could not be adduced for serial-range (i)(*a*) above) that Calisher & Terry, of Birmingham, were the makers of the pistols in serial-range (v).

Although Robert Adams's patent for the one-piece frame expired on February 24th, 1865, the value of this protection appears to have endured almost until expiry, for another interest in it was entered in the "*Index of Names, etc.*" during 1864. Only two parties were involved, it would seem, since the entries referred to "Tranter" and the "London Armoury Coy."

Nothing is known as to the terms of this interest, but it is assumed to have been an extension of Tranter's licence to make one-piece frames for use in his own revolvers (see Chapter XVIII), and not to have involved any manufacture of Beaumont-Adams revolvers for the London Armoury Company.

In sharp distinction to the history of Adams's self-cocking revolver there is little direct evidence that Tranter produced Beaumont-Adams, revolvers for others to retail, although reports from Tranter enthusiasts indicate that he produced this design for such retailers as Cogswell & Harrison or E. M. Reilly (of London), and Trulock & Harris (of Dublin); in some of these arms a separate barrel was reported to be screwed into the one-piece frame, but we have not been able to secure any specimens for examination.

Overseas, the Beaumont-Adams revolver was certainly imitated in Belgium, by A. Francotte of Liège, and Spanish-made copies have also been reported; one Belgian five-shot, 54-bore specimen examined had a 5½-in. barrel, measured 12 in. overall, and was stamped "ADAMS PATENT 1856" on the frame. A Spanish pistol of similar type was engraved "TRUBIA 1858 No. 378". It is believed that one or two superb examples were also made, by hand, at the Tula Arsenal in Russia, and R. Scurfield, "Early British Regulation Revolvers", *Journal of the Society for Army Historical Research*, Vol. XXXV, Nos. 143/4, Sept./Dec. 1957, both described and illustrated a 9 mm. version produced for the Prussian Navy; the source of manufacture for this last is still unknown.

In America, another excellent copy of the revolver was made, in two calibres, by the Massachusetts Arms Company, of Chicopee Falls. In Chapter VI, mention was made of Robert Adams's visit to the U.S.

in 1853, and the groundwork for licensed manufacture of his revolvers was presumably then laid. Apart from Adams's own U.S. Patent No. 9693, use was also made of Beaumont's U.S. Patent No. 15032 (dated June 3rd, 1856) for the double-action lock-mechanism, and Kerr's U.S. Patent No. 17044 (dated April 14th, 1857) for the lever-rammer; it can be assumed, therefore, that the Massachusetts Arms Company did not commence production of Beaumont-Adams revolvers until some time in 1857, despite Colonel B. R. Lewis's claim that such arms were supplied to the U.S. Army in 1856.

Specimens of two revolvers made by the Massachusetts Arms Company are shown in Plate 15; rumours of a ·44-in. calibre version have been heard for some years, amongst modern collectors, but no specimen has been shown to exist. In all the circumstances, it seems probable that confusion arose about the manufacture of this calibre, because a number of British-made ·44-in. revolvers were bought by Federal authorities, during the American Civil War.

In all conscience, there can have been few enough of these weapons. Colonel B. R. Lewis (*op. cit.* Appendix 5) gave a "Statement of ordnance and ordnance stores purchased by the (Federal) Ordnance Department from January 1st, 1861 to June 30th, 1866" which included 415 "Adam's revolvers" at a total cost of $7,526.38; no source or calibre was stated for these arms. Again, *U.S. House Executive Documents* v. 12, 40th Cong., 2 Sess. 1338, 954–956 (a citation for which we are obliged to the New York Public Library) gave the following details in relation to Federal "Purchases of cannon, ordnance, projectiles, and small arms since April 31st, 1861" in relation to British revolvers.

Contractor's or seller's name	*Date of Purchase*	*Quantity and kind of stores*	*Price* (each)
Schuyler, Hartley & Graham	Nov. 11th 1861	120 Adams's revolvers, double-action	$ 18.00
,,	Dec. 3rd 1861	93 Adams's pistols (Eng.) self-cocking	18.00
,,	Dec. 13th 1861	7 Adams's revolvers	18.00
,,	,,	12 Adams's revolvers, self-cocking	18.00
		16 Kerr's patent revolvers	18.00
		26 Beaumont revolvers	19.00
,,	Dec. 24th 1861	115 Adams's revolvers double-action	18.00
,,	Jan. 20th 1862	25 Adams's revolvers self-cocking	18.50

Contractor's or seller's name	*Date of Purchase*	*Quantity and kind of stores*	*Price* (each)
Schuyler, Hartley & Graham	Feb. 17th 1862	144 Adams's revolvers	$18.00
,,	Mar. 1st 1862	60 Adams's revolvers	18.00
,,	Mar. 11th 1862	22 Adams's Eng. revolvers	18.00
,,	Apr. 23rd 1862	24 Adams's revolvers	18.00

If the sixteen Kerr revolvers in this list are ignored, a total of 623 Adams or Beaumont revolvers were shown as bought from the New York merchant house of Schuyler, Hartley & Graham (between November, 1861 and April, 1862), and were presumably all weapons originally purchased, in Great Britain, for civilian sale. The total cannot be reconciled with Colonel Lewis's figure, but enough has been said to suggest that some ·44-in. revolvers (the most popular British calibre) were bought from sources other than the Massachusetts Arms Company, and that such purchases may have contributed to the legend of their manufacture by that Company.

The ·36-in. Massachusetts Arms Co. weapon, as most courteously described to us by Lt.-Col. R. C. Kuhn of Chicago, Illinois, and Mr. Roger C. Peterson of Portland, Maine (to whom we are deeply indebted for assistance here), was a five-shot revolver, with 6-in. barrel rifled in three grooves and a slow right twist, and had an overall length of 11½ in.; a blued finish overall was common, but plated specimens have been reported. Major differences in appearance from the English-made Beaumont-Adams revolvers will be clear from a comparison of Plates 15 and 13, but the reader's attention is drawn to a small plate, rivetted into the left side of the frame, just behind the cylinder on American weapons. This plate was not apparently intended to be removed readily, and can best be accounted for as evidence of differing methods, in manufacture or assembly, as between the English and American factories.

Lt.-Col. Kuhn's ·36-in. calibre revolver bore a U.S. Ordnance inspector's stamp, but otherwise had exactly the same markings as Mr. Peterson's "civilian" pistol. These markings comprised the words "Manufactured by Mass. Arms Co., Chicopee Falls" on the top strap of the frame, "Adams' Patent May 3, 1858" on the left side of the frame, "Patent June 3, 1856" on the right side of the frame, and "Kerr's Patent April 14th, 1857" in two lines on the rammer-lever. Both pistols had serial numbers, the Kuhn weapon being "609" and the Peterson specimen "12". It was observed that these legends did not agree with those U.S. Patent dates earlier recited, as to the "Adams Patent" date of 1858, and we were forced to the conclusion that frame-stamps prepared by the makers must have been defective. The error appeared on all ·36-in. Mass. Arms Co. revolvers reported to us, but the U.S. Patent Office confirmed that May 3rd, 1858 was not an issue date for *any* U.S. patent; it must therefore be accepted that the frame reference was intended to be to Robert Adams's U.S. Patent No. 9694 of 185*3*.

The smaller, ·31-in. calibre pistol was also five-chambered, and had a 3¼-in. octagonal barrel; it was generally (as may be seen on Plate 15) a smaller version of the ·36-in. revolver, but markings upon it differed in being concentrated upon the top strap of the frame, in a legend reading:

> "Made for Adams Revolving Arms Co., N.Y.
> By Mass. Arms Co. Chicopee Falls,
> Patent May 3, 1858, June 3, 1856, April 7, 1857".

The pistol in the Peterson Collection was serially numbered "40" on the left of the frame, and a double error can be observed in this patent legend, for Kerr's U.S. patent date was given incorrectly, whilst the error for Adams's patent was perpetuated. Courteous and most thorough inquiries (which we here gratefully acknowledge) were made by Mr. James J. Heslin of the New York Historical Society and by Mr. Rutherford D. Rogers of the New York Public Library (Reference Department), but no New York City or State directory (whether business or residential) gave any listing or address for this "Adams Revolving Arms Co., N.Y.".

The sliding safety-catch and captive-screw cylinder-dismount features of Robert Adams's British Patent No. 2645 of 1854 were fitted on the right-hand side of both ·36-in. and ·31-in. calibred weapons, although the safety-catch was here steeply angled, in line with the rake of the butt, instead of sliding horizontally as on English versions of the Beaumont-Adams revolver. Neither feature was ever patented in the U.S., by Robert Adams.

CHAPTER XI

SAMUEL COLT

SAMUEL COLT and his arms have an assured place in any book about British percussion revolvers, but much of their history is now so well known that there can be little point in reiterating it here. Accordingly, the reader is commended to H. L. Blackmore's admirable *British Military Firearms, 1650–1850* (Herbert Jenkins, 1961), for a summary of that history or to excellent specialist works by American authors, if detailed knowledge about this inventor is desired; these pages which follow are principally confined to the citation of contemporary material that is less familiar to an average reader.

As to the historical background to these pages, Samuel Colt was born at Hartford, Conn., on July 19th, 1814, and he died there on January 10th, 1862. Whilst young, he became interested in revolving firearms, and secured English Patent No. 6909 of October 22nd, 1835, and U.S. Patent No. 9430X of February 25th, 1836, in respect of such weapons, as soon as he was old enough to be granted protection.

The English patent was never worked, but Colt secured American financial backing for his ideas, and The Patent Arms Manufacturing Company (at Paterson, New Jersey) began to manufacture single-action percussion revolvers with folding triggers (and revolving carbines, rifles and shotguns) under Colt's U.S. patent, late in 1836. The venture was not successful, unfortunately, and the Company became insolvent in 1843.

In 1847, however, the Mexican War created a demand for revolving pistols, and Colt received an order from the U.S. Army for 1,000 heavy ·44-in. calibre holster arms, which are now very rare; in them the inventor improved upon his frail design of 1836, producing a robust six-chambered hand-gun with a powerful lever-ramrod, stout trigger guard, and one-piece frame. He did not, however, discard the use of a separate barrel, pinned to the frame by a cross-bolt or key passing transversely through the forward end of the cylinder arbor, nor his original single-action lock-mechanism.

In order to fulfil the United States Government order for his Model 1847 pistols, Colt had to use the services of Eli Whitney's Armoury (at Whitneyville, Conn.), but he had established his own factory at Pearl Street, Hartford, by 1848. In the years that followed 1847, Colt introduced the following variations upon his successful Model 1847:

1848. The six-shot "Dragoon" (later "Old Model Army") revolver, with 7½-in. barrel, in ·44-in. calibre.

1848. Five-shot pocket revolvers, with barrels of 3, 4, 5, or 6-in. lengths, and in ·31-in. calibre. No ramrod fitted.

1849. Pocket revolvers with lever-ramrods, in ·31-in. calibre. Five- and six-shot versions made, with various lengths of barrel.

1851. Six-shot belt or "Navy" revolvers, in ·36-in. calibre, with 7½-in. barrels, and a pocket arm in this calibre.

1860. Six-shot "Army" revolver in ·44-in. calibre, with 8-in. barrel. An improved ratchet-lever ramrod was fitted.

1861. Six-shot "Navy" revolver in ·36-in. calibre, with 7½-in. barrel and the improved rammer.

1862. Five-shot "New Police" revolver in ·36-in. calibre, with various barrel lengths available, and the improved rammer.

These single-action percussion arms were all built with separate barrels, connected to the frame by a key in the manner shown in Plate 16. In 1857, however, Colt collaborated with E. K. Root (the Superintendent of his factory) in manufacturing solid-framed revolvers, with side-hammers of the type shown on Plate 36 (*lower*); unlike Adams's pistols, these arms had a separate barrel screwed into their frame Small-calibre revolvers of this pattern were made, but it is principally remembered for application to carbines, muskets, rifles and shotguns.

Colt was undoubtedly fortunate in the period of American history which coincided with his career, and the revolvers mentioned above were sold in thousands, even before the American Civil War; in 1855, to meet this demand, a new factory was built on the banks of the Connecticut River, at Hartford, and remains the site of a modern plant today.

Endowed with great energy and perseverance, Colt was not content to rest upon the success that his arms enjoyed in America. Over the years, he traversed Russia and Europe, selling both his arms and the know-how for making them.

Although Colt secured English Patent No. 12668, of June 20th, 1849, covering the principal features of his improved arms, it was not until 1851 that he made any serious attempt to interest the British in his revolvers. The Great Exhibition, at London, served as a show-case for his wares, and something over 100 standard arms were displayed in ornamental panoplies; the effect, against competitors who exhibited only individual weapons (however decorative), was stunning.

It seems probable, moreover, that Colt also had in mind, from the start, a market considerably more profitable than that for revolvers alone.

The system used for manufacturing his arms was highly mechanized, by the standards of 1851, and he claimed for it the ability to produce revolvers with fully interchangeable parts; that system is more fully described in Chapter XIII, and we believe that Colt came to the Exhibition bent, at least in part, upon selling to the British Government his methods for the manufacture of infantry rifles with interchangeable parts.

As earlier mentioned, however, Colt's exhibition of revolvers, at the Crystal Palace, aroused considerable public interest in its own right, and he was quick to take advantage of this fact.

An elaborate importation procedure had been set up by H.M. Customs & Excise, to govern duty-free importation of all exhibits from overseas, and the majority of exhibitors appear to have been content to abide by that arrangement, and not to touch their exhibits. Colt on the other hand, apparently took the trouble to arrange withdrawal of twenty-five pistols from his show-cases, in July, 1851; these were sold to officers of the 12th Lancers (under orders for South Africa), and the publicity secured by his act must have been valuable to the American.

In October, when the Exhibition finally closed, Colt again made no bones about using the contents of his show-cases to advantage; a list (which still survives) was carefully compiled, and over eighty pistols were given to men or organizations likely to be helpful, in his plans to interest the British services in both his arms and manufacturing methods.

As a natural extension of these efforts, Colt also attempted a direct approach to the British armed forces, and something has already been said of these efforts, in Chapter V, and of his rivalry with the English revolver-manufacturer, Robert Adams.

As a result of these incidents, Colt decided that a substantial market for revolvers existed in the United Kingdom, and began preparations to open a factory for their manufacture here. Premises off Bessborough Place (in Pimlico, London) were leased in 1851, and production actually began in January, 1853, using the equipment and methods described in Chapter XIII.

Initially, there was little official interest in Colt revolvers, and although proof-figures indicate that they were probably made in some numbers during 1853 (see Chapter III), sales were to the public or to army officers buying their own weapons, and not for service issue. As later mentioned, it is believed that production was limited to the Model 1849 and Model 1851 pistols already described, and shown in Plate 16.

In 1854, however (see H. L. Blackmore, *op. cit.* pp. 254–5) events began to move in the direction desired by Colt, and on March 8th, the British Navy ordered 4,000 of his Model 1851 "Navy" revolvers; 5,500 more arms were ordered by the same source, later in that year.

More important, however, a Parliamentary Committee was set up in 1854, to determine "the cheapest and most expeditious and most efficient mode of providing Smallarms for Her Majesty's Service"; it was in this body that Samuel Colt's best chances lay for selling his whole system of manufacturing arms with interchangeable parts, as opposed to the sale of completed weapons. Financially, the opportunity was immense, for Service rifles were needed on a scale vastly exceeding that for revolvers, and it was to be hoped that royalties and other payments to those involved in rifle-manufacture for the British armed forces would be commensurate with the scale of the operation.

Deliberations by such a Committee were overdue, for H.M. Board of Ordnance had suffered long under the shortcomings of the British system of purchasing service small-arms from the gun trade at large. In themselves, such arms were as satisfactory as could be hoped for, from an industry that flourished on hand-work and used little machinery; however, even parts made by individual gunmakers would not interchange (a logistic weakness, when repairing damaged arms, whether recovered from battlefield or issue), and interminable delays were common in the delivery of weapons ordered by the Board, from Birmingham or London gunmakers.

The Committee (which had before it reports from a body sent to investigate American manufacture of interchangeable arms, in 1853) hoped to find solutions to these problems, and probably began its deliberations at least partly convinced of the need to set up a government factory for manufacturing service arms. Colt was unable to claim experience in producing rifles by his methods, but could demonstrate complete mastery of manufacturing revolving pistols in large quantities. The Committee therefore listened to his evidence respectfully, and also secured the opinions of both the Board of Ordnance and of various leading members of the Birmingham gun trade. Since the latter stood to lose an extremely profitable source of business, if Colt's methods were adopted, and were (almost to a man) temperamentally opposed to the idea of government manufacture in any field, their evidence tended to be carping or hypercritical.

The chief witnesses from the Ordnance were: Captain Sir Thomas Hastings (who was Principal Storekeeper), Lt.-Col. Tullock, (Inspector of the Royal Carriage Dept.), and John Anderson (Chief Engineer at Woolwich). All were very favourable towards Colt, and obviously thought that his methods should be adopted.

The following gave evidence for the gun trade, and their remarks were of sufficient interest (despite evident prejudice, in some cases) to be set out at length.

Mr. Richard Webb Gunner—Inspector of Small Arms.
Mr. Brazier—Gunmaker, Birmingham.
Mr. Charles Clark—Operative Gunmaker.
Mr. William Scott—Gunmaker, Birmingham.
Mr. Edward Philip Bond—Engineer in the Gun Trade.
Mr. Thomas Hunt—Supplier to the Board of Ordnance.
Mr. John Dent Goodman—Merchant.
Mr. Charles Millward—Gun-barrel Maker.
Mr. Tranter—Gunmaker.
Mr. Westley Richards—Gunmaker.
Mr. Gage Stickney—A worker for Colt.
Capt. Sir Thomas Hastings, R.N., C.B.
Lt.-Col. Alexander Gordon.
Major Robert Bruce.

Mr. Richard Webb Gunner, Inspector of Small Arms

He agreed that the barrels and cylinders were well made but thought the fittings, springs and internal work was inferior to our manufacture. The general finish of the Colt and the letting into the wood is careless and unsightly which he attributed to the want of skill in the artisan, who was a lower grade of man. Also the cock and tumbler are in one, which is an inferior mode of construction, the bents inferiorily cut and the little part where it catches on the sear, the sear nose did not fit. The work would be rejected here.

Ques. Is the internal work of Col. Colt's revolver so inferior as to impair the efficiency of the arm?
Ans. In some degree, I should say it is.
Ques. Have you seen the other revolving pistol made by Deane & Adams?
Ans. I have.

Ques. What is your opinion of that pistol?
Ans. The pistol I saw of Deane & Adams was very good work.
Ques. Would you pass the lock of such a pistol?
Ans. I should.

Mr. Brazier, of Birmingham

Ques. What is your opinion of the statement that the different parts might be thrown together into a basket and taken out indiscriminately and fitted together?
Ans. I do not believe it. I know that they would not go together again because I had two Colt pistols. I tried to reverse the rods and they would not act.

Mr. Charles Clark, an operative Gunmaker

Ques. Have you seen any of these new pistols by Col. Colt?
Ans. I have taken great pains to understand them.
Ques. Did you ever take any of them to pieces?
Ans. I have one here, a new one purchased last week and it is his highest-class pistol. This (pointing to same) has been dissected and it contains forty-seven pieces. Here is an English pistol (pointing to same) on an improved principle, a revolver and it contains thirty-one pieces only. This is a Deane & Adams. The workmanship of Col. Colt's is of a most inferior description, in fact it is opposed in the main parts to the leading mechanical principles, in fact it is almost laughable, the idea of this mainspring (pointing to same), this when striking the nipples loses power and when you raise it up to this extreme point (describing the same) it exerts a power of about eight or nine pounds and when it falls it falls about four, the cock is blown back with the power of the explosion but in the other the cock is stronger on the nipple losing power as it is raised up.

Mr. William Scott, Gunmaker of Birmingham

He said the Colt barrel and cylinder were good, but the lock inferior and the remainder perfectly common.

Mr. Edward Philip Bond, Engineer in the Gun Trade

Ques. Have you seen any of Col. Colt's pistols?
Ans. Yes. I am well acquainted with them.
Ques. Have you ever tried to interchange the different parts of the pistol?
Ans. I have tried it with a few and I have found the statement to be erroneous. The pistols cannot be said to interchange. I have tried to interchange two or three barrels and found difficulty. have tried to interchange two or three ramrods and have found

that they would not fit. The cylinders may occasionally fit because they are turned.

Mr. Thomas Hunt of Potts & Hunt, established in 1734—Suppliers to the East India Co. and the Board of Ordnance

Ques. Have you ever seen Col. Colt's revolvers?
Ans. Yes.
Ques. Do you consider his revolver to be a good and efficient weapon?
Ans. My impression is that all chambered revolvers are bad, but Colt's is not the best even of the bad ones. I consider it inferior to some which are now manufactured.
Ques. You think the revolving chamber is bad?
Ans. I do. There has never been a revolver produced yet by which the escape of gas is prevented between the chamber and the breech of the barrel.
Ques. Have you ever tried them?
Ans. Yes, but I have never shot out of one of Col. Colt's.
Ques. You stated with respect to revolving pistols that the escape of gas between the breech end of the barrel and the cylinder is very likely to make the other barrel explode, will you explain?
Ans. I read an account, I think furnished by an officer at the Cape that such had been the case. I had an order for some Colt revolvers from Vienna and I went to Col. Colt's to examine for myself. I could clearly see from the construction that it was impossible to be otherwise and the result was that I did not furnish our customer with the revolvers.
Ques. Does the fire make its way between the ball and the inner surface of the chamber down to the powder?
Ans. That would be the case if the bullet were not rammed in very tightly. I think Colt attempted to obviate that by the lever rammer which has its inconveniences as well as its advantages.

Mr. John Dent Goodman, Merchant trading between Birmingham and America

Ques. Have you ever seen any of Col. Colt's pistols?
Ans. I have had a few in my hand. I should say that the work in them is very inferior to what can be produced and is produced in Birmingham.

Mr. Millward, Gunmaker. (See also Chapter III.)

Ques. Yo you make pistols in your manufactory?
Ans. Des, that is a leading branch of our manufacture.
Ques. Have you ever seen any of Col. Colt's pistols?
Ans. I have had a few in my hand. I should say the work in them is very inferior to what can and is produced in Birmingham.

Mr. Tranter, Gun and Pistol Maker of Birmingham

Ques. Have you seen any of his pistols (Colt's)?
Ans. Yes.
Ques. Have you ever taken any of them to pieces?
Ans. Yes, I have taken one or two to pieces.
Ques. It has been stated that if they are taken to pieces and the different parts put into a basket they would fit together again indiscriminately; is that a fact?
Ans. I do not think they would.

Mr. Westley Richards, Gunmaker

Ques. Have you seen some of Col. Colt's pistols?
Ans. Yes.
Ques. Have you ever taken one to pieces?
Ans. Yes, a great many.
Ques. What do you think of the quality of the work?
Ans. I think it is inferior.
Ques. When you speak of the inferiority of the work in Col. Colt's pistol, do you apply that to the whole of the pistol or only to the lock?
Ans. To the arrangement of the lock, roller and mainspring. The whole of the barrel is very well done and I believe entirely by machinery. I have seen many of them not well done, they are not well done, they are not alike and different in bore.

Mr. Gage Stickney (an American who worked for Colt in that country)

Ques. Do you consider it is possible to interchange the parts of one pistol with another made by Col. Colt?
Ans. I never saw a case of the kind. I have heard of it but I defy any man to show me a case.
Ques. What do they mean when they say that the parts interchange?
Ans. I do not know. I know the American work was never gauged.
Ques. If they do not interchange in America, the Committee of Congress must have expressed a very wrong opinion.
Ans. They calculate that it is very near alike but it requires an armourer to file and fit the parts, the same as any other.

Capt. Sir Thomas Hastings, R.N., C.B.

Ques. Have you examined the interior of Col. Colt's pistol?
Ans. I have looked at parts of the lock.
Ques. Do you consider the parts to be good?
Ans. Yes, to be perfectly efficient and useful for the purpose intended.

Further questioned, he said that in his opinion the workmanship of the lock was not as good as that of the Minie rifle lock but "fully equal to its work" and in any case the comparison was between two dissimilar things.

Ques. How many pistols did you buy of Col. Colt?
Ans. We are to buy 4,000.
Ques. At what price?
Ans. 50*s*. each.

Lt.-Col. Alexander Gordon

Ques. Have you had any experience of Col. Colt's pistols?
Ans. I have known it for fourteen years. I have had one for fourteen years and I have bought another. I am going to the East and I have bought a larger one than I had before.
Ques. What is your opinion of the lock?
Ans. The lock answers its purpose perfectly well which is the only thing it is required to do.

Major Robert Bruce (who was in the Kaffir War)

Ques. How long had you these pistols of Col. Colt in use?
Ans. I think during the last six months of the war.
Ques. Have you ever made any experiments as to their qualities by firing them repeatedly for a long time?
Ans. Yes.
Ques. How many times did you ever see Col. Colt's pistols fired off successively?
Ans. I fired mine, 30, 40 or 50 shots. When I became tired I handed my pistol over to some of my brother officers. I do not know how many times but I dare say they went on firing as long as they had bullets, there was no complaint.
Ques. Will you state the grounds on which you formed the opinion (i.e. superior to any other)?
Ans. Its accuracy and in not failing in what you required. I have never seen an accident with it and I have never seen it in discharging become involved in any difficulty.
Ques. Does it wear as well as other pistols?
Ans. Quite. I have also seen it fall into the water and returned to its belt without being discharged and afterwards perhaps, after a lapse of some days, I have seen it discharged most perfectly without the least failure of any of the barrels (i.e. chambers).
Ques. Do you mean to say that the pistol was fully charged when it fell into the water?
Ans. Yes, fully charged and some days after it was discharged. The cap was not taken off and without discharging it, it would be difficult to draw the ball. It went off perfectly. Another great advantage is in riding. I have ridden hundreds of miles with a Colt revolver by my side and there is an impossibility of the

powder shaking out. I have found the ball in other pistols very frequently loosen but the advantage of Col. Colt's revolver is the great purchase of the ramrod. There is such a power on the ball, it forces it down a way no other ramrod could do.

Ques. Have you had an opportunity of seeing Deane's revolver tried?

Ans. I have seen it in the possession of other officers and it seemed to me to answer well enough in that respect. I have heard indeed friends of mine who had Deane's revolver rather complain of the ball loosening.

Ques. With regard to their misfiring, have you heard complaints made that those revolvers made by Deane frequently misfire, for some cause or other?

Ans. Yes, I have fired both frequently, but Deane's not so frequently as the other. I think that Col. Colt's is very superior in every way to Deane's. In my opinion it is very more simple, the simple fact of forcing the ball in so tightly is a great advantage and I consider that the ball has literally no windage.

Ques. Have you known any part fault after frequent use?

Ans. I have once or twice seen a little difficulty if the cap did not fit the nipple perfectly. I have seen once or twice the cap fall between the revolving cylinder and the breech but that has been because the cap was not perhaps a proper cap for the nipple, when the cap fitted accurately, I have not seen that.

Ques. Have you known the component parts of the Colt, a spring for instance, get out of order?

Ans. I have never seen that.

Ques. Did you find any advantage from the perfect adaptability of the different parts?

Ans. I have seen pistols of the same size taken to pieces and the various parts fitted on to another with perfect ease.

In examining this matter of interchange, the Committee also took evidence from the inventor himself, both as to the extent of his use of machinery, and the facility of interchange secured by such use.

Ques. Do you make your pistol chiefly by machinery?

Ans. Entirely except for taking the burr off and passing the parts from one machine to another, finishing and assembling the work.

Ques. Do you consider that you make your pistols better by machinery than you would by hand labour?

Ans. Most certainly.

Ques. And cheaper also?

Ans. Much cheaper.

The degree of interchange actually secured is more fully discussed in Chapter XIII, but Colt is on record as telling the Committee that:

"It is the constant habit of Government Officers who use my arms in Oregon or in California to order 100 mainsprings or 1,000 mainsprings or 1,000 hammers or 1,000 triggers . . . and I send them and one of the commonest operatives or soldiers will take an ordinary tool and put them in the place to which they belong. If the hammer is broken, he will put in a new one and so on."

For obvious reasons, the Committee was also concerned to discover the quality of work attainable by Colt's methods; in England, a long apprenticeship had to be served before a man was regarded as capable of making firearms, yet Colt said of the labour that he employed "all you want is interest and the faculty to manage machinery".

Lord Seymour (a member of the Committee) questioned one witness, who had been with the deputation sent to America, in 1853, to study production methods, "you stated that when you went to Col. Colt's you saw a young lad who had been a butcher managing four machines?" The witness replied in the affirmative, and added that he had commented upon this (which had no parallel in England) to Col. Colt, who replied "the most trustworthy men I have were originally agricultural labourers".

Mr. Clark (recalled)

Ques. Have you seen the machinery used there (i.e. at Colt's)?
Ans. Yes.
Ques. Did you notice to what extent he employed manual labour after he had done with the machinery?
Ans. I think I can state the exact amount of hand labour he uses. I should say it is at least 50 per cent. I can prove he has full 50 per cent of hand labour and a deal of it is high-class labour, first-rate mechanics.
Ques. Have you ever tried to fit together any pistol of his make?
Ans. Yes, that is an error, they will not reverse or interchange, you cannot get two to do so.

Mr. Gage Stickney (recalled)

Questioned on hand labour at Colt's he said: "Fifty per cent of the manufacture is by machinery and the rest hand-work, which was first-class labour."

Ques. What class of labour does Col. Colt use here?
Ans. About two-thirds medium class, the rest first class. Cheap labour 2*s.* 4*d.* per day, first class 6*s.* to 8*s.* per day, some 10*s.* per day and some 12*s.* per day.
Ques. Are the Committee to understand that all the different parts of the pistol made by Col. Colt, after having been prepared by machinery, are finished by hand labour?

Ans. Yes, by first-class mechanics.
Ques. Every part must be finished by skilled labour, must it not?
Ans. If it is done properly, every part will be so done.
Ques. Did he bring any skilled labour with him?
Ans. About twenty-five to thirty men. I consider them all skilful.

For reasons stated in Chapter XIII, malice can be imputed to Stickney's evidence, but it remains a fact that Colt's system for manufacturing revolvers depended heavily upon the skilled services of his supervisors, assemblers, and inspectors.

In the event, this Committee made no specific recommendations upon the use of Colt revolvers by H.M. forces, but early in 1855, the Board of Ordnance placed an order for 5,000 of his revolvers for use by the Army; the Board later bought 9,000 more pistols in August of that year.

The weapon chosen was again Colt's Model 1851 revolver (as earlier purchased by the British Navy), which is shown on Plate 16; the pistols differed very little from a version sold to private customers, but surviving specimens suggest that the standard roller-engraved ornamentation of the cylinders was turned off, for British service-contract arms. In addition, the "W.D. and broad arrow" mark shown in Plate 17 was struck upon their barrels, and most of the wooden butts carried an inspector's mark, lightly impressed upon them. Colt stamped the legend "Address Col. Colt London" on the top flat of each octagonal barrel, and this was not interfered with, on the service pistols; for commercial arms, London proof-marks were normally present, but service pistols can also be found which have, instead, official "VR and Crown" and "P with crossed flags" upon their barrels and cylinders.

The single-action lock-mechanism used in these arms was standard for every percussion revolver made by Colt after 1847, except as to the Model 1857 side-hammer design; see Figure 19.

The hammer (A) has a pawl (B) pivoted to it, and also actuates the cylinder-bolt (C), which see-saws upon a transverse frame-screw. When the hammer is thumbed back, (B) rises and turns the cylinder through part of a revolution, by engaging a ratchet cut at the rear of it; as soon as the hammer starts to move on its pivot, an actuating stud tips up the rear of (C) and rocks the front end of it out of engagement with locking-slots cut in the cylinder wall, to permit the cylinder to turn.

If the hammer is allowed to rest at half-cock, (C) is held out of engagement, and the cylinder can be manually revolved for loading and capping. If, however, the hammer is drawn back to full-cock, the bolt-tail (which is suitably split along part of its length) springs past the

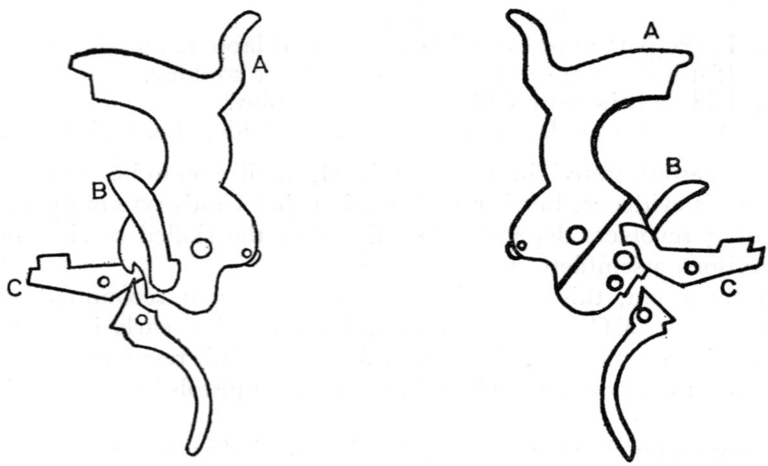

FIG. 19. *Colt revolver lock mechanism.*

actuating stud on the hammer, and a suitable spring rocks (C) into engagement with a cylinder locking-slot to lock chamber and barrel in alignment for discharge.

On the fall of the hammer, the actuating stud (which has a chamfered face) brushes past the tail of (C), and finishes at rest beneath it, ready to actuate (C) next time the hammer is cocked.

This classic single-action lock-mechanism may be compared with the competing, self-cocking, lock of Robert Adams, described in Chapter V and shown in Fig. 10; both designs had vulnerable points in them, but it is true to say that the Colt lock had springs of a less sophisticated shape than those used by Adams, and that it was therefore easier to repair after any spring failure.

The protagonists of each weapon also fell out over their respective frame-and-barrel constructions. Adams's one-piece frame and barrel was theoretically preferable to Colt's keyed assembly, as less liable to deform under abuse, or through wear. In fact, the advantage probably remained largely theoretical, for it is unusual, even today, for a Colt revolver to be found in such condition (as regards the union of barrel with frame) that it cannot be restored to shooting trim by an average collector.

Indeed, it is clear from the Committee evidence already cited that naval and military witnesses regarded Colt revolvers as good and efficient weapons, well able to carry out the tasks for which they were designed, and the report of an officer appointed by the Board of Ordnance to test both Colt and Adams revolvers bore out this view.

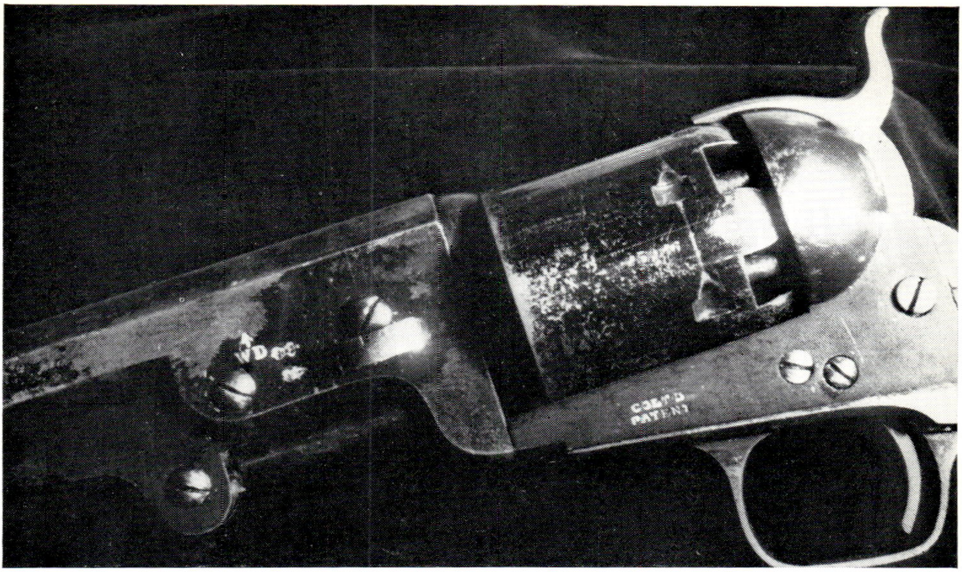

PLATE 17. COLT REVOLVERS – BRITISH SERVICE ISSUE

(*Top*) British "War Department" stamp upon barrel of Colt percussion revolver, Model of 1851. Note presence of commercial London proof-markings, and deliberate erasure of cylinder engraving. *Private Collection.*

(*Lower*) British War Department inspection stamp ("Crown over BR7"), on right butt-cheek of a Colt percussion revolver, Model of 1851. *Private Collection.*

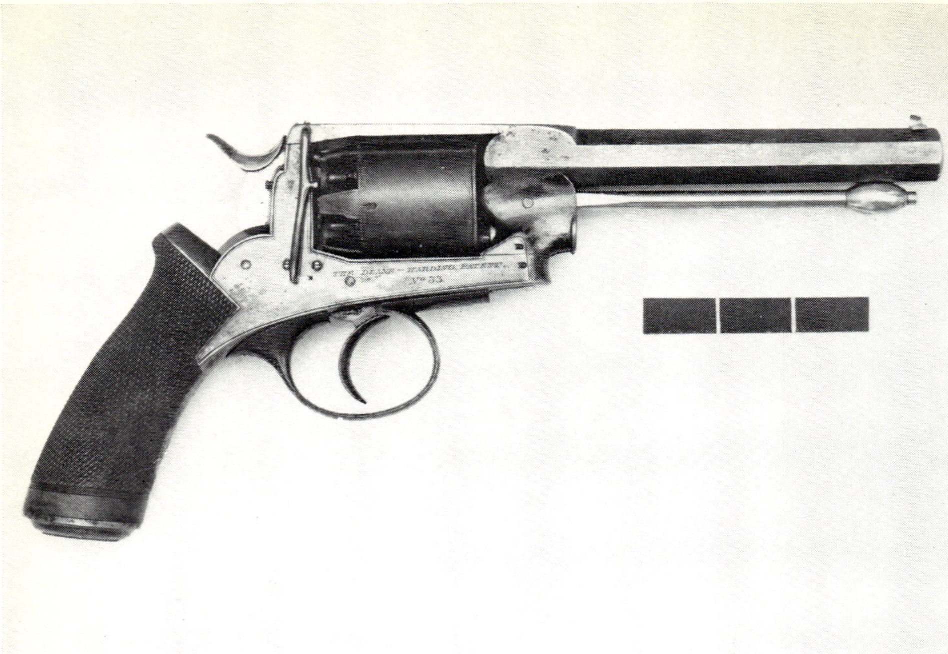

PLATE 18. DEANE-HARDING PERCUSSION REVOLVERS

(*Top*) First model of Harding's double-action percussion revolver, to Br. Pats. 669/1858 and 1159/1858. Made in 54 bore calibre, and serially-numbered 33. Compare frame and rammer-lever with later models shown in Plates below.

J. B. Bell Collection.

(*Lower*) Second model of Harding's double-action percussion revolver, in 54 bore calibre. Serially-numbered 6536L, but differs only in minor detail from versions with "P"-suffixed serial-numbers. *E. W. Taylerson Collection.*

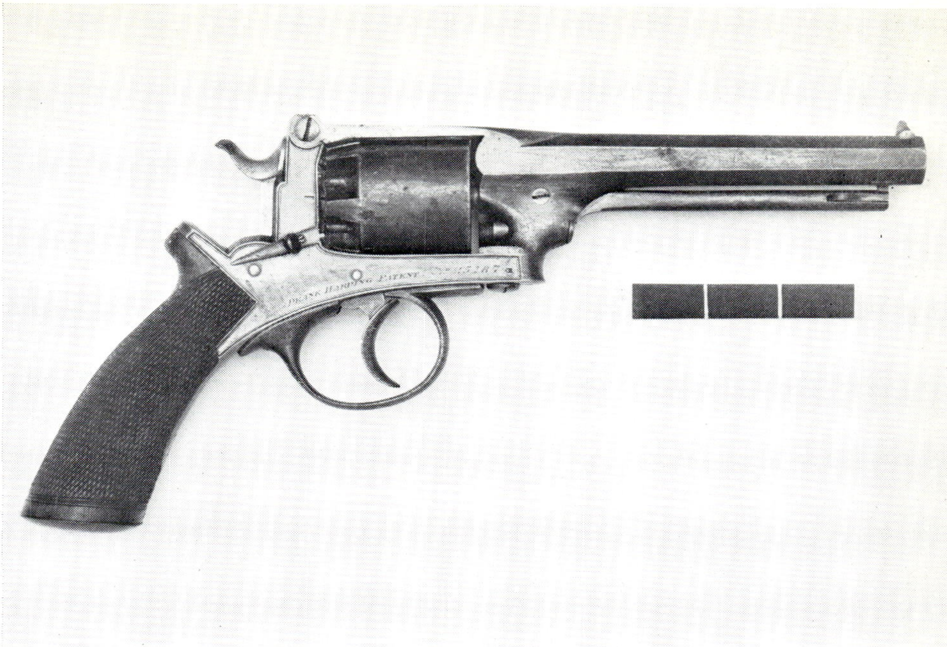

PLATE 19. DEANE-HARDING REVOLVERS
(*Top*) A double-action Deane-Harding percussion revolver, in 54 bore. This is the Third Model (with frame hinged to the top of the standing-breech), and is serially-numbered 25157. *J. B. Bell Collection.*

(*Lower*) A double-action Deane-Harding optional percussion rim-fire revolver, in ·442/54 bore calibre. The weapon is made under the Harding patents of 1858, and John Deane's patent of 1862. The serial-number is 93 DHD 15430P, and the rim-fire cylinder is shown in the Plate. *Darwent Collection.*

PLATE 20
CHARLES MANBY, F.R.S., M.INST., C.E.
Born: February 4th, 1804
Died: July 31st, 1884
Secretary of the Institution
of Civil Engineers
June 21st, 1839 – May 23rd, 1857
(Frontispiece from Volume LXXXI. 1884–1885, Part III, to the *Minutes of Proceedings of the Institution of Civil Engineers*. Reproduced here by kind permission of the Institution.)

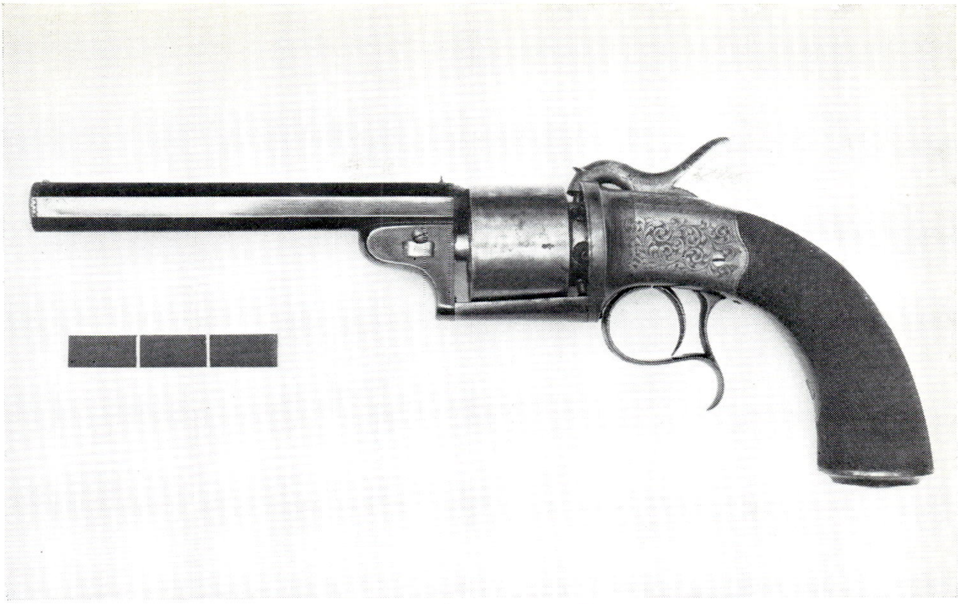

PLATE 21. DAW PERCUSSION REVOLVERS
(*Top*) Single-action gas-seal percussion revolver of the "Lang" type, in 54 bore calibre. Marked "Witton, Daw & Co., Improved. 57 Threadneedle St., London". Early revolver, without rammer. *Andrews Collection.*

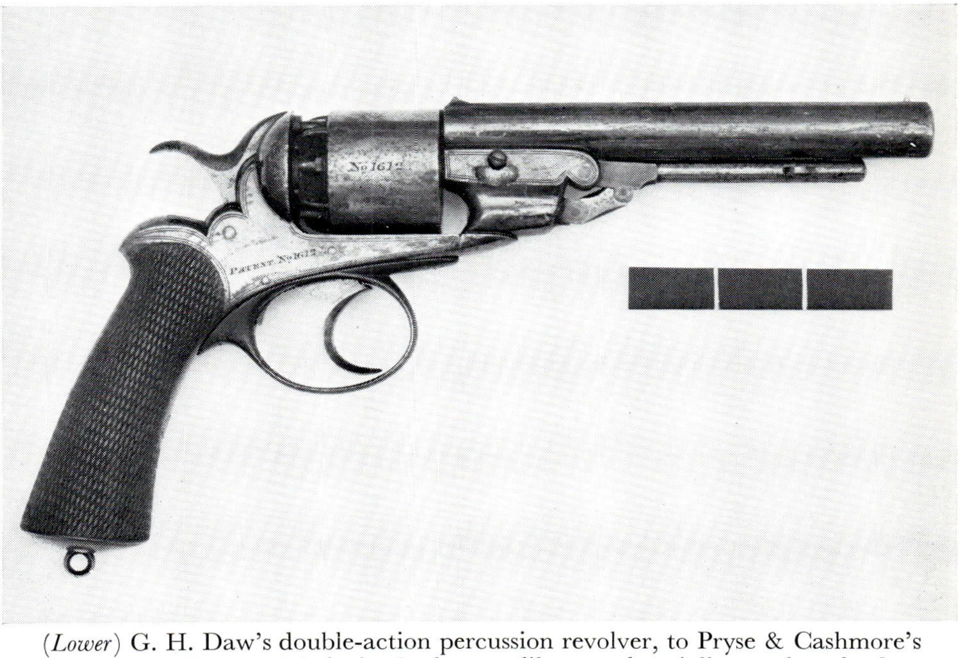

(*Lower*) G. H. Daw's double-action percussion revolver, to Pryse & Cashmore's Br. Pat. 2018/1855. Made in 80 bore calibre, and serially-numbered 1612, Retailed by G. H. Daw, from 57 Threadneedle St., London. *J. B. Bell Collection.*

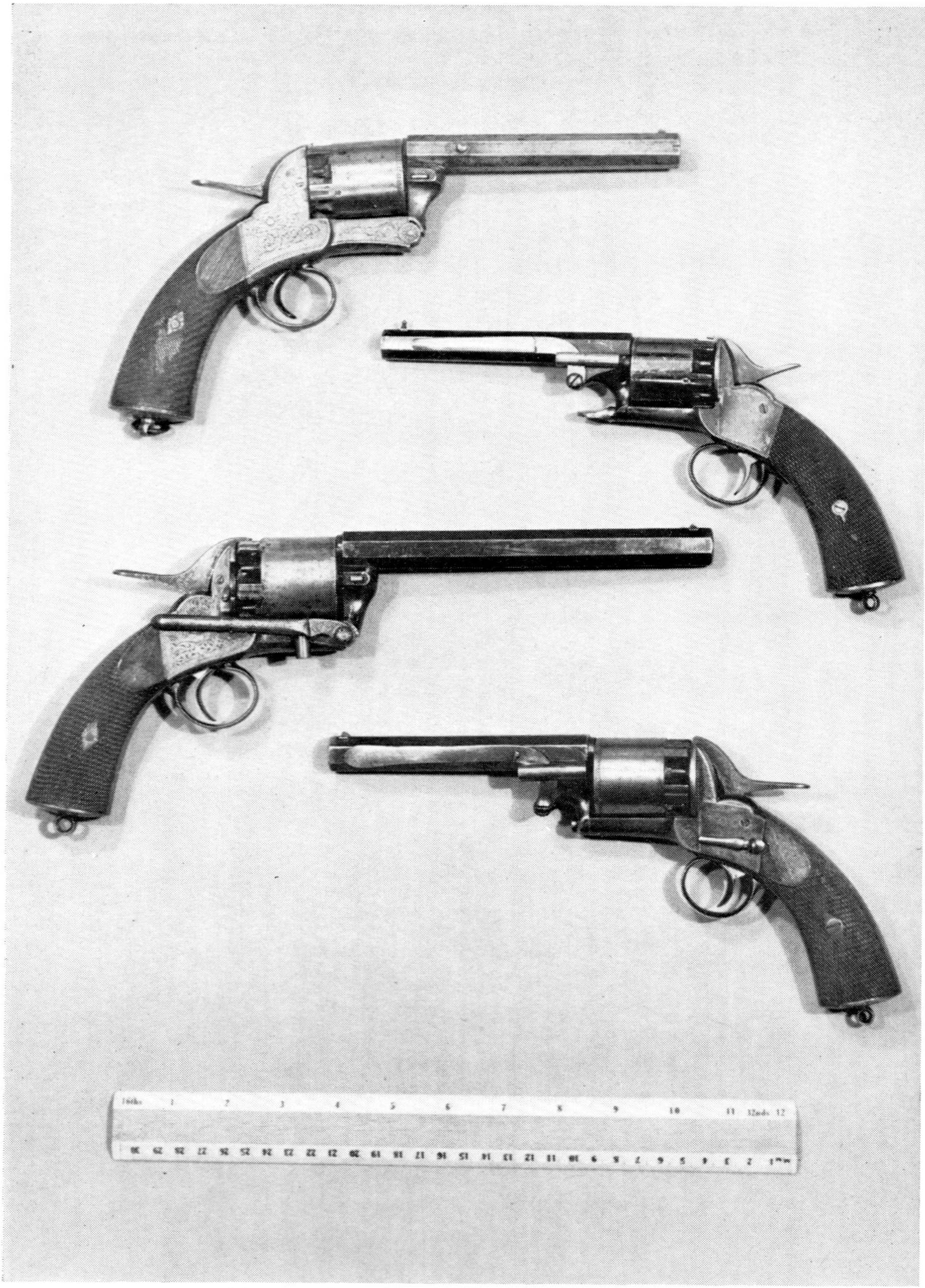

Plate 22. James Webley "Longspur" Revolvers

Single-action percussion revolver in 48 bore, made under James Webley's patent of 1853, and retailed by "Nobbs, Isle of Wight". Serial-number 332. The First Model.

Late single-action percussion revolver in 120 bore, made under James Webley's patent of 1853. No markings or retailer's name. No serial-number. The Third Model.

Single-action percussion revolver in 48 bore, made as above, and retailed by James Worrall, Chester. Serial-number 426. The Second Model.

Single-action percussion revolver in 64 bore, made as above, but un-marked. Serial-number 1025. An unusual Third Model, fitted with safety-bolt. *Private Collections.*

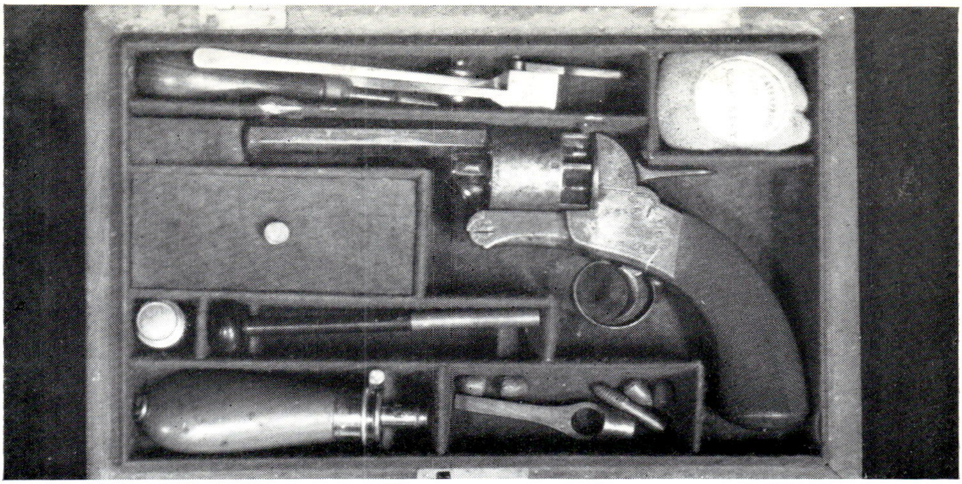

PLATE 23. CASED JAMES WEBLEY "LONGSPUR" REVOLVERS
(*Top*) Cased First Model single-action James Webley percussion revolver, retailed by Edwin Ladmore, Hereford. Serial-number 222. Made in 120 bore calibre. Note loading-rod in case. *Private Collection.*

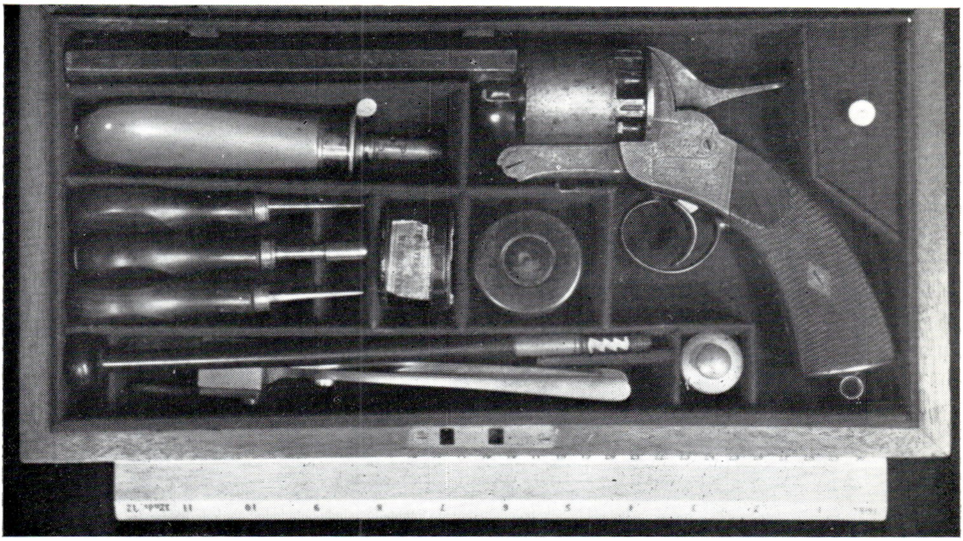

(*Lower*) Cased Second Model single-action James Webley percussion revolver, retailed by James Worrall, Chester. Serial-number 426. Made in 48 bore calibre. Flask may be associated. *Private Collection.*

PLATE 24. JAMES WEBLEY "LONGSPUR" REVOLVERS
(*Top*) Single-action percussion revolver by James Webley, of a type transitional between the Second and Third Models. Made in 48 bore calibre. Serial-number 985. Marked "Webley's Patent Improved" on barrel, and signed "James Webley" internally. *Private Collection.*

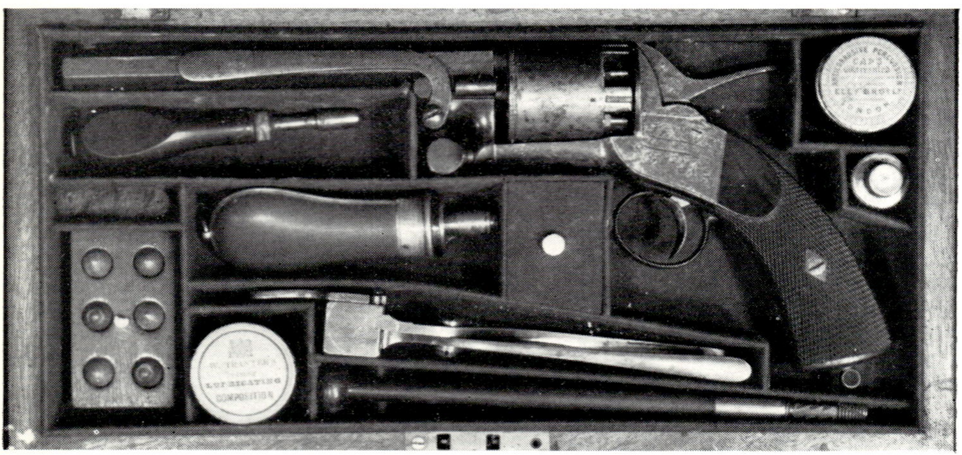

(*Lower*) Single-action cased percussion revolver by James Webley, retailed by Crew, Tetbury. Serial-number 1175. Made in 48 bore calibre. The Third Model. *Private Collection.*

Col. Chalmers, R.A., to the Secretary to the Honourable Board reporting upon Col. Colt's Revolving Pistols

Royal Arsenal, Woolwich.
28th March, 1854.

Sir,

I have the honour to report, for the information of the Honourable Board that three of Col. Colt's revolving pistols sent me for trial, have undergone the following tests:

No. 1	Large-size pistol	400 rounds
No. 2	Second ,, ,,	400 ,,
No. 3	also Second size	150 ,,

The whole of these pistols made good shooting at ranges of 50, 75 and 100 yds; but the accuracy of aim is of course dependent on the skill of the person who uses them.

Nos. 1 and 2 were carried in a holster by mounted N.C. Officers for an hour and a quarter, trotting and walking alternatively in order to try if the bullets were secure in the chambers. They were in no way displaced, though one of the pistols exploded in the holster (slightly injuring the leg of the N.C. Officer). This was occasioned by the pistol lifting up and down, there being no flounce on the holster, to confine the pistol securely in its position.

It is necessary, in order to see the chambers revolving freely, that the caps should fit the nipples accurately.

I have not had the advice or the assistance of the Sub-Committee on Small Arms upon these trials, as the officers are dispersed on foreign service, but Lts. Warlow and L'Estrange, R.A. have constantly attended.

It is our opinion upon the whole, that Col. Colt's revolving pistols are good, efficient, substantial and serviceable arms and with moderate care and attention would answer all the exigencies of service.

I have, etc.

Signed

J. A. Chalmers, Col., R.A.

Joseph Wood Esq.

This was not, of course, the first of such tests made by the British services (a competitive trial in September, 1851, has already been noticed in Chapter V), but it may well have led to those purchases of Colt revolvers for the British Army earlier mentioned as occurring in 1855.

It would appear that a comparative trial occurred later in the year, when Adams's self-cocking revolver was also tested at Woolwich. L. Jewitt (*Rifles and Volunteer Rifle Corps*, London, 1860) reported that:

"On the 5th Sept. 1854 the Foreman of the Gun Factory fired an

Adams revolver taken from stock at a target situated at a distance of 30 yards, 1,000 times in 4 hours, every shot hitting the target. There was not one misfire and no cleaning or oiling was necessary and no derangement whatsoever occurred to the mechanism. The test was conducted in the presence of Col. Chalmers, R.A. and other scientific officers who were full of admiration."

Jewitt also illustrated the foreman's last card (of fifty shots at a 6-in. square target), which exhibited a remarkable consistency in both man and weapon, after such a gigantic string with a self-cocking revolver.

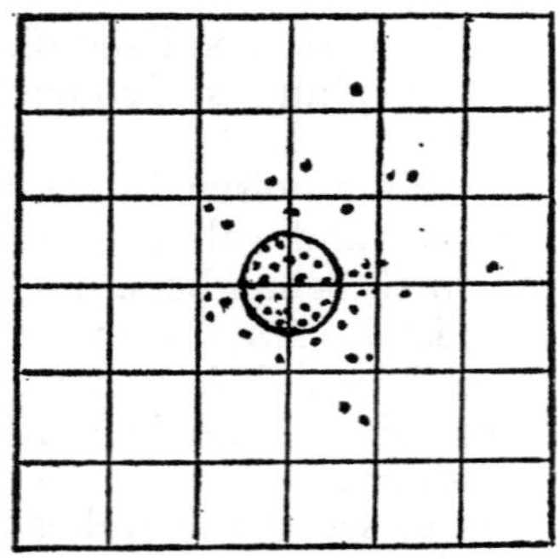

FIG. 20. *Last target at Woolwich, September 5th, 1854.*

At some time late in 1856 (or early in 1857) two cards were shot with a Beaumont-Adams revolver (see Chapters VIII and X), for the purposes of illustrating P. E. Dove's little book on "The Revolver", and we reproduce these as Fig. 21. It would appear that no service witnesses were present, and the two targets may simply be regarded as demonstrations of the standard of revolver shooting then regarded as creditable.

As a measure for comparison between the merits of the two weapons, we illustrate three targets shot at Woolwich, with Colt revolvers, in May and July of 1858; the reasons for these test cards (if such they were) are unknown, for by that date the British services had clearly accepted Beaumont-Adams revolvers as their preferred arm, although Colt revolvers already purchased remained in issue.

On May 28th, 1858 (a windy day) thirty-six shots were fired at 50 yards, against a target one foot square, with a central circle of 6 in. diameter, using "a small Colt revolver, No. 9066". This yielded the following pattern, from what we assume to have been a Model 1851 "Navy" pistol:

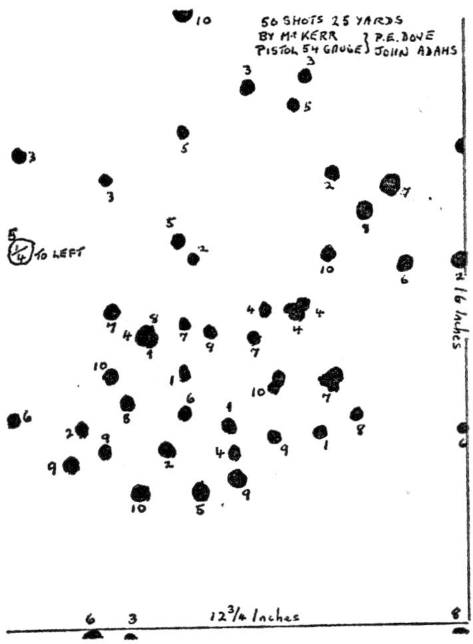

Fifty shots at twenty-five paces, by Mr. James Kerr. Target size 16 in. × 12¾ in. Date probably about 1857. Weapon, the Beaumont-Adams revolver.

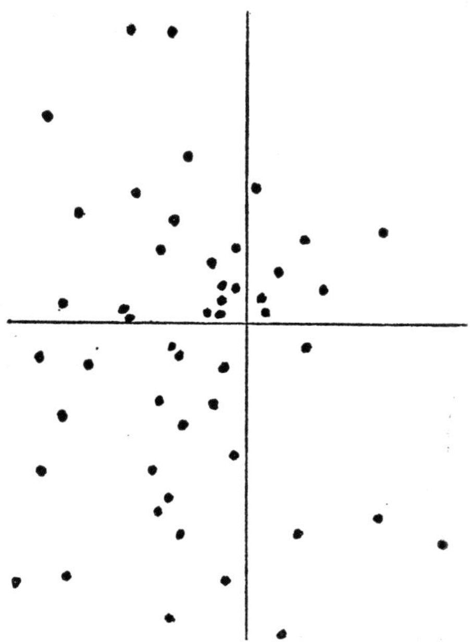

Fifty shots at same size of target and same distance, by Mr. John Adams. Date and weapon assumed to be the same as the above.

FIG. 21.

Targets from P. E. Dove's "The Revolver, etc." 1857.

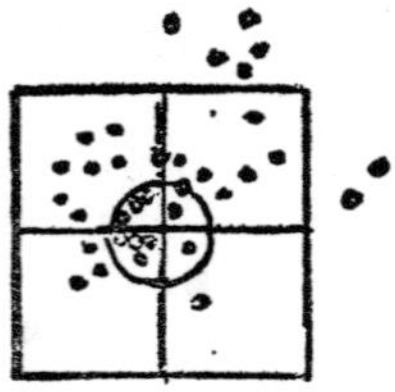

The test was repeated later, with the same revolver and the same number of rounds:

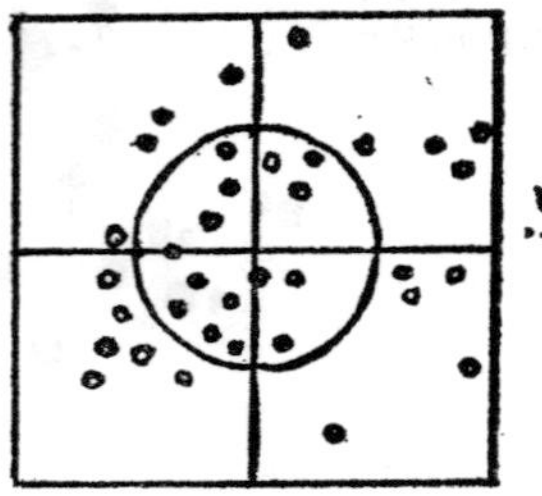

Finally, on July 21st, 1858, a target 6 ft. × 2ft. (with a 6-in. diameter bull) was set up at 200 yards, and ten shots were fired at it, using "a large Colt revolver" (presumably, a Model 1848 "Dragoon"), and this yielded a most creditable pattern, eight of the shots falling within the bull:

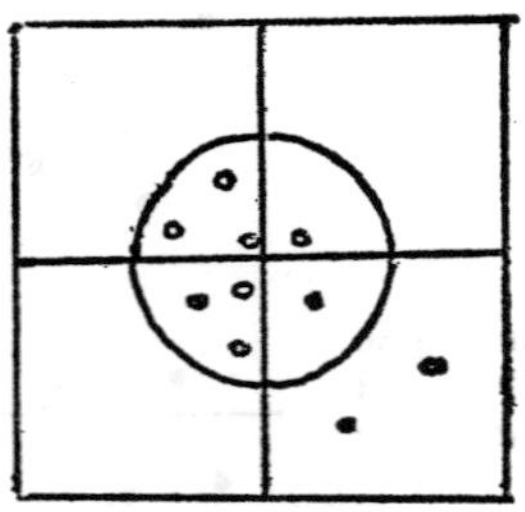

We must assume, however, that this test was either unofficial, or a belated attempt by Colt to re-awaken Service interest in single-action arms. As mentioned in Chapters VIII and X, H.M. Board of Ordnance had taken up the double-action Beaumont-Adams revolver during 1855, and had ordered substantial quantities of it, both in 1856 and in the years which followed.

H. L. Blackmore (*op. cit.* p. 256) related that Colt began to dismantle his London factory in December, 1856, and that he surrendered the premises in 1857. Conceding that his Hartford factory could have supplied any desired quantity of revolvers to H.M. Board of Ordnance thereafter, it seems unlikely that he would have shut down his London plant, if he had seriously anticipated such orders. The purpose of these 1858 tests therefore remains obscure.

Colt did not withdraw entirely from England, upon shutting down his factory, but maintained a depot and sales office at 14 Pall Mall, London. Presumably, this was first stocked with any arms left over from Bessborough Place manufacture, but thereafter Hartford arms were imported and sold.

It is possible to find almost every model of Colt percussion revolver marked as sold in London, but the Model 1849 pocket arm and the Model 1851 Navy pistol are the only weapons known to have been made here, in quantity.

CHAPTER XII

"DEANE & SON" AND HARDING'S REVOLVER

JOHN DEANE'S interest in revolving firearms did not end with dissolution of the "Deane, Adams & Deane" partnership (in August, 1856), although only the approximate date of his reappearance as a manufacturer of such arms is yet known.

It can be assumed that Deane continued in business as a warehouseman dealing generally in Birmingham-made merchandise after Adams had left, but 1859 was the first year in which a London directory listed the new firm of "Deane & Son", as gun-manufacturers at his old 30 King William Street premises, near London Bridge; the other partner in this undertaking was his own son, John Deane (jun.).

"Deane & Son" traded in Birmingham-made firearms for the remainder of our period, and the name can be found upon breech-loading arms dating well into the 1870's; however, the firm is best remembered (by collectors of revolving arms) for one percussion revolver, which was made under the patents of William Harding.

The Harding revolver was patented by its inventor on March 29th, and May 24th, 1858, as British Patents Nos. 669 and 1159 of that year; the first embraced most features of a lever-rammer used with the production revolvers, and the second protected their two-part frame construction, and double-action lock-mechanism.

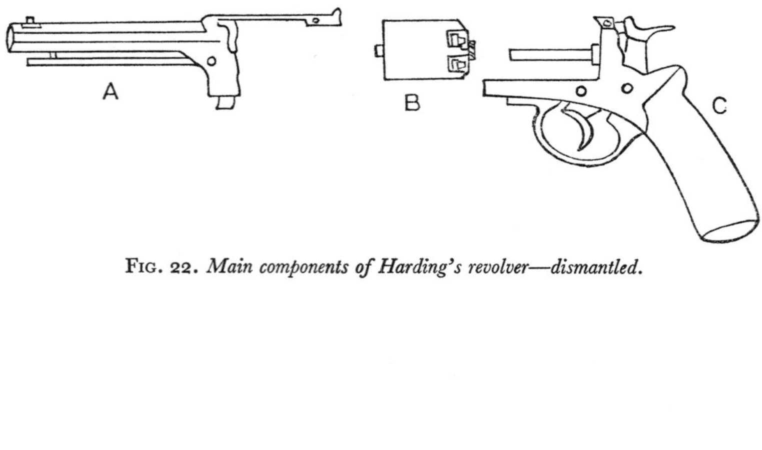

FIG. 22. *Main components of Harding's revolver—dismantled.*

The two-part build-up of the barrel and frame made Harding's revolvers particularly easy to clean, and some versions of the weapon could be dismantled and reassembled in seconds.

In Fig. 22, the barrel (A), cylinder (B), and frame (C) are shown separately:

With (C) gripped in the left hand, the cylinder was slipped onto its arbor, and the hook on the barrel-lug engaged with a recess in the front of (C): see Fig. 23.

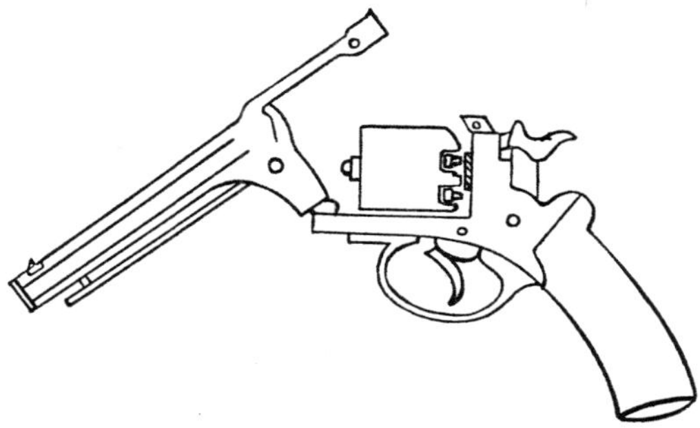

FIG. 23. ***Main components of Harding's*** *revolver—partly assembled.*

Finally (A) was tipped back until the top-strap engaged a lump on top of the standing breech, when the whole was locked together by one of the locking devices later to be mentioned; see Fig. 24.

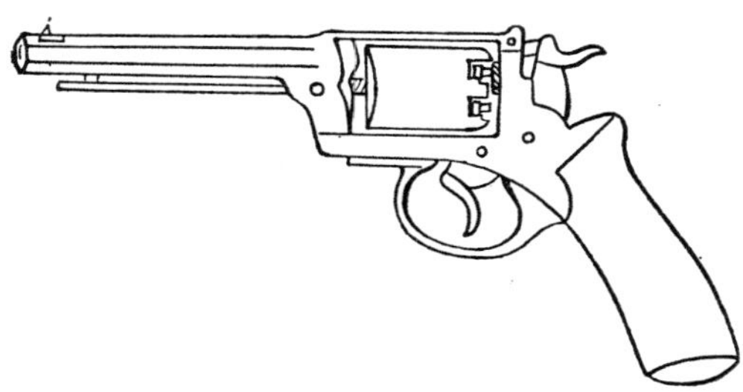

FIG. 24. *Harding's revolver—assembled.*

Opinions about the utility of Harding's lock mechanism are varied. J. N. George (*op. cit.* p. 220) cited Lord Roberts's statement that "it could always be depended upon to get out of order at the critical moment", whilst Captain Llewellyn Jewitt (*op. cit.* p. 64) regarded the mechanism as "so simplified as to place it in some respects no little in advance of some other makes".

Normal stripping of this lock yields four main components, namely, the hammer, the mainspring, the trigger, and the trigger-spring (pawl and cylinder-bolt being integral with the latter): see Fig. 25.

Pinned into a slot in the hammer-breast is a spring-loaded hammer-catch, which can be tilted in towards the hammer (against a light return-spring), but cannot be forced outwards when at rest. Rising from the top of the trigger is a nose, on the underside of which are cut half- and full-cock bents, and the hammer has a pronounced toe for engaging those bents.

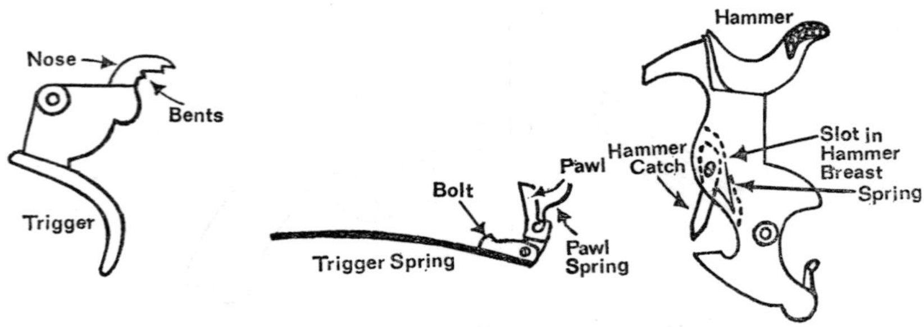

FIG. 25. *Harding's revolver lock-mechanism components.*

When assembled, the lock mechanism appears much as shown in Fig. 26, although, for clarity, we have shortened the trigger-spring in that figure; as made, cylinder-bolt and pawl are actually abreast of the trigger nose.

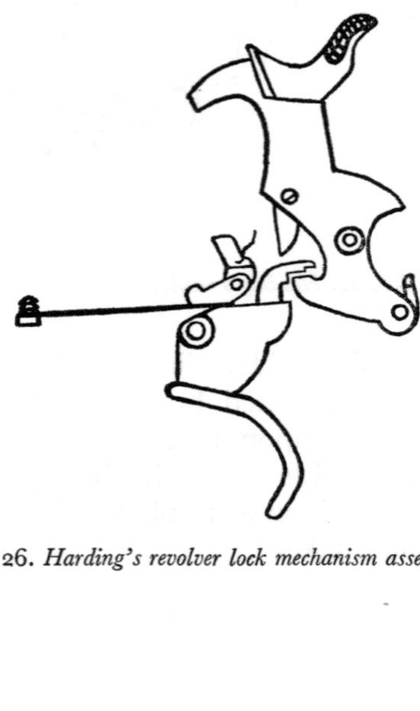

FIG. 26. *Harding's revolver lock mechanism assembled.*

In self-cocking operation, the trigger turns upon its pivot and forces the rear of the trigger-spring upwards, with its attached pawl and cylinder-bolt, which rotates the cylinder by the pawl bearing against ratchet teeth cut upon the rear of the latter. At the same time, the trigger nose engages the hammer-catch projecting from the hammer-breast, and forces the hammer back upon its pivot, until it has travelled far enough back for the hammer-catch to slip off the trigger nose, and release the hammer. When the trigger is released, the trigger-spring forces it back to its original position, the spring-loaded hammer-catch being pressed back temporarily, by the passing trigger nose, to allow this return.

If the weapon is cocked by pulling the hammer back with the thumb, the toe of the hammer catches the trigger nose, to pivot the trigger to the rear, and cylinder rotation occurs as above. When the parts reach their appropriate positions, the toe of the hammer slips into the half-cock bent on the trigger nose and, if cocking continues, ultimately catches in the full-cock bent on the nose to retain the hammer at full-cock. Upon the trigger being then pulled, the toe of the hammer escapes from the full-cock bent in the normal manner and the hammer falls. On releasing the trigger, the parts are returned to their former positions, as described above.

Whether in "self-cocking" or "thumb-cocking" this mechanism, the upwards pressure of the top of the trigger against the trigger-spring will always ensure that the bolt (which is an integral part of the rear of that spring) will protrude through a hole in the floor of the action-body to engage one of the nipple partitions on the cylinder, locking chamber in alignment with barrel at the moment the hammer begins to fall.

The principal and obvious defect of this mechanism is that if the trigger-spring breaks (which is a fairly common occurrence in all models of percussion revolver), then the weapon is completely crippled, with neither pawl nor cylinder-bolt in operation any more. This is undoubtedly a bad feature of the design, since many other percussion revolvers can be fired successfully, and safely, with a broken trigger-spring; in these cases, the trigger can be manually restored to its proper position, between shots. Deane & Son appear to have recognized this defect, and it was corrected by mounting the pawl upon the trigger in some late arms.

The rammer patented by Harding is shown in Fig. 27; although rather lightly constructed, and capable of being bent by rough use, it is unusual to find a totally unserviceable rammer. Towards the close of the service life for Harding's revolvers, an attempt was made to provide a more robust device, embodying a linked assembly, but the majority of Harding's arms have his device as patented.

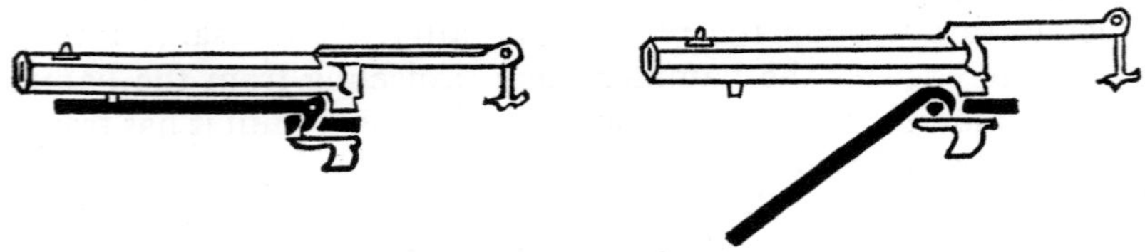

FIG. 27. *Harding's rammer.*

Over the course of manufacture later to be traced, Harding's revolver was (apart from breech-loading variants) always made as a five-chambered revolver, and in 54- or 120-bore embodiments only.

The approximate dimensions of the two models are given below, but will be found to vary slightly, even between weapons quite closely related in manufacturing date.

	"Army" revolvers	*"Pocket" revolvers*
Calibre:	54 bore	120 bore
Barrel length:	6 in.	$4\frac{1}{2}$ in.
Overall length:	12 in.	$9\frac{1}{2}$ in.
Cylinder length:	$2\frac{1}{16}$ in. overall (including ratchet and forward bearing)	$1\frac{9}{16}$ in. overall (including ratchet and forward bearing)
Cylinder diameter:	$1\frac{1}{2}$ in.	$1\frac{1}{4}$ in.
Weight (unloaded):	$2\frac{1}{4}$ lbs.	1 lb. 6 ozs.

Barrels were invariably octagonal, crowned at the muzzle, and normally rifled with three broad grooves and a right-hand twist of about one quarter-turn in the barrel's length; a late model (to be described) was unusual in having five-groove rifling. London proof-marks were always originally present on barrel and cylinder, but some pistols that have been converted to use metallic cartridges have Birmingham proof-marks also present.

Butts were one-piece chequered walnut, and occasionally had a cap-trap in their base. The normal finish was with a blued frame, cylinder, barrel, trigger guard, and butt-cap, but a bright case-hardened hammer, trigger and rammer; plated arms will occasionally be encountered.

Harding's patent specifications showed a bulbous rammer-lever tip, with a press-button in it for releasing the lever to use, but this type of rammer-lever is only found upon the first revolvers made. Most arms have a simple lever-latch at the side of a straight-sided rammer-lever, which is less comfortable to the hands, when loading, than the original bulbous tip.

Apart from its proof-marks, each weapon carried the retailer's name and address engraved upon the top of the barrel and top-strap. In a majority of cases, this name and address was that of "Deane & Son", but a few pistols have the names of country gunmakers engraved upon them instead. In addition, the right side of each pistol-frame carried either the engraved legend "The Deane-Harding Patent No. . . .", or (on most arms) "Deane-Harding Patent No. . . .", followed by a serial number.

A small stamp was usually impressed at the right forward end of each frame (on the side), and also beneath the hook on the barrel-lug, which engages the frame (see Fig. 23). This stamp appears as "ŒH", and may be an amalgam of the capital letters "D" and "H". In view of the methods later described for manufacturing Harding's revolvers, we suggest that this mark was an acceptance stamp, impressed on the weapons at their maker's factory, by an inspector acting for Deane & Son.

A few weapons will be found to have the name "T. Thacker" stamped internally upon their wooden butts. Thacker was a "Pistol Finisher", in business at No. 12 St. Mary's Row, Birmingham, during the 1860's, and was an outworker craftsman to whom various manufacturers passed arms for completion. Apart from making the butt, Thacker probably smoothed up the lock-mechanism, fitted sights, adjusted screw-heads, and arranged polishing, hardening, bluing, and chequering.

As might be expected, the serial numbers engraved upon Harding revolvers exhibit, on a smaller scale, a similar story of manufacture under contract, as was traced for some of the Adams and Beaumont-Adams pistols described in earlier chapters. When summarizing these numbers, it must again be emphasized that they do not indicate the scale of manufacture for Harding's revolvers; 25195 was the highest serial number traced, but it is clear from the number of pistols surviving today that total production nowhere approached this figure. As in other cases, blocks of serial numbers were allocated for use by contractors manufacturing under a patent licence (see Chapter II), and were deliberately dispersed (when practicable) to avoid confusion between the numbered products of different gunmakers.

The lowest serial numbers were found upon two rare 54-bore percussion revolvers of the type shown in Plate 18; one was No. 12, and the other No. 33. Both weapons had colour case-hardened cylinders, the frame-legend "The Deane-Harding Patent No. . . ." in two lines, and single-line barrel-legend "Deane & Son, 30 King William St., London Bridge". As can be seen by a comparison of Plate 18 (*top*) with the pistol shown below therein, the safety-catch, rammer-tip, and frame contours of these early arms differed markedly from those used on later weapons, and these are the only Harding revolvers in which the rammer-pivot was passed from left to right through the barrel-lug.

One 54-bore carbine (with 24¼-in. barrel) was marked "The Deane Harding Patent No. 1002", and may signal the use of a special 1000 serial-range for long-arms. In this weapon (the only specimen traced), a separate spring-loaded firing-pin was mounted in the frame, and the barrel (rifled like pistols 12 and 33) appeared to be screwed into the top-strap assembly, although stripping was accomplished in exactly the same manner as that demonstrated in Figs. 22, 23 and 24, for the hand-guns.

Following these early arms, came a more numerous series, with "L"-suffixed serial numbers in a 5000–7100+ range (see Plate 18). This series included the first 120-bore revolvers encountered, and more weapons were comprised in it, than in the aggregate of revolvers traced for two further serial-ranges that succeeded it.

The following arms were marked "Deane-Harding Patent No. . . .L" in this series, and were all retailed by Deane & Son, except as stated:

5076L	(a number pencilled upon a letter cited later in this Chapter)
5128L	A 54-bore percussion revolver.
5180L	ditto
5296L	ditto
5340L	A 54-bore percussion revolver, retailed by Dooley, of Liverpool.
5537L	A 54-bore percussion revolver.
5612L	ditto
5654L	ditto
5973L	A 120-bore percussion revolver.
6056L	A 54-bore percussion revolver, retailed by R. Freemantle, Fareham.
6062L	A 54-bore revolver, converted from percussion to use ·442-in. centre-fire cartridges, and cased as a pair with No. 6096L below. This conversion was of the split-cylinder type, in which a backing plate (carrying five separate firing-pins) has to be removed from the cylinder for insertion or ejection of the cartridge cases.
6070L	A 54-bore percussion revolver.
6094L	ditto
6096L	ditto (See remarks to 6062L, above.)

6107L A 120-bore percussion revolver.
6129L A 54-bore percussion revolver.
6137L ditto (Silver plated.)
6259L ditto
6288L ditto
6530L ditto
6536L ditto (See Plate 18.)
6541L A 54-bore percussion revolver. (This weapon was engraved "C. E. Condell. 93rd Highlands". Charles Edward Condell was an Ensign in the 93rd (Sutherland Highlanders) Regiment of Foot, who was commissioned in 1859, and sold his commission in 1863.)
6570L A 54-bore percussion revolver.
6790L ditto
6904L ditto (This weapon was engraved "Chelt. Coll. Sports. 1862. Ladies Prize", and was silver plated.)
6938L A 54-bore percussion revolver.
6950L ditto
6970L ditto
6985L A 54-bore revolver; split-cylinder conversion from percussion to ·442-in. centre-fire.
7004L A 120-bore percussion revolver.
7119L ditto

In the majority of these weapons, the letters "T. & L." were internally stamped on the lower butt-strap in minute characters, and at a position normally hidden by the lower limb of the mainspring; both mechanically and visually, this group most closely approached the weapon illustrated in the Complete Specification to William Harding's Br. Pat. No. 1159/1858.

During the course of their production, an interesting attempt was made to improve the latch which united barrel to frame, but it is not known whether the inventor was responsible for that attempt, or if it was made by the firm actually making these arms under licence.

As patented, Harding's revolver had an unthreaded cylindrical pin to secure barrel top-strap to standing breech, which passed right to left transversely through both the strap and a lump (visible in Fig. 23) on top of the breech, and which was formed with the safety-catch. This latter is the hook-shaped component visible in Plate 18, which can be turned forward or back (using the securing pin as a pivot) to engage with holes drilled in the nipple partitions and bolt the cylinder. In the early weapons previously described, a vertical retaining-screw had to be turned out before the screwing pin could be withdrawn, and the barrel tipped forward out of engagement with the frame.

For the first of these revolvers with "L"-suffixed serial numbers, a refined securing pin was introduced. In these arms, the safety-catch can be rotated backwards through about 180° and the pin pulled out sideways (to release the barrel) without having to turn out a retaining screw.

At about revolver No. 6070L, this arrangement was further improved (by grinding a flat on the retaining pin, and merely notching the lump on the standing breech) so that simply rotating the safety-catch rearwards freed barrel from action-body. No withdrawal of the pin is required.

Ingenious as this last idea was, it proved susceptible to wear, and weapons occasionally sprang apart on firing. Revolver No. 7119L had instead a simple threaded screw (which carried a rotatable safety-catch, and passed through barrel-strap and standing breech), and this arrangement was thereafter used on all Harding revolvers in the next group to be described; it had, in fact, been envisaged by Harding in the Complete Specification to his Br. Pat. No. 1159/1858.

To follow the "L"-suffixed serial numbers listed above, we traced a group exclusively retailed by Deane & Son in the 15000 range, but now with a "P"-suffix; not all of these revolvers were true percussion arms, having a small sub-group within the series which normally had a second serial number as well (with "D.H.D." suffix), and comprised weapons that could be used with metallic cartridges or as muzzle-loaders.

15054P		A 54-bore percussion revolver.
15105P		ditto
15143P		ditto
15177P		A 120-bore percussion revolver.
15206P		A 54-bore percussion revolver, fitted for a detachable carbine butt.
15207P		A 54-bore percussion revolver.
15254P		ditto
15260P		A cased 54-bore percussion revolver, with the case-lid escutcheon engraved "2nd Herts. Rifle Vols. Company Prize, 1865".
	14 D.H.D.	An optional 120-bore percussion or ·30 rim-fire revolver, made under John Deane Junior's Br. Pat. No. 626, of March 8th, 1862.
	18 D.H.D.	Number found upon a detached ·30-in. calibre rim-fire cylinder.

15283P	31 D.H.D.	A Deane 54-bore percussion/·442-in. rim-fire revolver.
15319P	47 D.H.D.	A Deane 120-bore percussion/·30-in. rim-fire revolver.
15342P	92 D.H.D.	A Deane 54-bore percussion/·442-in. rim-fire revolver.
15430P	93 D.H.D.	ditto (See Plate 19; *lower*.)
15440P		A Deane 120-bore percussion revolver.
15507P	132 D.H.D.	A Deane 120-bore percussion/·30-in. rim-fire revolver.
15563P		A ·442-in. centre-fire revolver, not made under John Deane Junior's patent above, but intended for use with an optional percussion cylinder.
15569P		A ·442-in. centre-fire revolver, with lever-ejector; not for optional use of percussion cylinder, and not a conversion.
15582P		ditto

A specimen of the optional percussion/rim-fire revolvers made under John Deane Junior's Br. Pat. No. 626/1862 is shown on Plate 19; these arms were originally sold in cases, which also contained a detachable recoil-plate and rim-fire cylinder, as well as percussion accessories. The hammer had the conventional percussion nose, but a rim-fire blade was formed above this; Harding's percussion rammer was fitted, and a separate rod-ejector was screwed into the butt-cap. The user could elect to assemble either percussion or rim-fire components into his revolver, dependent upon availability of the appropriate ammunition.

All optional percussion/rim-fire arms (together with metallic cartridge weapons originally made as such, like 15569P) were six-chambered, unlike the ordinary percussion or converted revolvers, and it can be seen in Plate 19 that cylinder locking-slots are cut or punched around the circumference of their cylinders. A simple see-saw cylinder-bolt was used, and tipped in or out of engagement with the locking-slots (against a weak spring) by engagement with the trigger. In optional percussion/rim-fire pistols, the old safety-catch was discarded, in favour of the sliding catch visible in Plate 19, which could be engaged with a slot in the trigger, to hold the hammer back off cartridges or nipples.

The last group of serial numbers traced had neither suffix nor prefix, and was in a 25000 range; weapons in this group were distinguished by the use of five-groove rifling, and some (of a type shown in Plate 19 *top*) had an entirely new form of two-piece frame construction. With one exception, they had been exclusively retailed by Deane & Son.

On the variant arms, the top-strap was permanently hinged at the top of the standing breech, and the barrel could be tipped up on that hinge for cleaning. The foot of the barrel-lug engaged a tenon cut in the forward end of the frame, as before, but was now vertically secured by a large-headed screw (turned into the barrel-lug) which had to be removed entirely, before the barrel could be tipped up on its hinge.

In such arms the old hook-type safety-catch was discarded, in favour of a sliding bolt on the right-hand side of the frame (see Plate 19); this was inclined at an angle to the horizontal, and could be slid forward to engage the nipple-partitions on the cylinder. Unlike the safety-catch used for John Deane's optional rim-fire/percussion revolvers, this safety-catch did not engage any part of the lock mechanism.

We traced the following revolvers in this 25000 group:

25070	A 54-bore percussion revolver, on the old frame construction, and with a simple threaded securing pin.
25137	ditto
25165	A top-hinged 54-bore percussion revolver.
25167	ditto (See Plate 19.)
25180	A top-hinged 54-bore percussion revolver, retailed by Fletcher, London.
25195	A top-hinged 54-bore percussion revolver.

It seemed reasonable to assume, from the use of four groups of serial numbers, that Harding's two patents of 1858 (Br. Pats. Nos. 669 and 1159) were developed with the help of other men, in a fashion similar to Adams's revolvers. In considering this assumption, we were fortunate enough to have access to *Index 24612* (*B.T. 79*), before the Patent Office destroyed it. This "*Index of Names of Assignees, etc., 1853–66*" showed only surnames of the parties to any agreements, the patents affected, and the year of each agreement, but the entries culled from it (taken with details noted from actual revolvers) were sufficient for us to suggest the probable course of events.

The *Index* showed that two agreements relating to Harding's revolver patents were filed at the Patent Office during 1858, in relation to "Deane" and "Thickbroom", and that a third was filed in the same year naming "Deane", "Thickbroom", and "Lawden".

We suggest that "Deane" was John Deane Senior, that "Thickbroom" (whom we could not trace in the available directories) was a bank nominee whose part in the affair was similar to that of T. A.

Hankey in relation to Robert Adams's patent, and that "Lawden" was Thomas Tipping Lawden, a noted Birmingham gun-manufacturer at 40/41 Constitution Hill and 18 Buckingham Street. The gravamen of these entries was, we further suggest, that Harding had leased or assigned shares in his patents to Deane and Thickbroom, and that these men then licensed Lawden (who traded as "Tipping & Lawden") to manufacture revolvers under the patents for sale by Deane & Son. For obvious reasons, the second group of revolvers described above with "L"-suffixed serial numbers are suggested as the product of this manufacturing licence (it will be recalled that many are stamped "T. & L." internally); the serial numbers would have governed royalty payments for each weapon manufactured under the licence, in a similar manner to that discussed in relation to Adams's patents.

Index 24612 contained a further entry in respect of Harding's patents during 1862, naming "Deane", "Pryse", "Redman" and "Thickbroom" as parties, and we suggest that this was a licence for manufacture of Harding revolvers by Charles Pryse and Richard Redman (noted gun-manufacturers at 84 Aston Street, Birmingham). Those revolvers with a "P"-suffixed serial number seem obvious candidates for such manufacture.

Although slightly different in dimension to their percussion counterparts (and six-chambered, to boot), we believe that the optional percussion/rim-fire revolvers bearing a "D.H.D." serial number were also made by Pryse & Redman under this licence; presumably with the consent of John Deane, Junior to the use of his Br. Pat. No. 626/1862, although we traced no Index entry in respect of it. These optional arms were numbered into the "P"-suffixed serial-range, it will be recalled, and consecutive numbering would have been difficult to accomplish in the hands of another manufacturer. Although not all of them have the "ŒH" acceptance stamp earlier mentioned, it seems improbable that manufacture of such arms was specially arranged out of the hands of existing manufacturing licencees.

Finally, for 1863, *Index 24612* contained another entry in relation to Harding's revolver patents, which named "Calisher", "Deane", "Terry" and "Thickbroom" as parties, and we suggest this as another manufacturing licence, under which those revolvers with serial numbers in the 25000 range were made. It is curious that Harding's Br. Pat. 669/1858 was mentioned, since it had been allowed to lapse in 1861 (see Appendix), but it seems likely that this defunct protection was again included in the licence because Harding's Br. Pat. 1159/1858 specifically referred to it in relation to the rammer proposed to be used. As to the parties licensed to manufacture revolvers by this agreement, Messrs.

Calisher & Terry (War Department contractors at 22/24 Whittall Street, Birmingham) would appear to be the firm involved.

To summarize our views, therefore, it is suggested that William Harding succeeded in interesting Deane & Son in his revolver patents, and that either shares in the protection were assigned to John Deane, Senior, and to Thickbroom (a financier), in 1858, or the patents were leased to them. Thereafter, Tipping & Lawden were first licensed to manufacture Harding's revolvers, and Pryse & Redman, and Calisher & Terry, were subsequently licensed to manufacture, in 1862 and 1863 respectively.

It is further suggested that the first arms described (i.e. revolvers numbered 12 and 33, and carbine No. 1002) were part of a small quantity made for promotional purposes, and that true production-arms begin at those serially-numbered with an "L"-suffix. It is unlikely that production under the Lawden licence overlapped with that by Pryse & Redman or Calisher & Terry, but it seems certain that manufacture under their respective licences, of 1862 and 1863, must have done so.

Harding's exact position in relation to the firm of Deane & Son remains uncertain. A manuscript letter in the author's possession, however, does suggest that he was an employee of the firm during at least part of his career.

Gun Manufactory,
30 King William Street,
London Bridge, E.C.
Feb'y 16th, 1860.

R. Westerman Esq.,
18 Reuben Terrace—Leeds.
Sir,

In answer to your enquiry we beg to give you the following instructions for loading your Revolver Pistol—First place a Cap on each Nipple & explode it to drive out any superfluous oil which may have been left in the cleaning the [*sic*] Pistol—then alter the Powder Flask to the middle charge—& having half cocked the Pistol to allow the Cylinder to revolve freely place a charge of Powder in each Chamber taking care to see that the top of the Powder Flask is duly emptied each time —prefs [*sic*] one of the greased Bullets into one of the Chambers & turn it round until it is brought under the Ram Rod which will force the Bullet down on the Powder—and so on until the five Chambers are loaded—then place on the Caps & bolt the cylinder until required to be fired—

There is no Wadding required with these Bullets which when rammed down fit the Chambers Air tight—

If this explanation be not sufficiently lucid we shall be happy to answer any other question you may wish to put, as, so far from thinking this a "trivial" matter, we are anxious that every one should know how to handle our Pistols properly—

We omitted to say above that the grease on the Bullets is for the purpose of lubricating the Barrel on the pafsage [*sic*] of the Bullet thro' it & thereby diminishing the friction.

We remain,
Sir
Your's mo. obediently
Deane & Son.
per William Harding.

The earliest reference to William Harding, that we could trace, was in the Deed of Settlement for the London Armoury Company (Ltd.), where "William Harding, Clerk, of Forest Hill, Kent" appeared as subscriber for five £5 shares (Nos. 1690–1695), on June 9th, 1856. Through the courtesy of *Crockford's Clerical Directory*, we were able to establish that Harding was not a Clerk in Holy Orders, and it therefore seems most likely that he was then a clerk in the employ of Deane, Adams & Deane; the letter cited above confirms his subsequent employment by Deane & Son, and presumably in the same position.

However, whilst a former "Patent Book. Mr. Solicitor General" (P.R.O. Ref. 74/4) gave his address in 1858 as at "1 Park Villas, Forest Hill, Surrey", an examination of *Kelly's Directory* for Essex, Herts, Kent, Middlesex, Surrey and Sussex (in 1859) yielded two William Hardings in the Forest Hill area; the private residents (and "Court") section showed "William Harding, Gentleman" at 4 Rutland Terrace, Forest Hill, and "Harding, William, plumber" also at Forest Hill.

Since John Deane had an excellent connection with contract-plumbing (see Chapter V), there is a temptation to infer some bond as stemming between the two men from this trade, by which Harding came to be connected with gunmaking. There was, however, a true gunsmith trading under that name, who retailed percussion revolvers, and in the present state of our knowledge, it would be unwise even to guess at the path by which the relevant William Harding first came into contact with the gun trade, and later designed and arranged manufacture for his revolver.

It may be assumed that the first promotional revolvers were made shortly after the granting of his two patents; the first of these patents also protected a revolver lock-mechanism distinct from that covered by the second and used in Harding revolvers. We have never seen a specimen of a revolver embodying it, but the Complete Specification clearly suggested that a two-part frame-construction would have been used in it.

Commencement of production (prior to that by Tipping & Lawden for Deane & Son) began around April 1859; on April 30th, *The Illustrated London News* (p. 426) featured an item stating that "Messrs. Deane & Son, of King William Street, London Bridge, have just brought out a new rifle and pistol, called the Deane Harding, which claims, from its extreme simplicity, the attention of all who are interested in the subject".

This story continued with various claims as to the simplicity and utility of Harding's pistol, and mentioned that "it can be made entirely by machinery; and so made as to interchange in its several parts . . ."

In view of the careful wording used in the claim to manufacture by machinery, and of the reference to "a rifle" version, we believe that this story related to arms of the type earlier described (as to a carbine and two revolvers) with very low serial numbers, and that production by Tipping & Lawden commenced later that year, or early in 1860. The tone of Harding's letter to Mr. Westerman is hardly consonant with answering any inquiry regarding a weapon already distributed in numbers, and was (it may be observed) dated February 16th, 1860.

CHAPTER XIII

MANUFACTURE OF PERCUSSION REVOLVERS

THIS chapter compares manufacturing methods used by Samuel Colt and Robert Adams. In the case of the American's factory, a good deal of contemporary documentation still exists to guide any comment; Charles Dickens wrote an article about it ("Revolvers". *Household Words*. Vol. IX. Ch. 15, May 27th, 1854), and a well-illustrated technical pamphlet was published (two years later) by A. V. Newton, under the formidable title "Colonel Colt's Small Arms Manufactory, with remarks on the system pursued by the English Manufacturers". In addition, various articles were published in American magazines of the period, which can be assumed, with reasonable safety, to describe processes also used by Colt in London. The factory of Deane, Adams & Deane, however, was not well-favoured in this respect, and we gratefully acknowledge our debt to Mr. Larry Wilson (of the Wadsworth Athenaeum) for discovering a contemporary newspaper article later cited, and to Mr. Robert C. Sale (State Librarian of the Connecticut State Library), who kindly provided a copy of it.

I. *Colt's London Factory*

As mentioned in Chapter XI, Colt leased a building off Bessborough Place, Pimlico, in 1851, and began to produce revolvers for sale in January 1853. This building was about 350 ft. long, and consisted of a basement and three floors, together with some separate out-buildings; a 30-h.p. steam-engine, in the basement, powered all of the machinery used in manufacture, *via* suitable shafting and belts. It would appear that demands upon this engine were frequently too much for it, and some acrid correspondence about power-failure still survives (in private hands) between Colt and Charles Manby, an English associate who had ordered the machine. Manby (whose portrait in age is shown as Plate

20) was Secretary of the Institution of Civil Engineers, between 1839 and 1857.

The work-force numbered about 200 British hands, with a strong leavening of between twenty-five and thirty American technicians, or supervisors; as was shown in Chapter XI, some doubt still exists as to how many of these latter were in fact necessary for successful operation of the factory.

Colt used unskilled British labour of either sex, but primarily as machine-minders rather than as craftsmen, and employed them in the true legal sense; if any proved unusually willing and skilful, they were given more responsible jobs and paid accordingly. Dickens stated that boys and girls could earn two or three shilling a day, men from three to eight shillings per day of ten hours, and one or two of them ("being quick, clever, and reliable") were paid twelve shillings a day. These were very good rates by British standards of the day; a decade later (Timmins. *Birmingham and the Midland Hardware District*, 1865, pp. 393–6) workers in the Birmingham gun trade were only paid about 30*s*. per week as skilled men, with 5*s*. to 10*s*. (for boys) or 15*s*. to 25*s*. (for adults) weekly paid to those assisting them.

It is possible that Colt used British employees in this fashion on the advice of Charles Manby. At his Hartford factory, piece-work was the rule, and almost all manual labour worked "by contract". Thus each part of the revolver had a price assigned to it, and the men making them contracted with the Colt Company as independent contractors, to supply the parts at that price. The Company provided room, power, machines, material, and so forth, to accomplish the contract, but it did not actually "employ" its labour in the legal sense, although the men or women were expected to observe certain company rules and regulations, and to make any assistants observe them.

It seems probable, too, that the skilled American work-force brought over to London as supervisors and assemblers (or to construct and maintain the special machinery) worked on that Hartford basis, as independent contractors, and not as employees.

If this were so, a fruitful source of friction between Briton and American would have existed, in the amounts paid to each for their labours. During a famous lecture to the Institution of Civil Engineers, in 1851, Colt stated that "in round numbers . . . supposing the cost of an arm to be 100, of this the wages of those who attended to, and passed the pieces through the machines was 10 per cent, and those of the best class of workmen engaged in putting together, finishing, and ornamenting the weapons, was also 10 per cent; thus leaving 80 per cent for the duty done by the machinery". It will be recalled (from Chapter XI) that Gage Stickney told a Parliamentary Committee that twenty-five to thirty skilled Americans came over to start up the factory; if that number of *élite* workers was maintained, and paid the same proportion of labour costs as was paid to the 200 or so unskilled workers, a certain jealousy amongst the latter could reasonably be expected.

The revolvers made at London (see Plate 16) had three basic components, which differed in size according to the model being made, but had always the same basic outline, and the same operations to be performed upon them. They were:

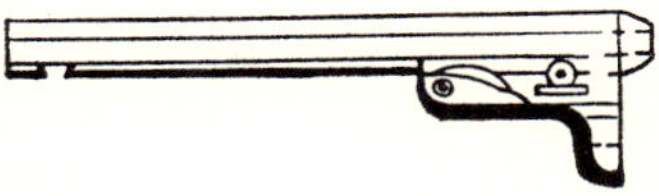

The Barrel

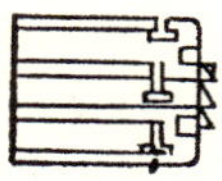

The Cylinder

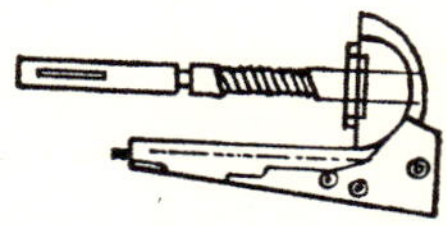

The Frame

(The frame component is shown with its arbor inserted, but this was actually a separate, turned component, screwed and pinned into the forged and machined frame.)

Basic frame forgings (together with those for hammer, rammer, and trigger) were formed under patented "American drop" hammers of the type shown in Fig. 29.

These stood to ceiling height in their workshop, and raised a forging hammer to the required level, whenever the workman engaged that hammer with the constantly rotating, threaded, central shaft. The hammer could be held until the workman released it manually, or allowed to fall automatically on reaching the agreed height, and it formed the heated metal inserted beneath it, by striking the latter against or between suitable dies.

Colt filed a "communicated" British Patent (see Chapter II), as No. 537, of March 3rd, 1853, for this mechanical hammer, but the application became void (with No. 536, of the same date, for an air compressor), because he failed to file the required Complete Specification relating to it. Subsequently, Gage Stickney (whose disaffected evidence has already been noticed in Chapter XI) secured British Patents No. 2039 and 2040, dated September 3rd, 1853, covering, respectively, the compressor and the "American drop" hammer. These two patents were later assigned to Samuel Colt, and it is not unreasonable to assume that the ill-feeling against Colt, shown by Stickney before the Parliamentary Small Arms Committee of 1854, stemmed somehow from this transaction.

The patented air-compressor was used for "blowing" the anthracite

COLT'S PATENT REPEATING POCKET, BELT, AND HOLSTER PISTOLS.

MANUFACTORY, THAMES BANK, NEAR VAUXHALL BRIDGE, LONDON.

OFFICES FOR SALE OF ARMS, BY SAMPLE.
1, SPRING GARDENS, COCKSPUR STREET.
LONDON, January 1st, 1854.

Colonel COLT, the Original Inventor and Patentee of the REVOLVING BREECH'D FIRE-ARMS, has the pleasure of informing the Trade and Dealers in Fire-Arms that he is now ready to supply his IMPROVED PATENT REPEATING PISTOLS, manufactured in London, of the purest refined Steel, on the most favorable terms: Orders which are now solicited, addressed to SAM. COLT, will receive prompt attention.

The tariff prices of the Weapons as sold at Retail will be found below, from which a liberal discount will be allowed to the Trade.

*** Terms Cash.

VARIOUS SIZES POCKET, BELT, AND HOLSTER PISTOLS.

	RETAIL PRICE.
N° 1.—FOUR-INCH RIFLE BARREL, POCKET PISTOL, FIVE SHOTS. Weight 23 oz., (84 Conical and 126 Round Balls to the lb.;) with Mould and Wrench.	£4 0 0
Same Weapon, in Case complete, with Caps, Flask, Moulds, and Wrench.	4 15 0
N° 2.—FIVE-INCH RIFLE BARREL, POCKET REVOLVER, FIVE SHOTS. Weight 24 oz., (84 Conical and 126 Round Balls to the lb.); with Mould and Wrench.	4 5 0
Same Weapon, in Case complete, with Caps, Flask, Moulds, and Wrench.	5 0 0
N° 3.—SIX-INCH RIFLE BARREL, POCKET REVOLVER, FIVE SHOTS. Weight 26 oz., (84 Conical and 126 Round Balls to the lb.); with Mould and Wrench.	4 10 0
Same Weapon, in Oak Case complete, with Caps, Flasks, Moulds, and Wrench.	5 5 0
N° 4.—PISTOLS FOR THE HOLSTER OR BODY BELT, RIFLED 7½ INCH BARREL, SIX SHOTS. Weight 2lb. 8 oz., (50 Conical and 84 Round Balls to the lb.); with Mould and Wrench.	5 10 0
Same Weapon, in Case complete, with Caps, Flask, Moulds, and Wrench.	6 10 0
N° 5.—HOLSTER OR CAVALRY PISTOL, 7½ INCH RIFLE BARREL, SIX SHOTS. Weight 4lb. 2 oz.; (32 Conical, 48 Round Balls to the lb.); with Mould and Wrench.	6 0 0
Same Weapon, in Case complete, with Caps, Flask, Mould, and Wrench.	7 0 0

N. B.—No Arms will be retailed by the Inventor except at these prices.
A few engraved and highly finished Arms, at prices proportionate to extra-work, always on hand.

FIG. 28. *Copy of original Trade Letter advising commencement of full-scale manufacture at the Colt, London factory.*

forges in which metal was heated before being forged, and also for purging debris from the forging dies. As patented, the "American drop" hammer was supposed to operate four hammers at once (i.e. one between each pair of its support legs), but Dickens's description of it suggested that only one hammer was used in each machine, at Bessborough Place. If this were so, it may have been that the steam-engine could not manage the four-hammer drop, or that the footings of these hammers could not support the irregular multiple impact of a four-hammer set-up.

Samuel Colt took out British Patent No. 654, of March 16th, 1853, for the heating forges, and also for the annealing furnaces in which

c. Hammers.
d. Dies.
e. Driving shaft.

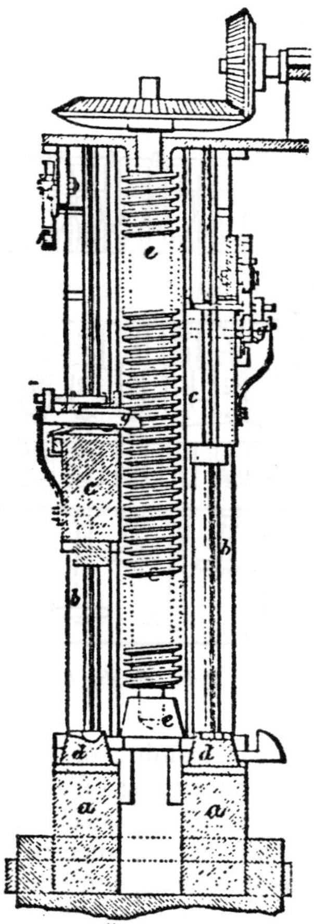

FIG. 29. *Stickney-Colt "American-drop" hammer.*

metal was case-hardened, stress-relieved, and annealed. He did not, however, hold any patent protection for another mechanical hammer (shown in Fig. 30), under which the barrel-forgings of his pistols were worked to shape at Bessborough Place.

The machine shown in Fig. 30 was known as a "Ryder Forging Engine", and had been patented on February 8th, 1841 (as Eng. Pat. No. 8835), by William Ryder, a "Roller and Spindle Maker", of Bolton, Lancs. It seems, from Dickens's remarks, that this forging engine was (apart from certain planing equipment) the only machine in Colt's London factory that had not been designed or imported by his own engineers.

Briefly, the operator heated one end of a bar of cast steel, and then inserted it, successively, beneath the five cylindrical hammers that

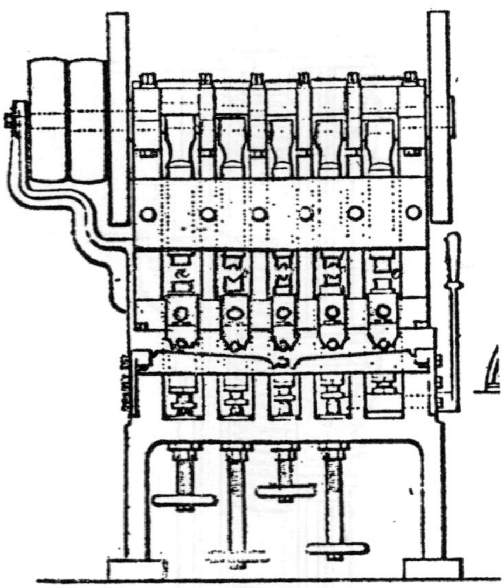

FIG. 30. *Ryder's Forging Engine.*

bobbed up and down in the framing of Ryder's machine. The last hammer chopped a completed forging off the end of the bar, and the cycle took about twelve seconds. In essence, the machine consisted only of a horizontal driving shaft, which successively cammed down the spring-loaded hammers as it rotated, but it appears to have operated most satisfactorily. The machinery used by Colt for barrel-forging at Hartford is not known.

As to the material used for barrels and cylinders, surviving correspondence suggests that the metal for Bessborough Place was secured from T. Firth & Sons, of Sheffield, as a British equivalent to the "best hammered cast steel" used at the Hartford factory. This was probably made by taking the so-called "Blistered Steel" (after its emergence from the closed, cementation furnace in which it had been first made), breaking it into small pieces, remelting in closed crucibles and recasting into ingots, then rolling or hammering this refined metal into the "Cast steel" bars delivered to Colt for use.

Disappointingly enough, Dickens restricted himself to generalities, when describing the "Armoury" or machine-shop at Bessborough Place; since this part of Colt's production methods contained the basis for his claims to manufacture pistols with fully interchangeable parts, it would have been interesting to have received a more detailed description, from this famous layman, of the processes shown to him.

"Across the yard strewn with chips of iron again," (Dickens wrote) "and through the tool room, where men are turning great screws and other bolts and portions of machinery, we mount to the first floor, and enter a long room filled with machines, and rather more redolent of hot rank oil. Considering that the floor supports a long vista of machinery in full action, the place looks clean and neat, and is not very noisy. Girls quietly attending to the boring and rifling of the barrels—having nothing to do but to watch the lathe narrowly, and drop a little oil upon the borer with a feather now and then—men drilling cylinders, holding locks to steam files, cutting triggers, slotting screws, treating cold iron everywhere as if it was soft wood, to be cut to any shape, without straining a muscle. It would be difficult and tedious to describe these machines minutely, although they are very interesting to a spectator, and cannot, I believe, be seen elsewhere. Every one of them is a simple lathe; but it is in the various cutters, borers, and riflers that the novelty and ingenuity exist. Where the thing to be made is of eccentric shape, the cutter is of eccentric shape also; and although the superintendent of each machine acquires more or less skill by practice, it is in the perfection of these cutters and borers that the guarantee for uniformity consists. The bores of barrels and cylinders must be mathematically straight, and every one of the many parts must be exactly a duplicate of another. No one part belongs, as a matter of course, to any other part of one pistol; but each piece may be taken at random from a heap, and fixed to and with the other pieces until a complete weapon is formed; that weapon being individualized by a number stamped upon many of its component parts. The advantage of these contrivances is obvious. In every case of revolvers are placed, when sold, a number of such parts of a pistol as are most liable to accident; and, with these, any soldier or sailor may, in a few minutes, repair his own weapon.

Seventy-odd out of a hundred of the injured revolvers picked up on the battlefield during the Mexican war were repaired with bits of other pistols on the spot."

Barrels and cylinders were bored in lathes that were standard in appearance, but ingeniously modified for revolver manufacture; a barrel-borer (shown in Fig. 31, with its ingenious quick-release for the tool) was protected in Great Britain by Colt's own Br. Pat. No. 1703, of July 16th, 1853; this patent also protected features of two "breech-cylinder borers", but one of these appears (from Newton's drawing of it) to have been a Fitch horizontal-axis turret lathe. Colt owed much of the modifications to these lathes to his Hartford factory Superintendent, E. K. Root; the latter held U.S. Patent No. 12002, of November 28th, 1854, covering two of these boring machines.

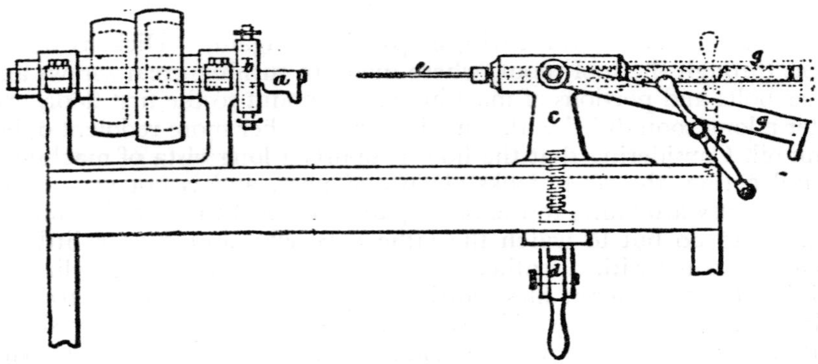

FIG. 31. *Colt's barrel-borer.*

Pistol barrels were draw-rifled, four at a time, in a power-operated machine, protected by Colt's Br. Pat. No. 429, of February 22nd, 1854; E. K. Root's U.S. Patent No. 12285 of January 23rd, 1855, appears to be the equivalent American protection.

A glance at the Colt pistol-frame, shown on page 199, will confirm that some quite advanced milling operations must have shaped it from the original forging. Colt held Br. Pat. No. 861, of April 12th, 1854, for the three machines designed for this purpose; one of them, known as the "Universal Milling Machine", was actually designed by another American, Frederick Howe. The latter was an enthusiastic designer of milling equipment, who appears to have collaborated with E. K. Root, in fitting up Colt's Hartford factory, and one of Newton's illustrations clearly established the machine mentioned above as one of Howe's index millers, and as probably made for Colt by the George S. Lincoln Company, of Hartford.

The same patent which protected this milling equipment also claimed the drill-box used in making revolver-hammers, together with a "screw-cutting machine" for making nipples and action-pins for the arms.

Oddly enough, Dickens mentioned the roller-engraving of pistol cylinders, in a context that suggested he was present in the factory during an all-out effort to meet the British Navy's order (mentioned in Chapter XI) for 4,000 "Navy" revolvers, Model of 1851.

It is impossible to be dogmatic on the point, since the Royal Navy may have been indifferent to the presence of decoration upon its weapons, but examination of British army-issue specimens of Model 1851 "Navy" Colt pistols, bearing the post-1855 "W.D." and Government broad arrow on their barrel-lugs, or with the Enfield Royal Manufactory stamp on their butts, certainly suggests that time spent on engraving service revolvers could have been wasted (see Plate 17).

Both the "War Department" pistols (principally iron-mounted weapons in serial-ranges below 26000 and with "London" barrel addresses) and the "R.M. Enfield"—stamped weapons (principally brass-mounted American-made revolvers with "New York" barrel addresses and serial numbers above 180000) were uncommon, but on what may be termed the majority of such pieces examined by us, the engraving had been deliberately removed from the cylinders.

This defacement was found on weapons in sufficiently good condition to be sure that it had *preceded* their passage through a partial refinishing, and that it was not the work of a collector or modern refinisher. Additionally, there were indications (no more) that electro-plating had been deliberately removed from the mounts of the brass-furnished revolvers.

It will also be noted, from Dickens's remarks, that an effort was made to sell him the idea that Colt's manufacturing methods yielded revolvers which were fully interchangeable as to their parts. In other words, a number of revolvers could be completely dismantled, all parts thrown into a heap, and an equivalent number of arms assembled without the need to match any component to the numbered frame or barrel from which it had originally been stripped.

However, it is clear from the British gun trade evidence cited in Chapter XI, that some contemporary observers were not satisfied that Colt could accomplish such a degree of interchangeability. Unfortunately, too, our own experience of "cannibalizing" weapons for parts to build up a complete revolver has not been on a scale sufficient for any firm decision as to the truth or otherwise of such a claim; where any interchange was attempted, however, no difficulty was experienced in relation to lock components.

On this point, however, it is worth recording that a Commission sent to America, in 1854, appears to have established fairly conclusively that the Springfield Arsenal there had been making fully interchangeable muskets for a decade. This Commission (sent over by the Parliamentary Small Arms Committee) satisfied itself that parts from any Springfield musket made between 1843 and 1853 would interchange with those of any other from the same source and period.

There may indeed be a case for argument as to exactly how much handwork was needed before Colt's machine-made parts could be assembled into a revolver, but this conceded, it seems past belief that Colt could not accomplish in 1854 the same degree of interchangeability as that practised in the U.S. Arsenal, at Springfield, ten years earlier.

II. *Deane, Adams & Deane's factory*

The *Daily News*, of Tuesday, July 18th, 1854, contained an article entitled "A Walk through a Revolver Factory", and this (apart from the first paragraph) is cited hereafter *in extenso*. The article was shown to be (and presumably paid for as) an advertisement, and it is possible that John Deane, Senior, wrote it; as earlier mentioned, his famous "History" was plagiarized from another work, but the turgid paragraphs which follow do seem to bear his literary stamp.

The *Daily News*—London, Tuesday, July 18th, 1854

"*The New Fashioned Fire Arms*

A visit to a 'revolver factory' is most interesting. The manufactory of the Messrs. Deane, for instance, is situated in Weston Street, Tooley-Street, close to the terminus of the Brighton Railway, and is well worth seeing. Here the visitor may see the whole process of manufacture, from the first forging of the rude masses of iron into something like the component parts of a rifle or pistol, to the final acts of polishing and adorning. The premises are of considerable extent, forming at the back four sides of a square yard; and employment is found for about 200 men. At the basement, the visitor may notice the raw materials of future rifles, fowling-pieces, pistols, etc.; and, if he be philosphic or imaginative, may speculate on the amount of death and bloodshed which is to be elicited from all that harmless-looking metal and timber. Huge bars of iron, raw from the iron districts, and looking as ugly as may be in their shapelessness and dusky asperity, lie side by side with heavy blocks of walnut-tree wood, equally shapeless and rough, almost colourless and with their beautiful grain, high surface, and warm, rich tint awaiting the development of the polisher. These are the bones and limbs and substance of uncreated weapons—the lifeless bodies into which the art and science of the workman is to put a kind of soul. In the furnace-room we may see the first process of organization going forward, and may behold the rude and sullen iron beginning to acquire something of form and feature. A bar of the metal which has been heated red hot is taken from the fire, and, being held firmly on the anvil by one of the workmen, is rapidly cloven in two lengthways by a second workman, and each piece becomes a rudimentary barrel, which, by a few blows from the sinewy Vulcans in attendance, is still further fashioned, the place for the trigger and for the cylinder which receives the bullets being cut out with as much apparent ease as though the substance were a ripe Stilton. The inexperienced spectator cannot fail to be astonished at the precision with which the men strike the bar at the right part, although the blow is given at arms' length, and with a circular swing of the long hammer employed, which, though of course augmenting the impetus and strength of the blow, would seem to the ignorant to render it of less certainty. But the skilled hand and arm possess a degree of judgement almost instinctive; and, the process goes on without a mistake."

The factory

As we earlier mentioned, in Chapter IX, the Weston Street factory was not owned by the Deane, Adams & Deane partnership, but merely leased from the South Eastern Railway Company. Since the lease was terminated within a year or so of the visit described here, however, and the buildings demolished to make way for the railway support arches which still cover so much of Weston Street, it is probable that the general appearance and layout of the factory will never now be known.

A study of the large-scale (10 ft. = 1 mile) "East Division of Southwark Pavements & Improvements Plan. 1853" preserved in the Central Library of the Metropolitan Borough of Bermondsey, showed that No. 2 Weston Street (in July, 1854) was the corner building shown in Fig. 32, at the site of the old Maze Gate and facing on to both Tooley Street and Weston Street, with its access or loading yard giving on to the latter.

The site appears on this map roughly as sketched, but the map did not establish exactly which part of the area actually constituted No. 2 Weston Street, and this fact remains a mystery, despite gallant efforts by the Bermondsey Central Library, the British Transport Commission's Archivist, the Regional Estate and Rating Surveyor for the Southern

Region, British Railways, and the Public Relations & Publicity Officer for that Region, to assist on the point.

Since the *Daily News* correspondent stated that the premises formed "at the back four sides of a square yard", it seems possible that "No. 2 Weston Street" covered only part of the site shown in Fig. 32. He could well have looked down into one of the enclosed yards shown, rather than into the long access yard, which can by no stretch of the imagination be termed "square".

The statement that a visitor to the factory might see "the whole process of manufacture" of the Deane, Adams & Deane arms is interesting,

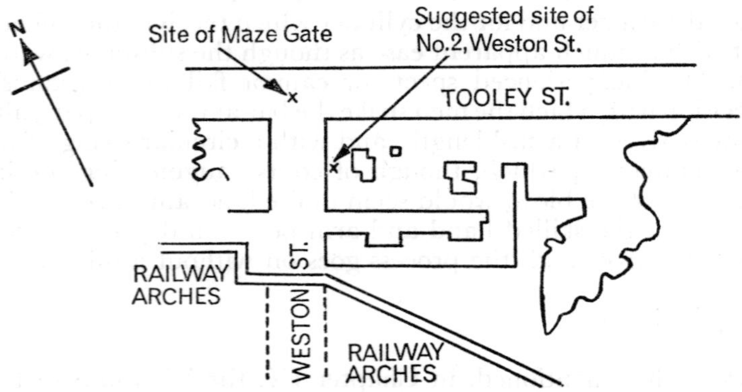

FIG. 32. *Suggested site of Deane, Adams & Deane's factory at No. 2 Weston Street.*

as it cuts flatly across a claim, made by Dickens, that this "all-under-one-roof" manufacturing process was peculiar to the Colt London Factory and not a method practised generally amongst United Kingdom gunmakers. A further parallel of interest, between the Colt and the Deane & Adams establishments, lies in the fact that their respective labour forces may both have numbered about 200 men. Dickens's words are not absolutely clear on this point, however, and he may have been excluding females and children from his total.

However, the labour force employed at No. 2 Weston Street would largely have comprised skilled workmen, trained or raised in the London gun trade, and would not have followed the variegated pattern permitted in Colt's factory, by the levelling influence of his machines.

Certainly, the methods of handiwork in the Weston Street forging department (see above) suggest long practice and a very high degree of skill amongst the smiths. To cut out even the most rudimentary barrel/frame forging for an Adams revolver by hand and eye alone, and within limits of accuracy permitting subsequent machining operations to be based on the work, seems an astonishing performance. Against it, and viewed in terms of sheer naked skill, the Colt smiths with their "drops" and mechanical hammers lose somewhat in stature.

The preference by Adams for the use of "iron" in manufacturing his revolvers may have stemmed either from technical advantages for this metal over steel, or from his past experience in making conventional firearms, and his known association with the Clive family of Birmingham barrel-makers. Certainly, some of the plates in which iron for rifle-barrels was supplied were temptingly close in dimensions to the sort of blank from which a revolver frame-forging might have been cut.

"Marshall's iron" appears to have been the favoured metal, and an unknown journalist writing upon military rifle-barrel manufacture in *The Illustrated London News* of April 26th, 1855; vol. 26, p. 410, (who formally acknowledged assistance from "Mr. Clive of Birmingham and Mr. Adams of the firm of Deane, Adams and Deane") stated of this particular material, in that context, that "the iron principally used for the Government rifles is called 'Marshall's' iron, and comes from Wednesbury, in Staffordshire". "Mr. Marshall" (he continued) "possesses some secret (not a patent), by which he is enabled to produce the purest iron, most free from what are technically termed 'greys' and this is found in all respects equal to the metal composed of twisted fragments".

"The iron is procured by the gunmaker in plates about twelve inches long by four inches wide and three quarters of an inch thick, varying in weight from nine to ten pounds . . ."

Adams's machinery

"The rude semblances or embryos of barrels beingthus produced they are taken into the upper chambers, and subjected to the processes of boring, shaping, and polishing, all of which is effected by the agency of steam, with the celerity and completeness only thus obtainable. In boring through the barrels—or rather in converting the solid bars of iron as they come from the forging room into hollow tubes—an admirable though very simple contrivance is resorted to for facilitating the operation. In the method usually employed, the boring-rod in a very short period produces so large an amount of filings that the newly-formed passage gets choked, and the rod has to be withdrawn several times in the course of forming one barrel, in order to allow of the tube being cleared, or the action would be stopped. In Messrs. Deane's factory this is prevented by directing a small stream of water side by side with the advancing rod; so that the passage is instantaneously and continuously cleaned, and the debris thrown out into a receptacle underneath. The whole of this portion of the work is very interesting. The piece of metal to be pierced is fixed horizontally point to point with the boring-rod, which is connected with elaborate machinery and moved by steam; and it is curious to observe it gradually eating its way into its stubborn antagonist. The process of 'rifling' is effected with the greatest exactness by a newly-invented and beautifully-constructed machine for which it is understood Messrs. Deane and Co. are taking out letters-patent. The planing of the barrels is also a beautiful process; in short, all the machinery employed in the production of these weapons appears so admirably adapted to its purpose, that it seems next to impossible to turn out anything short of a perfect weapon."

The barrel-boring and rifling machines described above were protected by Robert Adams's British Patent No. 2000 of 1854, dated September 14th of that year; the borer was admitted to be an adaptation of a standard lathe, but the rifling machine was original.

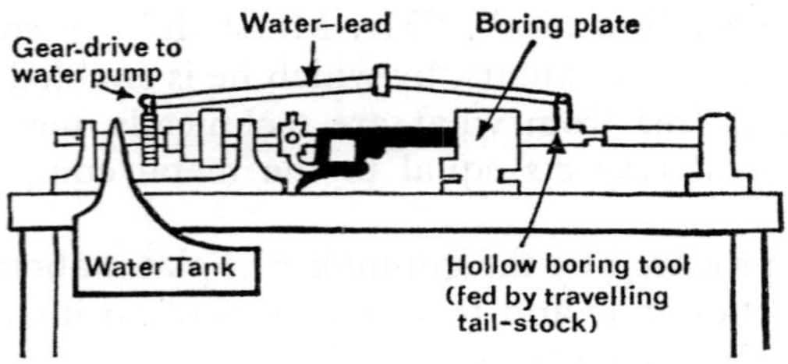

FIG. 33. *Robert Adams's barrel-boring lathe.*

Since this dating was, of course, that on which the actual application was filed (see Chapter II), it will be appreciated that Deane, Adams & Deane took something of a chance in permitting printed description of these machines (however brief) to appear in the previous July. The only real claim of novelty made for the barrel-boring machine, in Adams's patent, was the water feed purging the boring tool as it cut. To mention this improvement, therefore, in a national newspaper, and two months before applying for patent protection was a rash act, that could have jeopardized that patent claim as being no longer "novel". Strong indeed, are the claims of publicity.

Fig. 33 sketches Adams's barrel-boring lathe, and it is enough to point out here that the Specification drawing showed a virtually completed barrel/frame forging mounted in this machine, with all external planing and shaping apparently completed, with the frame cut out for the cylinder, and with the muzzle of the barrel already crowned.

It will be observed, from Fig. 33, that both lathe and technique were kept as simple as possible, for the barrel-boring process.

A gear-wheel was mounted on one end of the driving mandrel and operated a water-pump; the commonest of driver chucks was mounted on the nose of the driving mandrel, and centred on what would eventually be the frame-aperture for the pistol hammer-nose. The lower butt tang of the revolver served as the carrier. A water-tank for the pump, and a flexible water-lead from the pump to the head of the boring tool were also added. The lathe-saddle carried a boring plate to support the pistol muzzle (which had been crowned), and the boring tool was passed through this plate and fed forward into the work by a travelling tail-stock. This saddle also carried a receptacle for cuttings purged by the circulating water as it poured from the pistol muzzle, and a lead back to the main tank completed the water circuit.

The second claim of Adams's patent was for a "combination of machinery for rifling the barrels of firearms", and although the elaborate machine described and illustrated could be used for full-sized rifle barrels, it appears to have been designed primarily for Adams's revolvers, since it was an Adams revolver barrel/frame forging which was shown in the drawings as mounted in the machine for rifling. (See Fig. 34.)

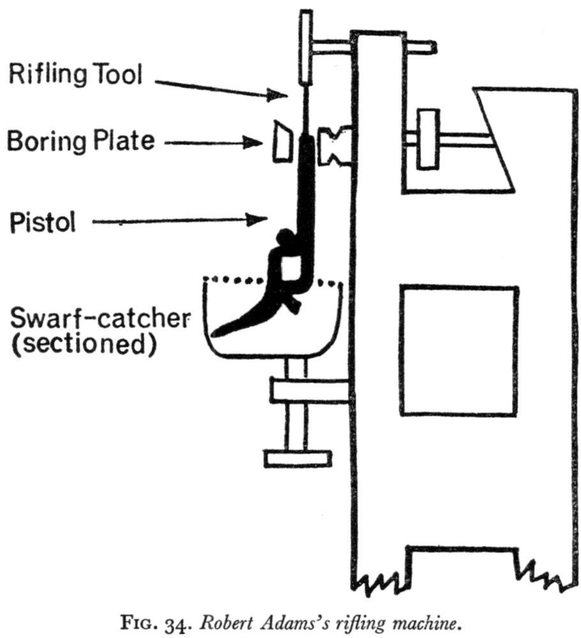

FIG. 34. *Robert Adams's rifling machine.*

It will be noted from the sketch that this machine (which was "self-acting") operated vertically, unlike its Colt counterpart, and was designed to stop or to cease cutting at a pre-determined point which was presumably on emergence of the rifling-tool head into the cylinder aperture of the pistol frame. Any number of rifling grooves could be cut (three-groove rifling is mentioned for the machine illustrated in the Specification) and the depth of the groove could be varied as the cut progressed, as could the degree of twist being given to the rifling. The cutting tool was chisel-edged (not serrated, as was more usual) and was pushed through the barrel, and not dragged—like that of the Colt equivalent.

"All the component parts of the rifles and revolvers are made by Messrs. Deane, upon their premises in Weston Street, and the visitor consequently can obtain a complete insight into the whole details. To connoisseurs in the Gun it would afford complete satisfaction to witness, in the different stages of progress towards completion, under the same roof, the manufacture of the best fowling-pieces and rifles. There are so few factories in London where this may be seen, that the genuine sportsman, who prides himself on his detailed acquaintance with his 'shooting-iron' would never regret an hour or two spent in the practical department of his favourite study at this factory of Messrs. Deane. The formation of the chambers which receive the shot is not the least interesting of these. A thick cylinder of steel is fastened in a similar way to that employed with the barrels, and five holes are simultaneously drilled through it by revolving wires. To an unaccustomed eye the elaborate construction of the machines, with their complex life of wheels, rods, shafts, and levers, and their swift and various movements, seems somewhat bewildering; but it will at once be admitted that the results are beautifully simple and direct. The giant steam moulds and pierces hard iron with the same confident mastery with which a clever cook manipulates a piece of soft pie-crust. Thus the solid cylinder is soon converted into a set of five chambers intended for the reception of the five bullets. When completed, these chambers are fitted into the pistol by a simple contrivance, which allows of their being taken out with the utmost ease; and, as any number of additional chambers may be had to one pistol, a man might carry about in his coat-pocket a small armoury, and having exhausted one set of five bullets, might take out the cylinder, rapidly fit in another, ready loaded, and deliver five more shots—and so on. The weapons are loaded, at the breech, and require no ramrod; and Messrs. Deane assert that the set of five chambers may be easily loaded and discharged twice in a minute by an expert hand."

The self-cocking weapons actually made at the time of this article have already been described in Chapter VII. It is as well to remind the reader, however, that this Weston Street factory may have accounted for only a small proportion of the Adams's self-cocking revolvers manufactured up to this date of July, 1854. As earlier mentioned, a majority of such weapons in serial-ranges below 14000 were probably made in Birmingham, and there is no record known of the machinery or methods there used for revolver manufacture.

The drilling machinery used, and too-briefly described in the article, appears to have been most unusual, for multi-spindle drilling equipment cannot have been common at the date in question. Colt on his side, and with all his love for speeding production by use of machines, had still to bore his chambers separately in several stages, albeit employing an ingenious chuck to simplify and accelerate the cylinder-manufacturing process. Yet, if our *Daily News* correspondent is to be believed, Robert Adams was boring five chambers as a single operation, and it is to be regretted that no light can be cast on the manner in which this was done; certainly, Adams's Br. Pat. No. 2000/1854 contained no claims relevant to such a technique.

"The observant have probably often noticed the fine, dark, lustrous blue which is sometimes presented by the barrels of pistols. This is effected by covering the iron with hot charcoal embers, which, by their chemical agency stain or dye the metal, not merely superficially, but to a considerable depth. This process, moreover, combines utility with ornament; since it hardens the iron, and renders it less liable to rust. A very beautiful grain, somewhat like that of wood, may also be observed in the barrels of some pieces, and which has the effect of rendering the metal more tough, by clenching, as it were, the different particles together. But this was produced naturally in the course of the original formation and is only indebted to art for heightening."

It seems fairly clear, both from the above remarks and from the off-hand dismissal of the whole subject by Dickens, that attempts by outsiders to pump "blueurs" as to what they were doing, and how they were doing it, met with little success. Like most craftsmen, these finishers guarded their secrets jealously, and whilst our *Daily News* correspondent appears to have been "blinded with science", Dickens probably received a flat and surly rebuff to his inquiries.

However the process was carried out, "heat blueing" aimed at achieving exactly the same object as by browning or blacking the firearm with a chemical paste or solution, namely, the formation on the surface of the metal of a suitable metallic oxide capable of being polished and of withstanding normal day-to-day handling.

Iron and steel happen to have the virtue that a particularly attractive bluish oxide forms naturally upon their surfaces in a fairly narrow temperature range between roughly 285° C. (545° F.) and 316° C. (601° F.), and can be "fixed" by then cooling or quenching the metal rapidly, before it passes the required temperature and begins changing to yet another colour. The oxide is not as durable as the protective coatings produced by some other methods, but serves well enough for the purposes stated.

Since the relationship between colour and temperature is not rigid (the desired oxide forming on short exposure to the higher temperatures or long exposure to the lower ranges), and the preparation and chemical composition of the weapon-metal can also be critical, then, plainly, long familiarity with the process and resultant "know-how" are probably more important than the actual methods or equipment used.

Basically, however, it appears that all "heat-blueing" methods are aimed at securing what may be termed the "cooking temperature" known to produce a desired colour, and at holding that temperature for the necessary period that the "blueurs" experience (or superstitions) fixes as necessary for the finest results.

Given sufficient skill or equipment, the desired temperature can be reached readily enough at the heat-source, and maintained there exactly, but the imparting of that heat evenly through a work-piece like Adams's revolver frame and barrel, can present a difficult problem, particularly to the amateur "blueur" using only equipment normally to hand in an average household. The massive action-body will absorb more heat than the lighter barrel, and if the heat source is at an even temperature throughout, but above that necessary for a blue oxide, then the barrel will reach the desired colour long before the frame, or will have overheated and ceased to be blue by the time the frame "arrives".

Amateur "blueurs" will have encountered this difficulty when trying to blue parts over a naked flame, and three methods are commonly used to try to secure an even heat-source which can be controlled and held at the necessary temperature whilst the work-piece is heating through.

A heavy metal plate of sufficient mass to absorb and disperse excess heat can be interposed between work-piece and flame. Alternatively, a sand bath, or a molten-metal bath composed of a lead/tin alloy known to melt at the temperature desired, can be used. By immersing the work-piece completely in the molten metal, and controlling the flame under it so as to keep the metal just liquid, the job can be heated uniformly throughout to the required temperature out of contact with the air. The oxide film is formed as the steel is withdrawn from the melt, when it has to be quenched in water, immediately, to fix it.

"Previous to 'blueing', and while the iron is yet in a comparatively soft state, the barrels are carried into the engravers' room, and elaborately embellished with fanciful designs. These are in accordance with the use to which the weapon is to be put, and are, therefore, rather various. If it be a fowling-piece, we have a representation of some pleasant English scene, with dogs, and pheasants, and heaths, and far-off hills. If the slaughter of tigers is contemplated, a tiger is represented; if an elephant, that fact also is pictorially shadowed forth upon the metal; and so on with the rest. Some of these designs are executed with great freedom and spirit. We do not know whether all who so desire may obtain the privilege of personally inspecting the various processes at Deane & Co's revolver factory, but this is certain, that such a visit is full of interest to those who care about improvements in arms or warfare."

All Deane, Adams & Deane revolvers were engraved, firstly with the barrel address, then with the cylinder- and frame-legends associated with manufacturing and patent royalties matters, and finally in various degrees and qualities of strictly decorative work. Despite the claims, implicit in this newspaper article, that the "barrels" of the revolving arms made at Weston Street were normally pictorially engraved, such pieces are the exception rather than the rule. The decorative engraving on a majority of Deane, Adams & Deane revolving weapons examined was foliate and not pictorial.

Outline graver or "burin"

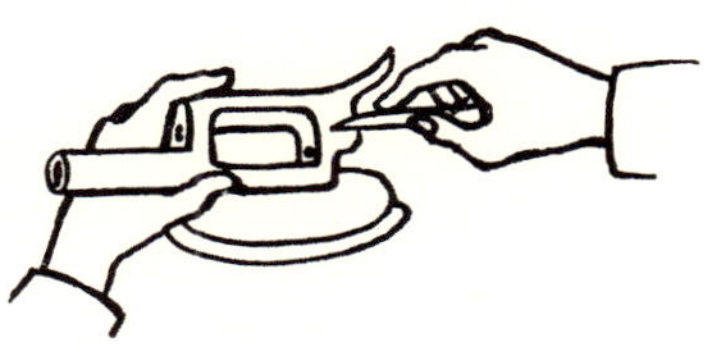

Suggested position of pistol-frame, for engraving.

FIG. 35. *Engraving.*

Certainly, a proportion of Adams's revolving rifles and carbines bore well-executed pictures of various sorts, but the remainder (and the majority of the revolving pistols of all sizes) carried the type of engraving shown, for example, in Plate 7, which we term "foliate".

This type of engraving was widely used on most English percussion revolvers (in association with lightly incised border lines cut around the periphery of the frames and cylinders or at the muzzles and on heads of screws, pins, rammer parts, etc.), and was presumably both inexpensive and in accord with contemporary taste.

We have not been able to establish exactly how the work was done, but would guess that it was executed broadly in the same manner as steel—or copper-plate engraving for the illustration of books. (See Fig. 35.)

The engraver would cut his decorations free-hand into the metal by means of a hardened steel "graver" or "burin" (a fine-pointed, triangular-sectioned chisel; see Fig. 35) which was pushed forward into the metal at a small angle of inclination to its surface. The work-piece would be held in the left hand, supported on a sandbag or small cushion, and for the standard foliate engraving as much of the work as was practicable would be executed with two sizes of graver only; one (extra-finely pointed and known as the "outline tool") being used to outline the proposed pattern and to form the transverse cut or nick which terminated each line, so that the wider-pointed graver actually used for cutting them would not overrun the end under the thrust of the engraver's hand.

A scraper (rather similar to a miniature modern triangular scraper) would be used to remove the burr in the metal raised at edges of the cuts, and a small hand burnisher would be employed to soften down lines cut too deeply or to remove small scratches inadvertently made during the course of the work.

CHAPTER XIV

THE LANG REVOLVER

IN Chapter IV, John Deane's opinion was cited that revolvers by Joseph Lang, J. Barnett, and Robert Adams, were exhibited in competition to Colt revolvers, at the Great Exhibition of 1851; Adams's, Barnett's, and Colt's pistols have now been dealt with, and Lang's weapon is the subject of this Chapter. It should be mentioned, too, that Joseph Lang was not the only British gunmaker to retail arms of the type to be described, and that reasonable doubt can be cast upon Deane's allegation that this man actually made them.

In describing Lang's pistol, Deane wrote (*op. cit.* p. 207) that "... compared 'as like with like' with the Colt revolver, it is considerably lighter; for the use of a ramrod and its concomitant lever being rejected, the metallic part adherent to the barrel is less massive. The rotating apparatus ... turns on a cylinder pin, and the barrel is maintained thereon by means of a key-bolt. ... The orifice of the chambers is widened, so as to permit the posterior orifice of the barrel to set into them, and thereby prevent the escape of flash and gases".

A revolver of this type is illustrated on Plate 21, but was actually retailed by "Witton, Daw & Co.", and it can be seen that Deane quite fairly described the type; it was a six-chambered, single-action, centre-hammer revolver, with cylinder moving to-and-fro upon its arbor, to mate chamber with barrel (in a gas-seal) at moment of discharge.

J. N. George (*op. cit.* p. 191) echoed Deane in stating that a version made by Joseph Lang was exhibited at the Great Exhibition of 1851, but went on to say that it was later produced in improved models by "Witton & Daw", and by "Parker, Field & Sons". He characterized Lang's pistol as being the rammerless weapon fitted with vertical nipples (unseparated by any partitions), and the improved models as retaining a primitive arrangement of nipples, but in association with Colt-type partitions, and mechanical or plunger-type rammers.

In this view, George was not wholly correct, for Joseph Lang also retailed the improved version, and other gunmakers sold the unimproved type. In view of the comments in Chapter IV, it is obviously impossible for us to accept George's claim (*op. cit.* p. 192) that Lang's design appeared in the late 1840's, but there can be no doubt that he was right in naming "Witton & Daw" and "Parker, Field & Sons" as retailers of improved models, and in stating that this design remained popular into the 1860's.

An examination of the Catalogue to the Great Exhibition of 1851 (with its Supplement) supplied no evidence that Joseph Lang had actually exhibited a revolver at all, amongst the beautiful arms that he did show. Certainly, his Catalogue entry (Vol. 1, Sect. ii, Class 8) for Stand 226, South Gallery West, at the Crystal Palace, listed revolving arms, but his wording, "the turnover 4-shot and *6-shot revolvers* of all sizes", would have been more applicable to his well-known (but fairly rare) single-action, mechanically-rotated pepperbox (by terminology then in use), than to any conventional revolver. It is also noteworthy that the Prize Medal secured by Lang was for the perfection in fabrication of his exhibits; it seems unlikely therefore (having regard to the interest in revolvers at this Exhibition) that Lang would not have attracted some contemporary comment or attention if a true revolver had been shown at his Stand.

It was also noted that "Witton, Daw & Co." exhibited rifles and duelling pistols (securing a Prize Medal for their guns and rifles) at Stand 203, whilst "Parker, Field & Sons" (Stand 224) exhibited a wide range of products, but that neither apparently showed any revolving arms, whether pepperbox or true revolver.

Indeed, it was the years from 1858 onwards which proved surprisingly rich in literary references to this type of "gas-seal" revolver, and it is submitted that the design was not even introduced until 1852 or 1853, and not produced in any appreciable numbers until 1857 or 1858.

Thus Deane's remarks upon Lang's pistol were published in 1858, and the reference noted by George (*op. cit.* p. 192) that Lieutenant Hans Busk (founder of the Rifle Volunteer Movement) regarded Witton & Daw's variation of the Lang "as being the best of the English revolvers then in existence", was based on *The Rifleman's Manual* published by Busk in that same year. His *Manual* prompted the appearance of a note in the *United Services Gazette* of December 25th, 1858:

> "Two or three weeks ago, we alluded incidentally to an improvement which Mr. Busk stated, in his little book on firearms, had been made in the action of revolvers by Mr. Daw. . . . It seemed, by the description, to have been a very material improvement, affording

protection against possible accidents from an escape of ignited powder between the cylinder and barrel. We have now had an opportunity of personally inspecting one of these pistols, at the moment before discharge, *the barrel enters to the extent of only one eighth of an inch into the chamber which contains the cartridge, and thus produces an uninterrupted prolongation of the tube*."

It must be conceded that the peculiar plunger-type rammer (with folding "T" head) found on Parker, Field & Son's variant of Lang's revolver was protected as a Registered Design on June 14th, 1852 (as No. 3300), but the accompanying drawing did not show that rammer as intended for any specific revolver, and it could as readily have been proposed for use upon an Adams self-cocking or T. K. Baker single-action weapon, when registered, as for use upon a pistol of the "Lang" type.

Finally, possessors of Captain L. Jewitt's book (*op. cit.*), in the 1865 edition, will find an advertisement in it by Parker, Field & Sons, at 233 High Holborn (Manufactory at 58 Mansell Street, Minories) announcing that they were "now manufacturing . . . Improved Revolving Pistols with their Patent Spring Ramrod and interlocking motion of the Chamber". This advertisement certainly suggests that the design enjoyed popularity over a number of years, and confirms that neither Lang nor Witton & Daw had a monopoly upon the gas-seal feature of it.

Two British patents for effecting gas-seals by thrusting forward a percussion revolver cylinder had already been in existence by 1858.

The first was that of William Moore and William Harris (No. 69 of October 1st, 1852), and was clearly (from its claims) not in point, whilst the second was that of Philip Webley (No. 2127 of September 14th, 1853); both had lapsed by 1858 (see Appendix), through failure of the patentees to pay renewal fees, but the Complete Specification to Philip Webley's patent claimed and illustrated the same method for effecting this gas-seal as that employed in Lang, Witton & Daw or Parker Field revolvers, namely, a simple ram within the action-body, which was cammed forward by the hammer-breast (as the hammer fell) to push forward the cylinder into engagement with the barrel.

Philip Webley restricted the claims for application of his reciprocating cylinder to "arms arranged for central fire", and by this he meant revolvers using nipples that were horizontally aligned with their barrel; the Lang-type revolver, on the other hand (and whether in its original version, or as subsequently improved), had nipples inserted pepperbox-style, either vertically to the bores or at a slight inclination to the vertical.

It is improbable that Philip Webley would have chosen such a limitation to his patent-claims, if the protection of his gas-seal cylinder for revolvers of Lang-type had also been available to him. It must follow, therefore, that the latter arms were on the market when he filed application for his patent, and his method for moving a revolver cylinder to-and-fro was thus already in the public domain (and unpatentable) by September 1853, so far as cylinders with vertical nipples were concerned.

The rarity of arms made under Philip Webley's patent also suggests that the public were well content with this arrangement of nipples, and saw no advantage in such revolvers "arranged for central fire".

As to the lock-mechanism used in Lang-type revolvers, and the device for operating their gas-seal, the design is best appreciated by reference to those sketches which follow.

Embedded in the rear face of the barrel lug is a flat spring which (when the weapon is assembled) presses against the front of the cylinder, and pushes that component back off the end of the barrel, except when the hammer is down. Since an arbor (and rammer, in some cases) must pass through this spring, it is of a peculiar shape, being either (A) *or* (B) as may be appropriate.

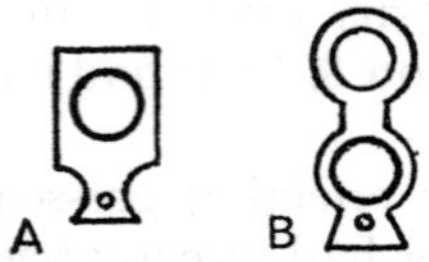

FIG. 36. *Lang's cylinder retracting spring.*

On sliding the cylinder off its arbor, it will be seen to be quite elaborately machined at the rear face, commonly into two "steps" if viewed from the side. A plain cylinder has been chosen for illustration in Fig. 37, but many of them were most attractively fluted.

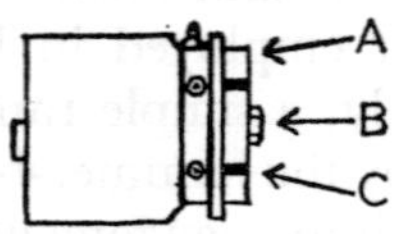

Front
a. Ram plate.
b. Ratchet.
c. Locking notches.

FIG. 37. *Lang revolver cylinder.*

The face of the standing breech, on the action-body, is in most cases very beautifully and exactly machined to fit these "steps".

Viewed from the rear, the cylinder usually appears as below, although more elaborate machinings have been found:

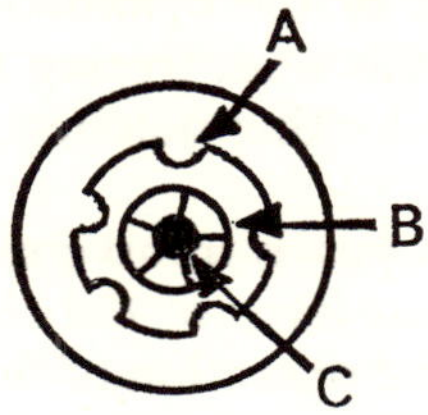

a. Locking notch.
b. Ram plate.
c. Ratchet.

FIG. 38. *Rear face of Lang revolver cylinder.*

Turning to the malleable iron (normally case-hardened) action-body, a large inspection plate on the left side may be removed, on turning out suitable securing screws, and a very heavy bearing-plate will be found beneath it, screwed to the floor of the body and taking the strain of the left end of the hammer axis-pin. This pin is turned (at its right end) into the wall of the action-body, which is also integral with the floor.

On removal of this bearing-plate, the cylinder-bolt will be found resting to the left side of the hammer, and pivoted at its forward end upon a small transverse pin in the front wall of the action-body.

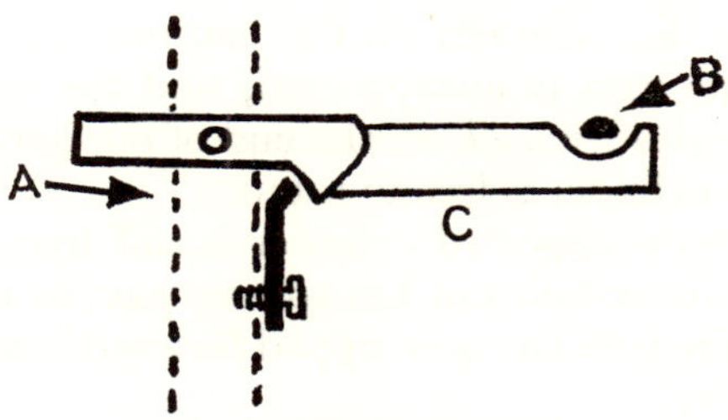

a. Front wall of action-body.
b. Actuating stud on hammer.
c. Cylinder-bolt.

FIG. 39. *Lang's cylinder-bolt.*

On removing this cylinder-bolt, the hammer (which is offset to the right side of the action to permit clear sight to be taken) can be removed; in some cases, it will be found that the hammer axis-pin has been partially cut through, vertically, to permit the tail of the cylinder-bolt to rise and fall as the weapon is cocked and fired.

The hammer is unremarkable in shape, except that it commonly has *three* bents cut in it to engage the trigger. A small stud on its left side rocks the cylinder-bolt out of engagement with locking notches cut on the periphery of the cylinder-ram plate (see Fig. 38) when the weapon is cocked; the tail of the cylinder-bolt has a spring temper, and eventually rides over this stud, freeing the bolt to tip down again (into engagement with the locking notches), under the pressure of the spring shown in Fig. 39.

Additionally, a small pin passes transversely through the belly of the hammer, to operate a thrust-lever of the old pepperbox type (earlier illustrated as Fig. 1), and the operation of this type of lever has already been described in Chapter III.

Embedded in the standing breech, in front of the hammer and directly in line with its breast, is the ram which thrusts the cylinder forward to seat over the end of the barrel, on firing. This is a stout block (square in section), which usually has a single retaining lip on top to prevent it from being pushed forward out of the action-body, and lost, and it can be removed to the rear, once the hammer has been taken out. This ram is operated solely by pressure from the hammer-breast, and is an easy sliding fit in the standing breech; this enables the flat cylinder retracting spring shown in Fig. 36 to push both cylinder and ram to the rear (when not supported by the hammer-breast), and thus allow the pawl to rotate the cylinder without barrel and chamber fouling each other.

On cocking these weapons, the hammer rocks the cylinder-bolt out of a notch on the edge of the ram plate at half-cock, and also operates the pawl to rotate the cylinder. As the hammer travels back through its cocking arc, the ram is unsupported, and the cylinder retracting spring pushes the cylinder back off the end of the barrel, so that (being also unbolted) the cylinder is free to rotate.

At full-cock, the unsupported cylinder is still free of the end of the barrel, but the cylinder-bolt tail has sprung past its operating stud on the hammer and the bolt has now tipped forward into a locking notch, to index the cylinder.

When the trigger is pressed, the hammer falls and its breast strikes the ram (which in turn hits the ram plate on the rear of the cylinder), forcing the cylinder forward about $\frac{1}{8}$ in. to effect a "gas-seal" of chamber-mouth to barrel-breech. As it falls, the hammer (by its protruding screw) drags the loosely-mouthed pawl down to a position where it can pick up the cylinder ratchet for the next rotation, and the hammer also picks up the tail of the cylinder-bolt under the bolt stud on its left side.

Examination of these revolvers suggested the following gunmakers as principal retailers of the type, whether as introduced or as later improved:

T. K. Baker, 88 Fleet Street, London.
J. Beattie, 205 Regent Street, London.
J. Blanch & Son, Gracechurch Street, London.
J. Blissett, 316 & 321 High Holborn, London.
J. Lang, 22 Cockspur Street, London.
Parker Field & Sons, 233 High Holborn, London.
Witton, Daw & Co., 57 Threadneedle Street, London.

Specimens were occasionally found with provincial names upon them ("Tipping & Lawden" of Birmingham, was a case in point), and any of these names could be found upon revolvers without the gas-seal cylinder refinement. Upon occasion, pistols were found to have had the gas-seal components deliberately removed (or unserviceable through rust and congealed oil), but enough specimens were traced to establish that a model without reciprocating cylinder was also offered by the gun trade.

It was once hoped, by collating our notes, that some pattern of details would emerge to establish that one or more versions of this revolver were made and sold as the standard gun-trade line for retail by others.

In fact, these notes yielded little or nothing to support such a view.

Serial numbers

It appeared that the gas-seal revolvers were often numbered, but we could not establish whether these numbers were those of the retailing gun makers, or serial numbers used by the manufacturers. The lowest number recorded was 44 (on a 58-bore weapon retailed by T. K. Baker) and the highest was 9575 (upon a 50-bore pistol retailed by Parker, Field & Sons).

Calibres

In general, a difference between the gauge of each chamber and that of their barrel was remarked: a 70-bore barrel served 60-bore chambers, a 58-bore barrel served 52-bore chambers, and so on. This sort of variation is to be found in most percussion revolvers, but did appear to be unusual in degree amongst these arms.

The smallest calibre found was in two revolvers with 120-bore barrels, by Joseph Lang and T. K. Baker; several arms (with and without gas-seal cylinders) had 70-bore barrels; most specimens encountered had barrels of about 50-bore calibre; one or two unusual specimens were larger, a 44-bore pistol being the largest calibre.

Rammers

As may be seen from Plate 21, many of these arms had no rammers fitted to them, the owner loading his pistol by using a separate loading rod.

In rare cases, a straight-line "plunger" rammer was fitted into the barrel-lug, complete with a coiled return-spring to withdraw it from a chamber after use. The Parker Field rammer (under Registered Design No. 3300, of June 14th, 1852) was of plunger-type, but mounted in a tube brazed or soldered on to the right-hand side of the barrel-lug; this rammer had a folding "T" head, which could be turned flat against the barrel when not in use.

A majority of these arms (irrespective of their retailer) were fitted with quite powerful lever-rammers, mounted Colt-style beneath their barrels. Some (but not all) of these rammers were marked "VR" and Crown, "Patent No. . . ."; in our own experience, these numbers were in a 2000 range (rammers numbered 2015, 2251, 2633, 2772 and 2815 being noted), which numbers did not reflect true production-quantities for the device.

It has been suggested that these rammers were made under John Blissett's Br. Pat. No. 2069, of September 13th, 1855, but we were unable definitely to establish such a parentage from Blissett's provisional patent specification. In fact, we found the use of rammers taken from Charles Pryse and Paul Cashmore's Br. Pat. No. 2018, of 6th September, 1855, to have been preferred, and the use of the "Patent" marking and number strongly suggested that rammers so marked were made before that patent was allowed to lapse in September, 1858. The appearance of such rammers bearing no markings at all was consonant with the continuance of manufacture after that date.

Rifling

In the matter of rifling used in these weapons, only one model appeared to exhibit any consistency; where Joseph Lang had retailed a weapon lacking the gas-seal cylinder, it appeared that the maker of those arms preferred to use a five-groove rifling, twisted to the right.

In other cases (whatever the retailer, and regardless of the presence or otherwise of a reciprocating cylinder) confusion reigned; five-groove rifling was used on some pistols, whilst others had twelve, fourteen, or twenty grooves in their barrels. Even weapons bearing the same retailer's name were found to have been rifled differently.

Safety-catches

Safety-catches were not fitted to every weapon, but some had a catch on top of the action-body, which could be slid forward to bolt the hammer; in principle, such catches were identical with those used for good-quality centre-hammer pocket pistols, or for pepperboxes.

Dimensions

The largest pistol found was over 14 in. in overall length, with a 7-in. barrel; below this unusually heavy pistol, the overall lengths ranged between 10 and 12½ in., the shortest barrel being 4⅕ in. long for an 80-bore pistol.

One or two revolving rifles were also traced, with 28-in. barrels in 54-bore calibre, but they were sufficiently rare to suggest that only a handful were made. An unusually heavy specimen is shown as Plate 35.

In the light of what has already been written, it is obviously impossible to offer the reader any firm propositions about the history of these arms; however, it is reasonable to suggest a likely course of events, pending publication of the researches of others.

It does seem probable that Joseph Lang introduced a single-action revolver not long after the 1851 Exhibition, since John Deane mentioned him by name (in preference to any other maker), and because he is known to have sold single-action, mechanically-rotated pepperboxes of the type mentioned in Chapter III. These pepperboxes were also retailed by others, but specimens signed by Lang are known, and may have been the "turnover . . . 6-shot revolvers" shown as exhibited by Lang in the 1851 Exhibition Catalogue; mechanically similar to the revolvers, these small-calibred (90-bore) six-barrelled pistols measured about 8 in. overall in length, and had the same centrally-hung hammer. If Lang made them (or had access to supplies for retail), it would have been a reasonable step for him to introduce a true revolver based on that design, to meet the demand for such arms aroused by the Great Exhibition. This could have been accomplished in 1852, or early in 1853, and it will be recalled that Chapter IV proposed a similar course of events in relation to the self-cocking "transition-type" revolver.

We cannot confirm the source from which the first manufacturers drew their design for a reciprocating gas-seal cylinder; such an idea may have been suggested by Moore & Harris's patent of 1852, or even noted from surviving Collier arms, but it was clearly well established in the gun trade by 1853, or Philip Webley's patent would certainly not have been limited in the fashion earlier described.

As to the source from which this gas-seal feature came, Hans Busk (in the 1860 edition of his *Manual*) wrote that it was ". . . originally the contrivance of a Birmingham lock-filer", and although this was a trade to which Philip Webley was originally apprenticed (see Chapter XVI), we are unable to accept that man as the inventor, by reason of the limited claims to his patent of 1853. It is suggested, therefore, that if Busk's reference was, in fact, to Webley, then it was prompted by either a defective memory or misunderstanding of the claims to Webley's Specification.

Demand for the Lang design obviously encouraged other makers to take it up, and the addition of rammers and nipple-partitions (the latter cutting down risk of multiple discharges) was a logical improvement. On the matter of dating such improvements, we traced a revolver by T. K. Baker (numbered "105"), which had a lever-rammer but no nipple-partitions; engraved upon the lock-case was the legend "Thomas K. Baker. 1855", and this was a year in which no patents or Registered Designs were secured by Baker. It seems probable that this dating referred to an improvement in Lang's original design, which would logically have been the rammer on Baker's revolver; whether that rammer was made under the patents of Blissett or Pryse & Cashmore, in that year, could not be established. However, the latter's Br. Pat. No. 2018/1855 contained claims for both single-action revolvers and lever-rammers (amongst other matters), and they seem at least possible suppliers to the gun trade for these arms, in 1855.

CHAPTER XV

GEORGE HENRY DAW
and
"DAW'S REVOLVER"

GEORGE HENRY DAW is probably best-known for his work on centre-fire cartridges (under two patents of François Eugene Schneider—Nos. 1487 and 2203 of 1861—initially, but later to his own designs), and many collectors of nineteenth-century firearms literature will have studied *Daw's Gun Patents* (London, 1864) and *The central fire cartridge before the Law Courts, the Government and the Public* (London, 1867) in which he recorded his prowess both as gunmaker and as litigant. He was, however, active in the promotion of revolving arms, one of which (retailed by "Witton & Daw") has been described in Chapter XIV.

Despite his trade-label claims to represent a line established in 1780, Daw was not traceably located in London until 1851, when "Witton, Daw & Co." appeared at 57 Threadneedle St., E.C. His associate, Witton may have come from the merchant house of "D. W. Witton & Co." (at 63 Fenchurch Street, London), or have been J. Witton (an arms contractor, at 82 Old Broad Street, from 1844 to 1850); one David William Witton seems to have merged with "Lacy & Co." (at 21 Great St. Helen's St., E.C.) in 1855 to create the firm of "Witton Brothers", in 1857, and was presumably a member of the same Witton family.

The trading style "Witton, Daw & Co." endured until 1855, when it became "Witton & Daw"; this new name lasted until 1860, when "George Henry Daw" became the trading name at 57 Threadneedle St. At this time, Witton's interests may have switched to the "Witton Bros." business at gun-making and arms-contracting addresses in Great St. Helen's St., and at Dunning's Alley, Bishopsgate, or he may simply have died or retired.

As earlier mentioned, in September, 1855, Charles Pryse, of Birmingham, and Paul Cashmore, of West Bromwich, took out British Patent

No. 2018, covering revolvers (with their locks) and a variety of lever-rammers and rammer-latches; some of the rammers had been used on "Witton & Daw" gas-seal revolvers dealt with in a previous chapter, and upon achieving a trading style of his own, George Daw turned his attention to other features covered by this protection.

The claims of Pryse & Cashmore's provisional Specification to this patent mentioned, amongst other matters:

"... an improved construction of repeating fire-arm, having a double action, and self-acting or non-self-acting, that is to say, that may be cocked either by means of the trigger or by the use of a thumb piece on the hammer, neither of these actions interfering with the utility of the weapon or its rapidity of fire as a repeating arm, ... making the body or frame of this arm in one piece of metal.

. . . . .

"And, lastly, in improved forms of double-acting bolt, which will prevent the chambers from revolving and secure the hammer or striker at half cock at the same time."

Pryse and Cashmore's patented double-action lock-mechanism and one-piece frame were taken up enthusiastically by G. H. Daw, and were either manufactured entirely for him by the patentees, or else supplied "in the white" for locking and stocking at Daw's London premises.

The date for commencement of manufacture is uncertain. Daw himself claimed that these arms were successfully used in India, during the Mutiny of 1857–8, but, if this statement was true, it seems unlikely that Daw himself sold the weapons; most specimens encountered are marked "George H. Daw, 57 Threadneedle Street, London", and there is genuine doubt that Daw was actually trading under this style prior to 1860.

It can be assumed, however, that the necessary arrangements between Daw and the two patentees were completed before British Patent No. 2018/1855 was allowed to become void, in September, 1858. Messrs. Pryse and Cashmore were experienced manufacturers, and well aware of the value in their patent; since they allowed the renewal-date for their patent to pass without payment of the necessary fees, it is certain that some manufacturing agreement signed with George H. Daw ensured that the protection had served its purpose.

It is always possible that Daw's reference to use of these revolvers during the Indian Mutiny related to arms supplied by other gun-makers, for whilst it is a fact that a large majority of all Pryse & Cash-more double-action revolvers made under this patent were marked and retailed by Daw, a few early arms (with low serial numbers) can be seen as retailed by other men; Henry Holland (of King St., Holborn) and Alexander Henry (of London and Edinburgh) respectively signed pistols numbered 215 and 1084.

Daw devoted some space to describing this weapon in his *Gun Patents* pamphlet, and certain of his comments regarding the purpose in features of the design cannot be improved upon for clarity, when read in association with a study of the specimen arm shown on Plate 21 (*lower*).

"The Daw revolver" (he wrote) "has a boss in front of the lock action that rises hemispherically so as to completely cover the nipples of the chamber cylinder with its flat side. . . . The upper part of the boss . . . is cut away like the lower end of a U, to the extent of the breech part of the space of one of the chambers. This answers two purposes; it gives the opening through which the hammer falls on the nipple, and it supplies the only place where a cap can either be put on, taken off, or otherwise meddled with.

"In some revolving arms having bosses, these have been cut away on each side. The result has been that caps have been accidentally struck and exploded, producing both fatal and serious accidents. In the Daw revolver the impossibility of such accidents is unalterably secured. The nipple once capped it is passed round in front of the flat side of the boss, so close that there is not room enough for a loose cap to get off the nipple; it rests against the boss until it comes up in front of the hammer. The caps are put on when the hammer is at half-cock, because at this position the chamber cylinder turns round without stop from right to left.

"I spoke, just now, of the upper part of the boss being cut out in the form of the lower part of the letter U so as to allow the hammer to come down on the nipple. Now, the crown of the hammer is shaped exactly in the same way, so that when it is down it fills up to a mechanical nicety the space cut out in the boss, fitting so neatly that it all looks like one piece. The value of this shape is that by being part of the boss itself when down, it follows that upon exploding a cap on the nipple it acts as a shield to the eyes of the shooter.

"Between the nipples are stop-studs, upon any one of which the hammer can be let down to rest, the latter has a notch undercut in its nose into which the stop fits. By this contrivance the revolving breech is held firmly immovable and each nipple may be capped and the pistol carried about with the greatest confidence in its safety."

"Double-action Trigger"

"When the hammer is at half-cock the cylinder is free to revolve in one direction only, but when the full cock is set, the act of setting also sets the trigger, upon which a strong flat bolt shoots through the centre of the under side of the stock against the projecting "fence" that divides each chamber breech and encloses each nipple. This bolt wedges the cylinder tight and holds it so until the trigger is let go after firing, it then sinks into its slot and remains until it is re-set by the hammer being again placed on full-cock."

In addition, Daw also secured (or arranged) laudatory comments about the weapon, from contemporary writers. W. H. Russell (special campaign correspondent to *The Times*, during the Indian Mutiny) wrote of it approvingly, whilst Hans Busk remarked of it that "since my last notice of this arm, Mr. Daw has improved it still further. The position and arrangement of the sights, the action of the lock, the very handy form of the grip and the attention obviously given to various minute but important parts, fairly justify one in bestowing upon it the most unqualified commendation."

Another contemporary, Henry Downes Miles (author of *The Book of Field Sports*) wrote fulsomely of this revolver that "its form is of the handsomest, its finish exquisite and it is one of the pleasantest in the hand of the many varieties of revolver we have examined", and Daw also succeeded in securing a similar (if more practical) commendation in Captain L. Jewitt's book, when this was published in 1860.

Finally, in 1864, this indefatigable promoter achieved national Press-coverage again, and took care to note the fact in *Daw's Gun Patents*:

GENERAL GARIBALDI

Partaking of the national feeling of admiration for the noble character of the Great Italian Patriot Soldier, I obtained an interview with him, which the following paragraph from the *Daily Telegraph* of April 22nd, will sufficiently explain:

"Yesterday Mr. Daw, of Threadneedle Street, waited by special appointment on General Garibaldi, and presented him with a Patent Breech-Loading Gun of novel and simple construction, together with a handsome Double-Action Revolver. Upon each weapon, on a gold plate, was engraved the following inscription: 'Presented to General Garibaldi, from the maker, George H. Daw, April, 1864.' The General, on receiving the presents, said: 'This is very kind. I thank you, my friend, for your handsome and useful presents.' Referring to the Breech-Loader, he said: 'I am grateful to you for it, and will use it for your memory—the system is so simple and so good.'"

Although Daw promoted this pistol so hard, it was in fact an excellent weapon in its own right; the methods used in its manufacture probably involved a good deal of handwork, so that parts were not interchangeable, but this was a criticism that could be levelled at revolvers considerably less reliable than Pryse and Cashmore's product.

Daw offered his pistol as a five-chambered arm with horizontal nipples, and in a range of calibres (some of which are extremely rare today); the 6th edition (1860) of Hans Busk's *The Rifle and How to Use It* included an advertisement referring to "Army, Navy and Pocket sizes in Holster case, with extra nipples, Bullet Mould &c. from £6.10.0. Ditto in Oak case and apparatus complete £7.10.0. Ditto highly finished £10.10.0. Revolving rifles from £15.0.0.", but did not define the calibres of weapons thus described.

J. N. George (*op. cit.* Plate XXVI) illustrated a 38-bore version, with 8-in. octagonal barrel and an overall weight of 3 lb. 10 oz., but the average collector is more likely to encounter round-barrelled 54- or 80-bore pistols, and these may have been Daw's "Army" and "Navy" weapons; arms in 120-bore calibre are also believed to exist, as suggested by the above, but no specimens could be secured for description here. A five-shot 46-bore rifle (with 26¾-in. barrel) passed through the salerooms of Messrs. Wallis & Wallis, at Lewes, in November, 1961, and may have been the version advertised; G. H. Daw's salesmanship was evident again in this arm, with a range of sights up to 1,000 yards!

The version of Daw's double-action revolver most likely to be encountered is shown on Plate 21, as a five-chambered 80-bore weapon, with 5½-in. barrel and an overall length of 10¼ in. No safety-catch is fitted (the hammer-nose engaging safety-pins on the cylinder base, instead), and the lever-rammer is clearly based upon one of the devices claimed in Pryse and Cashmore's patent. Rifled with five grooves, twisted to the right, the round barrel is London-proved, and attached to the arbor by a Colt-type transverse key.

The hammer on this version of the weapon strikes through a "U"-shaped aperture in the frame boss, just as G. H. Daw described it, but has pronounced "ears" formed at either side of the nose. When the hammer is down, these ears fill the aperture in the boss completely, and it is impossible for any cap to fall off the nipple, and to jam the cylinder; when the hammer is at half-cock to permit capping, however, this eared formation of the hammer and boss leaves a comfortable aperture through which thumb and forefinger can accomplish the necessary capping.

Pryse and Cashmore numbered their revolvers consecutively, the lowest serial number traced being 215 and the highest 5461; however, below serial number 1000, the arms differ in minor details from the standard specimen shown on Plate 21, and described by Daw himself.

In these first revolvers, the butt lacked a comfortable prawl visible in Plate 21, and no boss was formed behind the cylinder; the "U"-shaped frame-aperture for the hammer-nose had not been evolved, and the hammer (which had no ears) struck the nipples through a hole drilled in the standing breech. The rammer conformed, as to both linkage and latch, to one device claimed in Pryse and Cashmore's patent, and some of these arms were fitted with an externally sliding safety-bolt claimed in that patent, which locked both cylinder and hammer at half-cock.

In addition to these double-action pistols, Daw also retailed a self-cocking revolver of quite similar appearance. J. N. George (*op. cit.* p. 224) mentioned a 120-bore six-chambered version, but we have only encountered this calibre as retailed by James Beattie; such weapons signed by Daw were found as 80-bore pistols (with $5\frac{1}{2}$-in. barrels) and in 54-bore, with $6\frac{1}{2}$-in. barrels. Unlike the double-action arms, these self-cocking pistols were all six-chambered.

There has been some dispute amongst collectors, as to whether or not the lock-mechanism used in these self-cocking revolvers was originally designed to "hesitate" in a similar manner to that described in Chapter VII as patented by Robert Adams; "hesitating" lock mechanisms, it will be recalled, permit the user to cock his weapon by pulling the trigger, to take deliberate aim, and to discharge the arm by a second pull at the trigger.

Whilst it is a fact that the lock components used in Daw's self-cocking revolvers can be deliberately adjusted to operate in this "hesitating" manner, there is doubt as to whether their designer ever intended his mechanism to function in such a manner. R. Bedford (in the article cited in Chapter VII) dismissed, out of hand, any idea that such an effect was deliberately built into these revolvers, and it is a fact that not only is extremely careful "tuning" of the lock mechanism needed to achieve hesitation, but also that although such arms were marked "Patent", there is no known British patent for achieving hesitation in the manner thus effected. A toggle-lock would be the medium through which the revolver-hammer was held at full-cock, by the components provided, and such a device should be readily discernable in any patent specification drawings. Careful examination of such drawings has failed to reveal any inventor as claiming such a lock mechanism, and until (if ever) such a patent is discovered, it seems best to accept Mr. Bedford's view.

CHAPTER XVI

JAMES AND PHILIP WEBLEY

ALTHOUGH the Webley family did not make its undoubted contribution to any history of British revolvers until after 1865, the groundwork for that contribution was certainly laid in our period.

Before dealing with the known facts about James and Philip Webley, however, two points have to be made; unless these are grasped at the outset, it may be difficult to make clear why the two men were apparently so slow to advance (prior to 1867) in an expanding field, which the family came later to monopolize.

Firstly, it is a fact that the gun trade never dominated Birmingham's economy, even in boom periods, and this was the town in which the Webley fortunes were built. Nail-making or coal-mining were considerably more important industries than gun-making during the period, whilst the manufacture of jewellery, or "toys" (i.e. buttons, buckles and the like), and the brass trade were of equal size and importance to the gun trade. It follows, therefore, that Birmingham "gunmakers" were not necessarily either important tradesmen or rich men.

Secondly, it was almost unknown for any Birmingham arms-manufacturer to have a factory, as we now understand the term. Broadly, a "gunmaker" arranged and financed the manufacture of his guns at the hands of craftsmen who were rarely employed by him, in the strict legal sense of employment. Such men might work in their own homes, or in galleries where they rented space, and they worked as independent contractors, to make parts or assemble guns using materials or components drawn from the "gunmaker"; they were paid no wage, but received an agreed price for each item. In varying degrees, each workman was a specialist, the "material suppliers" severally concentrating upon the manufacture of a particular gun part, and the "setters-up" building those components into complete guns.

It is little exaggeration to say that an average "gunmaker" was virtually a warehouseman, and that if he did own working premises then most of the men using them for work on his behalf paid him for light, heat, power, bench space, and so forth (albeit by deduction from the price that he paid for their completed work), in the same manner as any other tenants to whom he stood as landlord.

James and Philip Webley were the sons of Thomas Webley (a white button turner), and both were apprenticed for seven years to craftsmen of the type mentioned above. By working for very little money, between their fourteenth and twenty-first birthdays, such boys learned their master's craft, and became entitled thereafter to a craftsman's rate of payment.

James Webley was born in 1807, and the craft to which he was apprenticed in the gun trade is still not known. In the light of later events, it seems probable that he became a lock-filer or a "percussioner", and that he was apprenticed in 1821, and completed his training in 1828. Philip (born in 1813) is known to have been apprenticed to Benjamin Watson the Younger, a gun-lock filer, for a period of seven years from June 26th, 1827.

A Directory of Birmingham, 1835 lists the brothers in business together as "percussioners, gun-lock, etc., makers", in Weaman Street. A percussioner did not simply file up the breech and run in the nipple, but finished the nipple-seat, put in the nipple, filed up the percussion fences, adjusted the fall of hammer to nipple, and handed over the gun to his principal with lock standing firm in its seating.

This sort of Directory listing was a necessary means for advising the gun trade at large that the Webley brothers could accept work in their respective crafts, but no indications of prosperity should be inferred from it. There is no real evidence one way or the other, but it is likely that the two young men were illiterate, worked extremely hard at their trades, and were dependent for daily bread entirely upon their own labours.

On January 5th, 1838, however, Philip married (stating his trade as "percussioner"), when he was living at Tower Street, and this event apparently terminated the joint venture with his brother. The bride, Caroline Davis, lived at 84 Weaman Street, and when a son (Thomas William Webley) was born later in that year, the birth certificate showed Philip established there as a "Gun Smith".

Caroline Davis was the eldest daughter of William Davis, who had died before her marriage, and her mother controlled the thriving business which her father had built up as a "Manufacturer of Bullet, Shot, and Fishing Net Moulds, Gun Implements, Gun Makers' Tools, etc.," at 84 Weaman Street. Under the marriage laws then obtaining Caroline's share in this business effectively vested in Philip Webley, and by 1840 the latter had changed its trading name to "Philip Webley (late

William Davis)" and added gun-lock making as another line to be offered by the firm.

Even at this stage, Philip Webley was still only a small manufacturer; his orders would principally have come from the factors or merchants who supplied wholesale houses outside Birmingham, and in bad times (or for orders of larger value) his principal probably advanced money or materials to him, with which to carry out an order. This value would have been deducted from his price for the goods, on delivery, in exactly the same manner as he would have deducted charges (at the end of each week) for bench space, hire of tools, or heat and light, from what he had paid to any independent workmen temporarily employed at 84 Weaman Street.

The progress of James Webley, after the brothers had split up, in 1838, is not clear. However, it is known that he remarried on October 14th, 1849, at St. Martin Parish Church (when he described himself as "42, Widower, Gun Maker"), and that his bride was Louisa Green. The lady was a thirty-four-year-old widow, whose father (Thomas Smallwood) had a grocery in St. Martin parish.

However, taking into account the very sound quality of the few signed "James Webley" revolvers examined, coupled with the perfectly practical (if rather fragile) design of such arms, it is suggested that James was employed in pistol manufacture during these missing years, 1838–49, and that he probably had experience in making revolving weapons of some kind.

Admittedly, *Slater's Commercial Directory, 1852*, listed him simply as a "gun and pistol manufacturer", at 14 St. Mary's Row, in the year immediately preceding that in which he filed application for a British revolver patent, but it seems only reasonable to assume that he was engaged in revolver-manufacture by that date, even if he did not advertise the fact. His neighbours were Thomas Thacker and William Tranter (both specializing in that type of weapon), and it is past belief that he could have had no knowledge of a subject in 1852, but have been capable of patenting a viable design in it twelve months later.

At about this time, Philip either added a "kick-stamp" to his existing equipment at 84 Weaman Street, or had access to a subcontractor using one; gun-lock and gun-furniture forgings were added to the products that he offered after 1851, and a light mechanical hammer of this type would have been useful in such manufacture.

For reasons already made plain, 1851 had also been the year in which Birmingham "gunmakers" and workshop masters decided that revolvers (whether as pepperboxes or as cylinder arms) were weapons for which a serious market either existed or shortly would exist, and further grasped that the much-touted revisions occurring in British patent law would enable them to secure necessary protection to seek a place in this new field, and to offer to the factors or merchants who were still their most important market something which could only be secured from the one source, i.e. themselves, as patentees of it.

Clearly, however, neither of the Webley brothers (nor most of the other Birmingham gunsmiths and factors who took out revolver patents in 1852 or 1853) seriously regarded the revolver as a product upon which to base their business. Such weapons were simply a useful sideline to meet largely civilian demands, which could (as to most parts and accessories) be as well and as cheaply made by hand, in the traditional Birmingham manner, as by the expensive machinery of specialists like Samuel Colt or Robert Adams.

To this end, James Webley secured British Patent No. 743 (of March 29th, 1853), and his brother took out protection on February 4th, 1853 (Br. Pat. No. 305), and September 14th, 1853 (Br. Pat. No. 2127); the fact that James could find the fee for this protection (see Chapter II) suggests that his affairs, too, were prospering by 1853.

James Webley's patent protected the lock-mechanism and frame-construction of a single-action percussion revolver more particularly described in the next chapter; this weapon enjoyed a modest success (see Fig. 40), and it is probable that the manufacture of it (which involved a lot of handwork) occupied a good deal of the pantetee's time and energy. A self-cocking revolver was also protected.

Philip Webley's Br. Pat. No. 305/1853 covered a self-cocking revolver, with separate barrel, and claimed the lock-mechanism, a safety device, and a barrel-hinge assisting in rapid removal and replacement of the cylinder. His lock-mechanism was simple and ingenious, involving use of a "hammer catch" of the type later used by William Harding (see Fig. 25), but no specimens of revolvers made under this patent are now known to have survived. Enough has already been said of his gas-seal cylinder patent (No. 2127/1853) in Chapter XIV, and the device requires no description here.

An examination of the Specifications to the three patents clearly shows that both patentees had a real familiarity with revolvers, with their manufacturing problems, and with desirable design features like horizontal nipples, "gas-seal" cylinders, and quick reloading. However, these designs show neither serious grasp nor study of the revolver as a fighting weapon, and were (with the possible exception of James Webley's "Long Spur" revolver lock-mechanism) devoid of any characteristic that would ever have been adopted by the armed forces.

In truth, the years which saw both Crimean War and bitter Colt/Adams rivalry to supply an official Service revolver were probably of greater importance to Philip Webley in other fields. This may have been the period in which he became contractor to H.M. Board of Ordnance, or the Honourable East India Company to whom he claimed vaguely (in one advertisement covering his full range of firearms and related products) to be the "original" manufacturer, rather than for any success with his two revolver patents. The exact nature of the articles supplied to these bodies is not known, but it may have included the prosaic (but profitable) army boot; the reader's attention is drawn to English Patent No. 13611 of April 30th, 1851 (in the name of Philip Webley, "Manufacturer" of Warwick), which protected an artificial leather, suitable for such a product by nineteenth-century standards.

By 1856, however, demand and interest was sufficient to warrant the appearance of an advertisement by James Webley shown as Fig. 40, together with a separate Directory-listing as "Gun and Pistol Maker (patent revolving)" for him; the statement that this inventor was contractor to the Board of Ordnance referred, however, to products other than revolving arms.

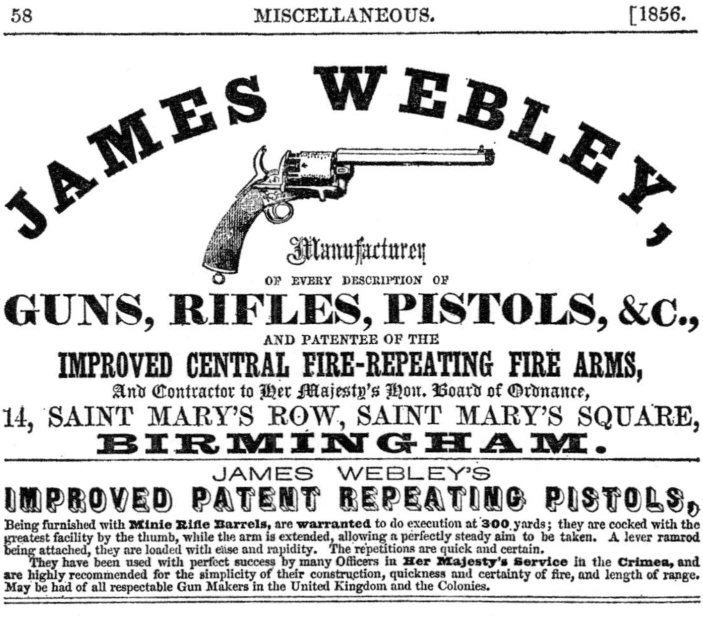
58 MISCELLANEOUS. [1856.

JAMES WEBLEY,

Manufacturer

OF EVERY DESCRIPTION OF

GUNS, RIFLES, PISTOLS, &C.,

AND PATENTEE OF THE

IMPROVED CENTRAL FIRE-REPEATING FIRE ARMS,

And Contractor to Her Majesty's Hon. Board of Ordnance,

14, SAINT MARY'S ROW, SAINT MARY'S SQUARE,

BIRMINGHAM.

JAMES WEBLEY'S

IMPROVED PATENT REPEATING PISTOLS,

Being furnished with **Minie Rifle Barrels**, are **warranted** to do execution at **300** yards; they are cocked with the greatest facility by the thumb, while the arm is extended, allowing a perfectly steady aim to be taken. A lever ramrod being attached, they are loaded with ease and rapidity. The repetitions are quick and certain.

They have been used with perfect success by many Officers in **Her Majesty's Service** in the **Crimea**, and are highly recommended for the simplicity of their construction, quickness and certainty of fire, and length of range. May be had of all respectable Gun Makers in the United Kingdom and the Colonies.

FIG. 40. *Advertisement by James Webley.*

Kelly's Directory of Birmingham & District, 1856.

On balance, had actual events seen a continuance from success to success of Philip and James (with their separate interests and businesses), until death or retirement in old age, there is a real possibility that the Webley name might never have gained its later prominence in the field of revolver-making. James, the revolver-maker, had no sons but two daughters; the latter would presumably have passed on their inherited shares in his business to their husbands (if they had married), at which time the name would certainly have changed. Philip, on the other hand, had the necessary sons to perpetuate his name, but (as a study of one of his later advertisements will show; see Fig. 42) a business later so well established in other fields, that concentration upon revolver-manufacture might never have been seriously considered by him.

However, on March 28th, 1856, James Webley died suddenly, at 14 St. Mary's Row, through an internal haemorrhage, and his "Long Spur" revolver patent lapsed officially on the following day; the lapse occurred by reason of his failure (whether intentional or not) to pay the renewal fees. He was forty-eight years old, and was survived by two daughters (Louisa and Hannah Amelia) and by his widow, who took out Letters of Administration for the comfortable estate of £2,000.

Although the "James Webley" business was carried on under that name until 1863, it is unlikely that his widow (the child of a grocer) accomplished this feat. There is no reason for stating flatly that she did not do so, for some Birmingham gun trades were either practised or supervised by women, but it is believed that this business was, in fact, integrated with that of "P. Webley" (as an embryo revolver department) as soon as possible, and that the trading name "James Webley" was used thereafter only so long as any goodwill attached to it. Certainly, the implication of "P. Webley" advertisements appearing as Fig. 42 and Plate 40 support such a view, since they offer arms developed by James, without reference to him.

Louisa Webley (James's widow) died at Potters Hill, Aston, on August 26th, 1868, leaving her daughters in the charge of Philip until they should marry or reach their majority. The will by which she placed this charge upon him was dated February 17th, 1863; it shows that she was still living at 14 St. Mary's Row at the time of making it, and it also reveals that Philip Webley still regarded himself primarily as a "Gun Implement Manufacturer" at the time of her death.

If 1856 saw the conclusion of the Webley family's first successful entry into revolver-making, it must also be regarded as a year which marked the true beginnings of their assault on this market, and as the start of a battle which eventually saw them established as British monopolists in it. Left to his own devices, Philip might reasonably have had little interest in developing revolvers (for he allowed his own revolver patents to lapse in 1856), but given the existence of his late brother's successful revolver business, he was not the man to let any advantage go by default, and this was the first year in which his own advertisements specifically included "Pistols and Revolvers".

The same year had probably also seen the arrival of "Gibson's Steam Mills" in Weaman Street, and Philip may have seen in these steam engines (which were at the same address as that for the gun-making side of his own business) a means by which serious entry into revolver-manufacture might be effected, possibly as a supplier of barrels and cylinders to "the trade".

Generally speaking, Birmingham gunmakers still used little mechanization at that time, unless employed in manufacturing military barrels (which required rolling and grinding), and even twelve years later the *Reports of the Inspectors of Factories*, October 1868 (pp. 116–25) recorded the use of only a bare ¼ h.p. per worker in the Staffordshire gun and pistol "factories" (i.e. establishments employing more than fifty persons). Philip's activities, however, had always included items in which the basic steps of manufacture either required or would benefit from mechanized stamping and forging equipment, and we believe that his enforced re-entry into the revolver business, on his brother's death, coincided with earlier plans for mechanizing more of these processes in his business, and that the weapons dealt with in Chapter XVII may have benefited from that decision.

On April 15th, 1859, Thomas William Webley (Philip's eldest son) came of age, and the name of the firm changed to "P. Webley & Son (late William Davis)" shortly afterwards. It has been suggested by that firm's successors, Messrs. Webley & Scott Ltd. (*Webley, 1790–1953*) that Thomas William had visited the United States in search of revolver-making machinery even before this date, and their booklet noted that he eventually placed his order in Birmingham. The tale is interesting, but may be incorrectly dated; Thomas William Webley was only a teenager at the date suggested for this trip, and any contract signed by him would have been legally unenforceable. It is improbable, therefore, that his father entrusted him with a task that no American manufacturer would have allowed him to fulfil.

However, the coming-of-age of this young man may also have been the year in which plans were first considered for the introduction of a double-action revolver by his firm, although surviving evidence on this development actually dated after 1859.

This evidence for Webley activity in the double-action revolver market was two-fold, consisting of an advertisement in *Kelly's Directory for London, 1862* (Birmingham advertising section), which specifically mentioned this type of weapon, together with the entry of a Webley interest in Joseph Bentley's double-action revolver patent (No. 2657 of 1857) in the *Index of Names of Assignees, etc., of Patents*, earlier mentioned.

The advertisement in *Kelly's Directory* was presumably compiled and proofed in 1861, so that double-action percussion revolvers were obviously available to (if not actually manufactured by) the firm during that year. The type of revolver offered is discussed in Chapter XVII.

As to the entry in the *Index of Names of Assignees, etc., of Patents*, this was one of two such agreements filed at the Patent Office to register an interest in Joseph Bentley's patent; the other entry did not refer to the Webley name at all, but both entries were filed late in 1860.

These *Index* entries (which carried no initials for the names given) showed that the first agreement involved "Hargrove, Perry, Redman and Turner", whilst the second related to "Hargrove, Redman, Turner and Webley", and consideration of these names suggested that the men signing or party to the agreements were probably:

(1) Samuel Hargrove, of "Saml. Hargrove & Sons", malleable iron and brass founders at 44 Princip Street, Birmingham.

(2) Richard Redman, of "Pryse & Redman", gun, rifle and pistol makers at 84 Aston Street, Birmingham.

(3) Thomas Turner, gunmaker at 8 Fisher Street, Birmingham, and gun-barrel manufacturer under the style "J. & J. Turner" at Holdford Mill, Perry Bar, and "T. & J. Turner" at 6 Fisher Street, Birmingham.

(4) Philip Webley, of "P. Webley & Son" at 84 Weaman Street, Birmingham, and "James Webley" at 14 St. Mary's Row, Birmingham.

(5) James Bracey Perry, merchant, of Birmingham and Lea Hall, Handsworth, who died suddenly on October 28th, 1860, but whose business seems to have had South American and other overseas connections in the hardware trade.

Certainty is impossible now, but it seems likely that Bentley approached Hargrove, Perry, Redman and Turner, and arranged with them to manufacture and sell his patented double-action revolver under licence. Shortly after this licence had been noted in the *Index of Assignees, etc.*, Perry (who may have been interested as a distributor only) died, and Philip Webley stepped into his place. Another licence was executed, and manufacture and sale arranged by the four men, through the usual Birmingham gun-trade channels.

It will be noticed (from the Appendix), that Bentley abandoned his patent in 1861, but earlier remarks upon this sort of practice are applicable here. Once an enforceable manufacturing agreement had been

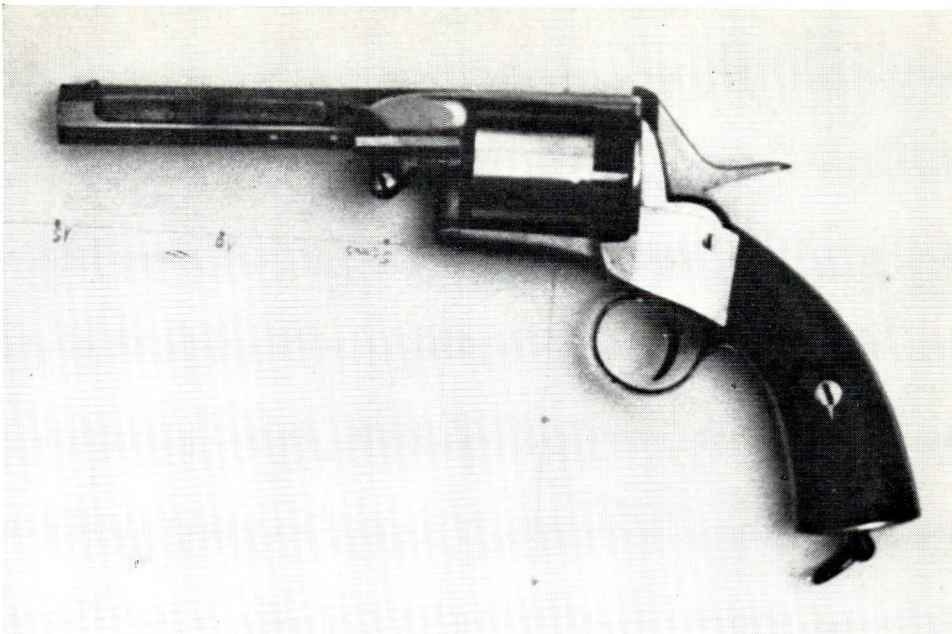

Plate 25. Webley Single-action Revolvers
(*Top*) A single-action Webley percussion revolver, with a top-strap between barrel and frame, in 54 bore calibre. No maker or retailer markings. The Fourth Model "Longspur" pistol. *Musee d'Armes Collection. Liege.*

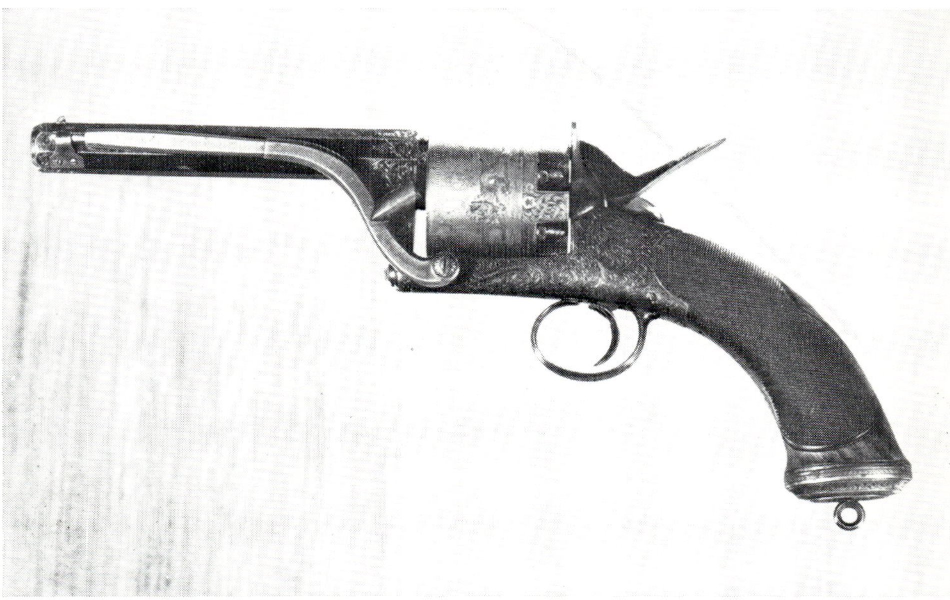

(*Lower*) A P. Webley single-action percussion revolver, made to Philips' patent of 1853. Retailed by "Bentley. Liverpool". Marked "P. Webley's Patent", but unnumbered. *Private Collection.*

PLATE 26. WEBLEY CARTRIDGE REVOLVERS
(*Top*) Six-shot, single-action breech-loading revolver, retailed by "Walter Scott. Birmingham". A British copy (possibly by P. Webley & Son) of the American "Smith & Wesson No. 2" revolver, in ·32 rim-fire calibre. No serial number. *Tower Collection.*

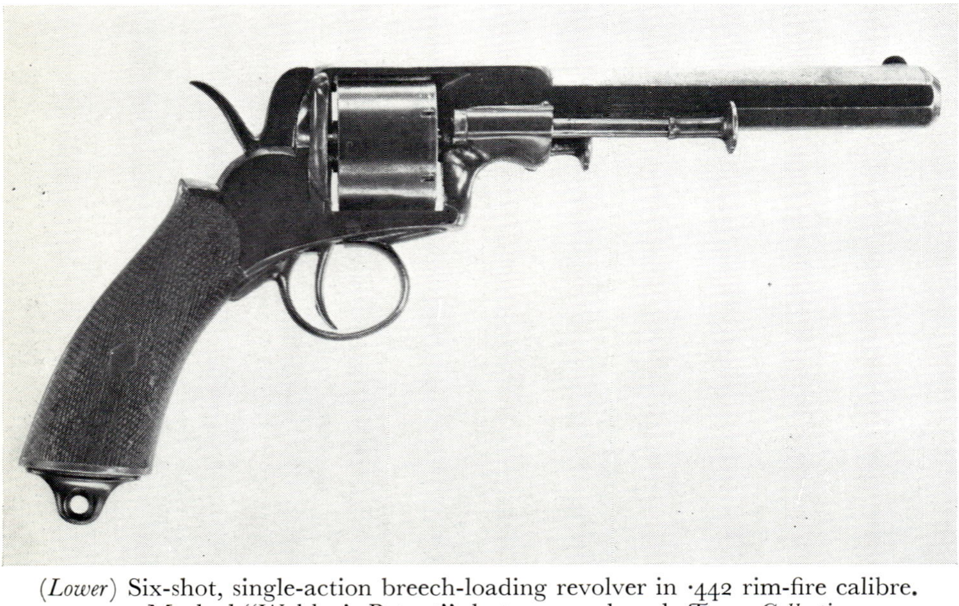

(*Lower*) Six-shot, single-action breech-loading revolver in ·442 rim-fire calibre. Marked "Webley's Patent", but un-numbered. *Tower Collection.*

PLATE 27. WEBLEY PERCUSSION REVOLVERS

(*Top*) Five-chambered, self-cocking percussion revolver, in 60 bore calibre, and signed "Osborne". Similar in appearance to the so-called Webley-Bentley arms, but lacks the characteristic hammer-nose safety. *Andrews Collection.*

(*Lower*) Five-chambered "wedge-frame" Webley double-action percussion revolver, in 54 bore calibre. Engraved "Dan'l Leonard. Patent No. 4079". *J. B. Bell Collection.*

(*Bottom*) Five-chambered solid-framed Webley double-action percussion revolver, with separate barrel screwed into frame. No serial-number or retailer's name. Like the wedge-frame pistol above, has multi-groove rifling, and is in 54 bore calibre. *J. B. Bell Collection.*

PLATE 28. TRANTER DOUBLE-TRIGGER REVOLVERS
(*Top*) Two Tranter double-trigger percussion revolvers, made under Br. Pat. No. 212/1853 and (under licence) Eng. Pat. No. 13,527/1851. Pistol at top is in 54 bore calibre and serially-numbered 13702R (retailed by Deane, Adams & Deane); lower weapon is in 80 bore calibre, serially-numbered 20944Y, and retailed by Owen Powell, Sheffield. *Darwent Collection.*

(*Lower*) A Tranter double-trigger percussion revolver in 54 bore calibre. Engraved "Pattern 1856", and not serially-numbered. *Darwent Collection.*

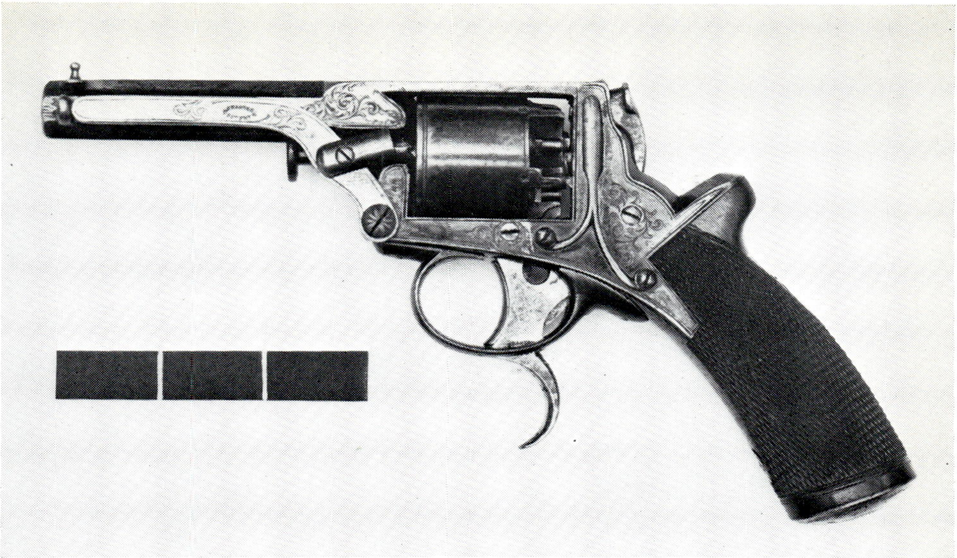

PLATE 29. TRANTER PERCUSSION REVOLVERS

(*Top*) Late Tranter double-trigger percussion revolver, in 120 bore. Serially-numbered 13149T, and retailed by "H. Richards, Old Hall St., Liverpool". *Andrews Collection.*

(*Lower*) Tranter's single-trigger double-action percussion revolver, in 54 bore calibre. Made under Br. Pat. No. 1913/1856 and (under licence) Eng. Pat. No. 13,527/1851. Serially-numbered 9980T, and retailed by "Thomas Jackson, Edward St., Portman Sq., London". *Darwent Collection.*

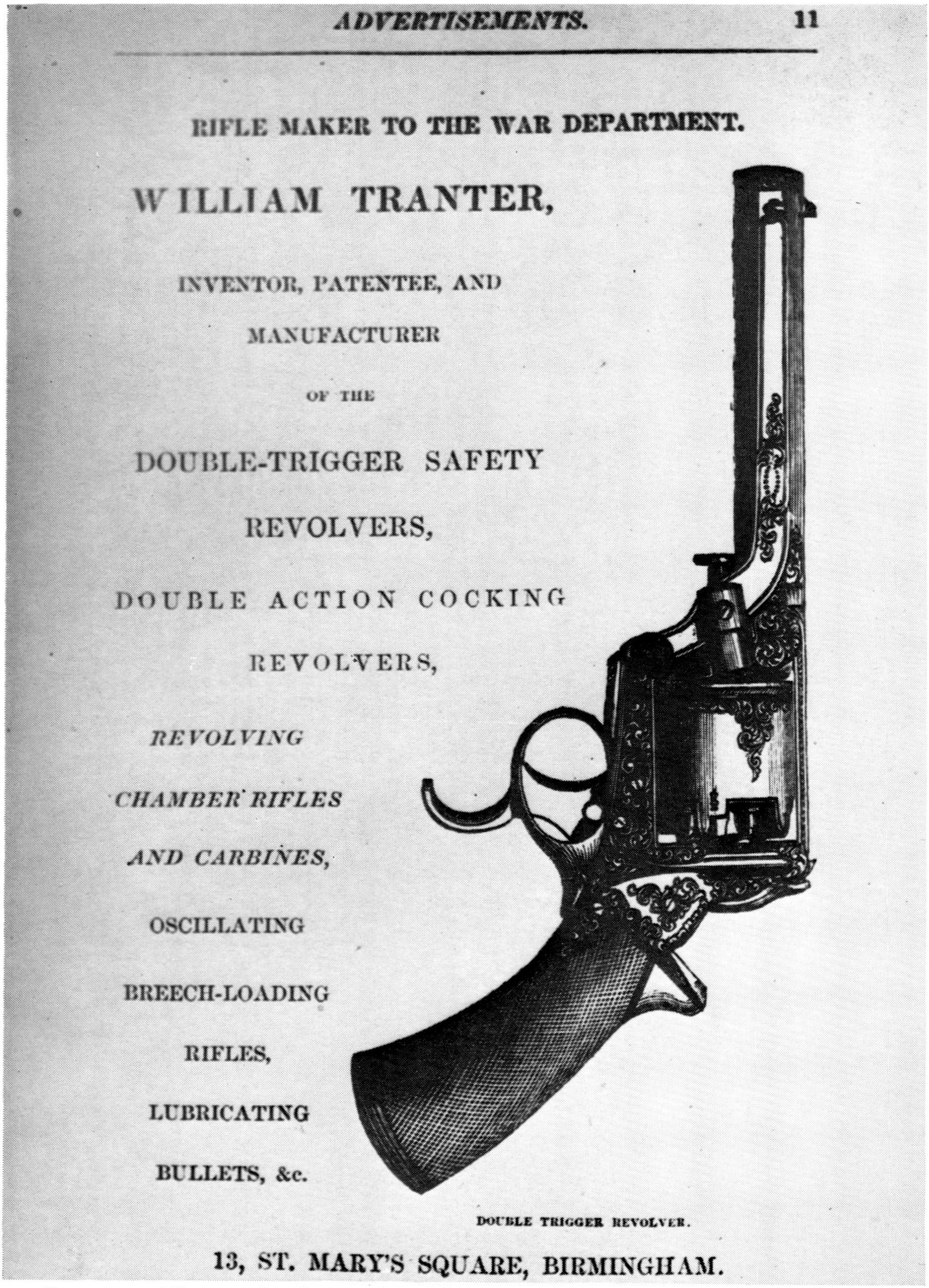

PLATE 30. *Tranter advertisement from W. W. Greener's "Gunnery in 1858", illustrating a "Third Model" double-trigger Tranter revolver.*

PLATE 31. TRANTER CARTRIDGE REVOLVERS

(*Top*) Seven-shot, single-action breech-loading revolver, in ·32 rim-fire calibre. Made under Br. Pat. No. 2067/1862, and serially-numbered 3959. Gold-plated frame and ivory butt-cheeks; retailed by "Henry Beckwith, Skinner St., London". *Darwent Collection.*

(*Lower*) Six-shot, double-action breech-loading revolver, in ·442 rim-fire calibre. Made under Br. Pats. Nos. 1913/1856 and 1862/1863; serially-numbered 5244. No retailer's name. *Darwent Collection.*

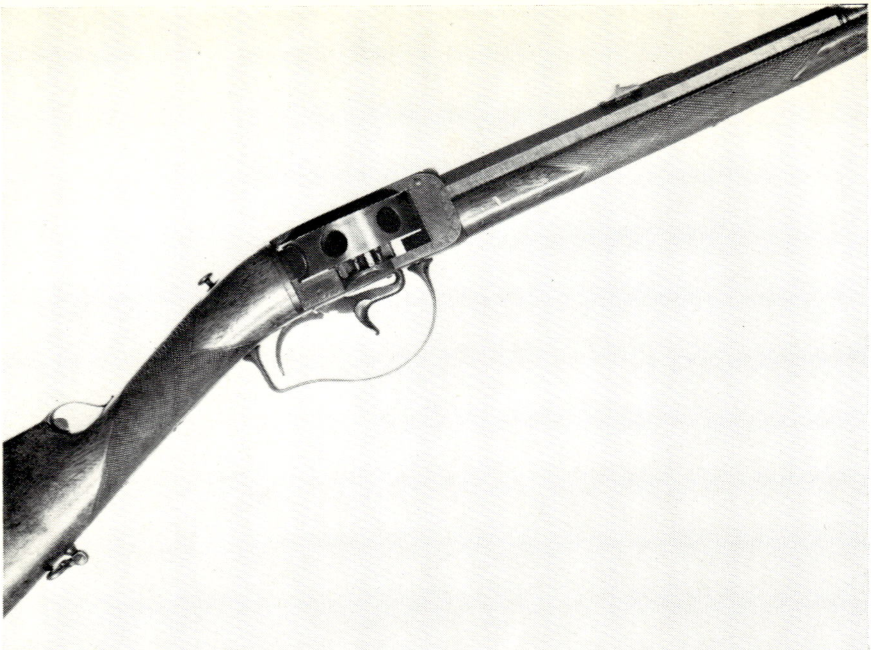

PLATE 32. A RADIAL PERCUSSION REVOLVING RIFLE
(*Top*) a side view of a manually-rotated under-hammer radial percussion rifle, retailed by "Wilkinson & Son, Pall Mall, London", but made to M. Poole's Eng. Pat. No. 7286/1837.

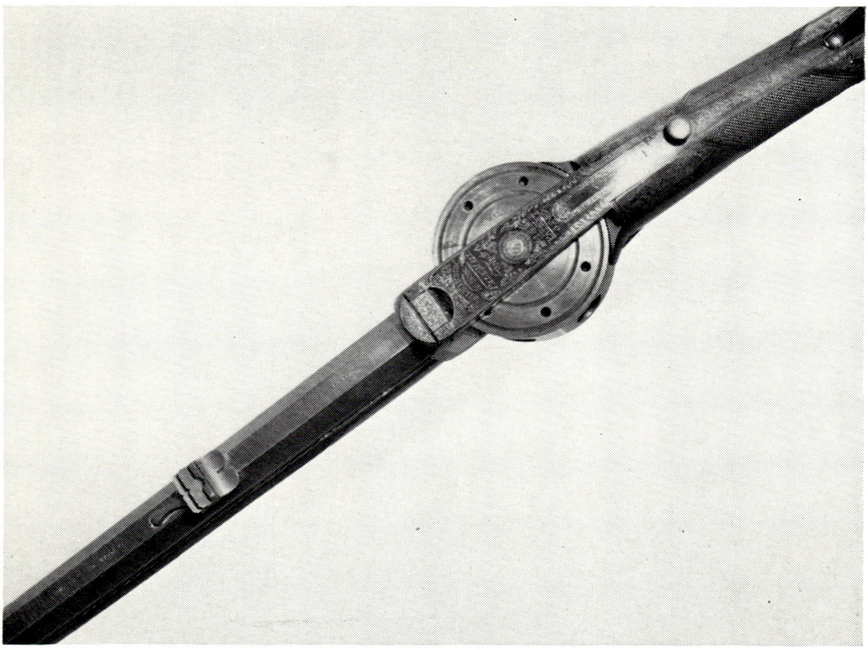

(*Lower*) Top view of the Wilkinson rifle above referred to. The hinged top-strap may be raised for removal of the charge-carrier. *Private Collection.*

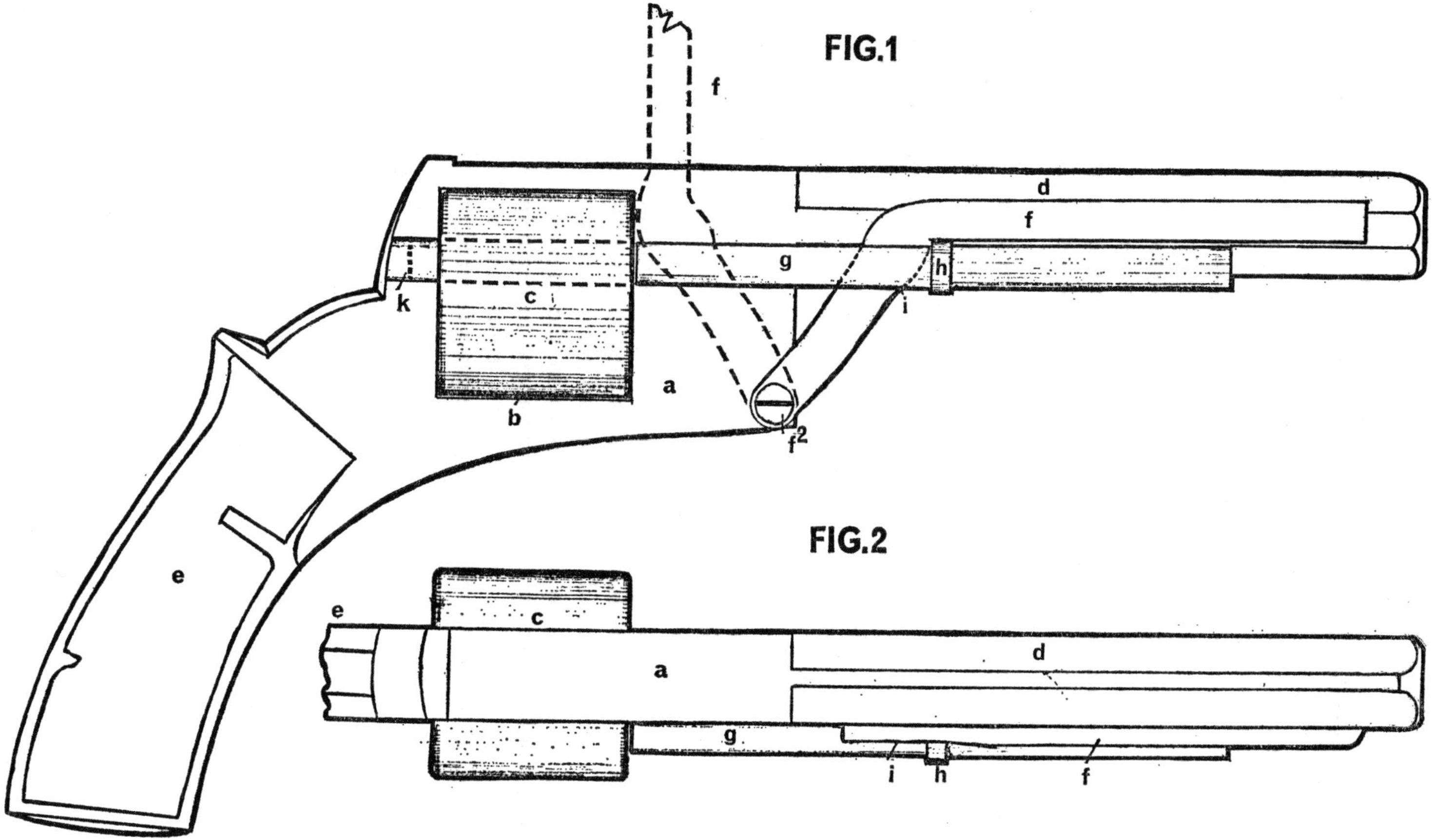

FIG. 41. *Philip Webley & Sons' "Design for Lever Rod for expelling exploded cartridge cases from Repeating fire arms".*
(Registered Design No. 4634 of May 7th, 1864)
The shape or configuration claimed as new related only to the lever rod and its parts.

arranged, the need for expensive patent protection upon a modestly successful invention was at an end. Piracy of Bentley's revolver, by other manufacturers, was unlikely in the teeth of production by men with exclusive access to all the special tools, patterns and so forth needed to produce it.

As will be demonstrated in Chapter XVII, these Bentley revolvers went through a major revision in the design of their frames, and the dating for this change is not known. By 1864, however, P. Webley & Son were advertising "Smith & Wesson and other breech-loading revolvers", in preference to muzzle-loading percussion arms, and it is therefore possible that the two types of frame-construction for the Bentley weapon overlapped prior to 1864.

On May 7th of the same year, the firm also took out Registered Design No. 4634 for a lever-ejector shown in Fig. 41, and began to offer heavy ·44-in. single-action rim-fire revolvers to use it (see Chapter XVII), whilst a pin-fire revolver patent was secured sixteen months later. This latter protection (British Patent No. 2030, of August 4th, 1865) was filed by Thomas William Webley, and proposed the use of a revolver-hammer that would fire either pin-fire or centre-fire cartridges in the same weapon.

By this date, however, "P. Webley & Son" stood at the threshold of its first major success in revolver manufacture, for plans must by then have been in hand for production of the famous ·442-in. Model 1867 centre-fire revolver, chosen for issue to the Royal Irish Constabulary in that year. This weapon was deservedly successful, and was to be followed by even more famous designs as the century progressed.

Philip Webley's second son, Henry, had been born in 1846; it is not known if he served any apprenticeship in the gun trade, but such training would probably have been arranged to begin in 1860 and to terminate in 1867. He could, therefore, contribute nothing to the history of "P. Webley & Son" during the years covered by this book.

CHAPTER XVII

WEBLEY REVOLVERS

IN the course of events outlined by Chapter XVI, the two Webley families were involved with the production of revolvers using any one of our three major lock-mechanism classifications; single-action arms were sold by both James and Philip, whilst the latter offered double-action and self-cocking weapons as well.

James Webley's untimely death makes it difficult to split our remarks into the two chapters apparently required, and all Webley revolvers appropriate to our period are therefore dealt with in this Chapter XVII, by reference to the type of lock-mechanism used in them.

Single-action Revolvers

(1) *James Webley's "Longspur" pistol*

The "Longspur" title for this weapon is modern, and describes the spoon-shaped thumb-piece to its hammer; James Webley (who patented the arm as Br. Pat. No. 743/1853) preferred to advertise it as his "Improved Patent Repeating Pistol", see Fig. 40, and specimens are shown on Plates 22 to 24. The long thumb-spur was not featured in his patent drawings, and may have been taken from T. K. Baker's design. (See Fig. 8.)

James Webley's patent contained four claims, of which the first related to a self-cocking lock-mechanism that we have never encountered; Claim Two described a single-action revolver lock-mechanism, in which a deep half-cock bent was formed so low on the hammer-belly that it was theoretically impossible for the trigger-nose to be released from that bent by trigger-pressure alone; Claim Three described two methods for hinging a revolver barrel to its frame, so that it might be readily tipped down for removal of the cylinder, and also a method for retaining the cross-bolt (of Colt-type) by which a hinged barrel was united to the arbor; Claim Four described making the middle portion of the arbor "of a square or prismatic figure", so that oil and dirt from the ends of the arbor worked inwards to this square section, and (on accumulating there) would not foul the rotation of the revolver cylinder.

Claims Two, Three (in part) and Four were all used, at one time or another, in the manufacture of these Longspur revolvers.

Within the model variations described below, this was a single-action percussion revolver of the type shown on Plates 22 and 23, normally without top-strap connecting barrel to frame, and usually five-chambered although six-chambered versions do exist. The frame was of malleable cast iron, the barrel octagonal, and the butt normally consisted of two chequered wooden grips secured by a cross pin in the normal Birmingham fashion; occasionally, one-piece butts were fitted.

Apart from a model hereafter described as the "Fourth Model" (which had a top-strap to the barrel; see Plate 25) sight was taken through a notch in the hammer nose after the style of Colt percussion revolvers, and there was normally no safety-device fitted other than the patented half-cock bent on the hammer (but see Plate 22). The removal of a securing-screw permitted access to the action through an inspection plate, normally on the left of the frame, and four action-springs were of suitably formed flat strip.

On most models, a simple see-saw cylinder-bolt (operating when the trigger was pressed) engaged in shallow slots cut in the nipple partitions (see Plate 24) to lock the cylinder at the moment of discharge; this bolt was unpatented, but was used in conjunction with a pin-stop, linked to the hammer, which had been claimed in the patent. Cylinder rotation was normally, but not invariably, in a clockwise direction, and the chambers were often numbered anti-clockwise. Only four specimens have so far come to light bearing London proof-marks, as opposed to those of the Birmingham Proof House, and a very wide range of retailers signed the various pieces examined; usually such names were those of country gunsmiths, but some famous gun-making surnames (as well as those of saddlers and military uniform suppliers) were found.

Nearly all the specimens were decorated with light foliate engraving, and the finish was originally with blued barrel and frame, and colour case-hardened cylinder, hammer, trigger and rammer, although blued cylinders and a bright case-hardened finish for the rammer and some smaller parts were found, and a few fully-plated specimens are known.

Examination of some sixty of these quite rare weapons (or the study of detailed reports on them) have disclosed the existence of four distinct models, and although there is a possibility of simultaneous model-production in some cases, we have adopted "First", "Second", "Third" and "Fourth Model" designations for easy reference to the successive versions of the arm.

The majority of First, Second and Third Models were numbered at the top of the butt-strap, and this number was usually also stamped upon most principal parts of the pistol. Unnumbered revolvers bore other identification marks (stamps, file-cuts, etc.), by which their components were similarly guided and reunited, through the hands of various workshops and craftsmen forming James Webley's "production line".

Our First Model is shown in Plates 22 (*top*) and 23 (*top*) from which it can be seen that no rammer was fitted to the arm. Instead, a detached rammer (of Adams type, but with swivelling head) could be slipped on to a peg on the side of the barrel, for use in ramming home the charges. Lock-mechanism and hinged barrel followed the claims of James Webley's patent, which fact was normally recognized in the legends "By Her Majesty's Royal Letters Patent" (engraved upon the top butt-strap), and "James Webley, Patentee" (engraved upon the removable side-plate). This version was six-chambered, and offered in 48- and 54-bore holster sizes (with barrel lengths between $5\frac{1}{4}$ in. and 7 in.) or as a 120-bore pocket arm (with barrel lengths between $2\frac{3}{10}$ in. and 4 in.); a sixteen-groove rifling, twisted to the right, was used, and we found weapons numbered 131; 218; 222; 241; 319; 332; and 338 in this group.

Our Second Model appears in Plates 22 and 23, and now had a permanently attached rammer of Adams type, with fixed head, pivoted upon the screw that formed the barrel hinge; when not in use, this rammer was turned back along the frame, and secured by a spring clip.

In this version of the arm, the butt-strap legend remained unaltered, but (part way through the series) a change occurred in the side-plate engraving; thus revolver No. 573 had the old wording "James Webley, Patentee", whilst No. 615 had the style "Webley's Patent", which was thereafter maintained for the remainder of our Second Model series.

This version was usually five-chambered, although six-chambered pocket arms were also encountered, and it was offered in several calibres. We traced the following revolvers within it:

Number	*Retailer*	*Bore*	*Rifling*	*Barrel length*
*199	Veisey & Son	120	3-groove right twist	4 in.
293	James Webley, St. Mary's Square, Birmingham	48	16-groove	7 in.
315	None stated	48	16-groove	6 in.
414	J. D. Dougall	48	3-groove	Not known
426	James Worrall, Foregate Street, Chester	48	3-groove	7 in.
435	J. Sampson, Liverpool	48	3-groove	7 in.
442	S. Isaac Campbell & Co., St. James St., London	48	3-groove	$6\frac{13}{16}$ in.
450	Moore & Woodward, 64 St. James Street, London	48	3-groove	$6\frac{5}{8}$ in.
497	Tipping & Lawden, London	54	3-groove	6 in.
*504	J. Colson, Stowmarket	120	3-groove	3 in.
*546	James Webley, 14 St. Mary's Square, Birmingham	120	3-groove	$3\frac{1}{2}$ in.
549	Scholefield & Sons	90(?)	3-groove	$3\frac{1}{2}$ in.
555	Alex. Martin, Royal Exchange Square, Glasgow	48	3-groove	$7\frac{1}{8}$ in.
573	James Webb, Salisbury	60	3-groove	4 in.
590	J. H. Clark	54	3-groove	7 in.
615	Victor Harold & Co., 9 Green Street, Leicester Square, London	60	3-groove	$4\frac{7}{8}$ in.
620	None stated	60	3-groove	$4\frac{7}{8}$ in.
629	Hy. Hook, Tenterden, Kent	60	3-groove	$4\frac{7}{8}$ in.
642	Walker, Newbury	60	3-groove	4 in.
*662	Powell, Sheffield	120	3-groove	$3\frac{3}{8}$ in.
*693	Hy. Hook, Tenterden, Kent	120	Not known	4 in.
710	J. Lofley, Brigg	60	Not known	$4\frac{3}{4}$ in.
773	J. W. Browne & Co.	48	3-groove	4 in.
813	T. E. Mortimer, Edinburgh	48	3-groove	7 in.
817	Not known	48	Not known	7 in.
884	Not known	48	Not known	$6\frac{1}{2}$ in.
922	Thos. Blissett, Liverpool	60	3-groove	5 in.
934	John Blissett, 322 High Holborn, London	60	3-groove	5 in.
969	G. B. Bales, Ipswich and Colchester	60	3-groove	$4\frac{3}{4}$ in.

N.B. Those arms marked with an asterisk * were six-chambered.

In the majority of these Second Model pistols, the rammer was fitted to the right side of the frame. However, revolvers numbered 922, 934 and 969 had their rammers mounted on the left; this alteration must have required changes to the pattern from which frames were cast (since it involved transposing the inspection aperture from one side of the frame to the other), and may therefore indicate an attempt to provide a stock of arms for left-handed customers.

The revolver shown as Plate 24 (*top*) was a transitional type, between our Second and Third Models. It will be seen that a compound rammer, of Kerr-type, was now fitted, and that, although a cross-bolt was still used, the hinged-barrel configuration had been abandoned. This 48-bore weapon was numbered 985, and marked only "Webley's Patent Improved" upon its 6-in. barrel; internally, the side-plate was stamped "James Webley", but no external legend was engraved upon the plate. It is as well to record here that the patent protecting James Kerr's lever-rammer did not lapse until July, 1862.

The true Third Model is shown on Plate 22 and was apparently made in similar quantities to the Second Model. We found pistols (and calibres) numbered 1013 (60); 1020 (60); 1025 (60); 1149 (48); 1173 (48); 1186 (48); 1297 (48); 1307 (48); 1310 (48); 1371 (54); 1420 (60); 1555 (48); 1602 (60); 2509 (54); 6967 (86); and 6974 (120) within this group. In addition, seven unnumbered revolvers were traced, some of which lacked both of the engraved legends.

With this Third Model the patented cross-bolt and hinging of barrel to end of the frame were abandoned, and the barrel was now simply screwed on to the threaded end of the arbor, and generally secured by a "butterfly" or other large-headed screw passing through the front of the barrel-lug and into the front end of the frame; this method of securing or mounting a revolver barrel was specifically claimed in an earlier British patent by Joseph Bentley, being No. 960 of 1852, which did not lapse until December, 1859. Another interesting variant pistol, numbered 243, passed through the saleroom of Messrs. Wallis & Wallis of Lewes in September, 1956. Originally a conventional 60-bore First Model, it had been converted to the Third Model by fitting a new barrel and rammer and removing the frame hinge; a small filler-piece had been inserted to make up the required length of frame. The new barrel was not of a type fitted by James Webley, being rifled with five grooves in a right twist, and it was attached in a "wedge-frame" Webley style described later in this chapter. The conversion was probably done quite late in this weapon's working-life, after breakage of the frame-hinge.

The existence of a 54-bore weapon shown at Plate 25 (in the Musée d'Armes at Liège), coupled with the existence of another known specimen, has (in conjunction with Philip Webley's advertisement on p. 6 of William Greener's *Gunnery in 1858*; see Plate 40) encouraged us to designate a "Fourth Model" for this type of Webley percussion revolver, although its rarity in collections indicates that very few specimens were actually made; both of the known examples are unsigned and unnumbered, and have 54-bore barrels (rifled with three grooves twisted to the right), which are 4 $\frac{7}{10}$in. long.

The Musée d'Armes weapon (bearing Birmingham proof-marks) is "in the white", i.e. it has not been finished. The distinctive hammer-nose and overall appearance of the known specimens faithfully reproduce those in Greener's plate, although his picture of the advertised weapon shows a "Webley's Patent" side-plate legend missing on the actual embodiments; as a further possible argument for the existence of a Fourth Model, we would also refer the reader to an unillustrated "P. Webley & Son" advertisement (in the 1860 edition of *Kelly's Directory* for Birmingham) which offered "Webley's Gunlock Action Safety Revolvers and Rifles". This was, at least, a palpable reference to the single-action feature. (See Fig. 42.)

Admitting the slenderness of the evidence, we feel safe in assuming that the makers of Longspur revolvers recognized that the absence of a strengthening top-strap to the barrel was a defect in an otherwise admirable design, and a Fourth Model (by our designation) was accordingly prepared around 1858. In the event, it is suggested, stocks of parts for the Third Model outlasted the useful life of the single-action design, and this top-strap was therefore developed upon a double-action pistol described later in this chapter.

The reason for numbering a majority of these revolvers is not yet known. As Chapter II should have made clear, this practice in relation to a patented weapon normally meant that a royalty was payable to someone on each weapon made under the patent, yet we found no evidence that James Webley ever granted a manufacturing licence under Patent No. 743 of 1853, at least, not in terms which he saw fit to record in the Patent Office register intended for the minuting of such arrangements.

This does not mean, of course, that some informal or unrecorded bargain was not made, but the two likeliest explanations for these numbers would seem to be either that James used a technical contribution from someone else in the production-version of his revolver (and paid that person a royalty on each weapon) or, alternatively, that James himself received the royalty from a factor or merchant who had financed the whole development and patenting of the weapon, controlled all sales of it, and took the lion's share of profits from such sales.

The first suggestion seems the most likely, for it is worth noting that not only did the Third and Fourth Models embody one feature from Joseph Bentley's earlier patent No. 960 of 1852 (as already noted), but all models used a frame-anchored pawl spring and single-leaf trigger spring, which were specific claims of Philip Webley's own patent No. 305 of 1853.

A case can therefore be postulated for the view that James Webley (until his death) paid a small royalty to his brother, for using patented construction details from the latter's Br. Pat. 305/1853, and numbered the Longspur revolvers in respect of this royalty. Upon his death, Philip chose either to use or to continue to use the barrel-anchorage system patented by Joseph Bentley (as Br. Pat. 960/1852), and maintained the numbering in respect of royalty payments for such use.

It is to be hoped that the point will one day be settled. Meantime, the reader may accept or reject the proposals made above, as to the reason for numbering Longspur revolvers at all.

As to possible datings for our four Models of this pistol, it may be assumed that (for reasons made clear in Chapter II) no First Models were sold prior to the date of filing for James Webley's Complete Specification to Br. Pat. No. 743, which is to say, September 29th, 1853.

FIG. 42. *"P. Webley & Son" advertisement from Kelly's Post Office Directory of Birmingham, etc., 1860.*

(Reproduced by permission of the publishers. Copyright reserved)

Our Second Model pistol appeared in the advertisement shown as Fig. 40, which was presumably proofed in 1855, and the cased Second Model Holster weapon on Plate 23 has an undocumented Indian Mutiny history, which may either indicate production in 1857, or else that the owner was satisfied with a pistol purchased a year or two earlier.

Turning to the other end of the tale, we have the Fourth Model featured in the *Gunnery in 1858* advertisement for which an exact publication date is unknown, and we know that the basic patent lapsed (and the patentee died) in 1856; P. Webley & Son actually ceased to advertise "Gun Lock Action Safety Revolvers and Rifles" in the 1863/64 edition of *Kelly's London Directory* (in favour of Smith & Wesson and "other breech loading revolvers"), although these were still featured in the 1862 edition.

Bearing in mind that Joseph Bentley's patent No. 960 of 1852 (which covered the barrel-mounting system used in most of the Third Models) could have "lived" until December, 1866, but was deliberately allowed to lapse in December, 1859, we suggest an intermittent production life for all models of some eight or nine years in all, between 1853 and 1862 with a carry-over of sales to the special order of satisfied (or highly conservative) customers, and the usual problem of disposal for any parts or completed weapons left in stock when active demand tailed off.

On the basis of numbered arms, and the roughly established proportion of unnumbered weapons to them, we are disposed to regard J. N. George's suggestion that about 2,000 specimens were produced in all as erring on the high side; despite the existence of pistols numbered in a 6000 range, a total output of about 1,500 revolvers seems more probable.

Since the change-over in side-plate legend (from "James Webley Patentee", to "Webley's Patent") can be established as occurring quite early in the Second Model series, it seems logical to suggest that this event signalled the intervention of Philip Webley in his deceased brother's affairs; revolver No. 573 had the old legend, whilst No. 615 had the new one. As might be expected, however, the division was by no means neat and tidy, for pistol No. 549 had a "Webley's Patent" side-plate, whilst the two pistols recorded on either side of it in our listing above had "James Webley, Patentee" plates.

However, in the present state of our knowledge, we do not regard this discrepancy as invalidating the suggestion, which concerns, after all, a change in "get-up" of a weapon produced by methods not materially different from those of a century before. By the law of averages, some pistols were bound to be caught in the "production line", by any change, and would reflect that fact.

Whether Philip's intervention occurred (as earlier suggested) on the death of James in March, 1856, or later, when the widow found that she could not manage the business alone, or even whether it was arranged between the two brothers to meet an unexpected demand for the Longspur revolver, must be left to the reader's judgement and to the researches of others, or it may seem that an undue proportion of space is being allotted to a weapon whose rarity should limit the time devoted to it here.

However, no study of these revolvers would be complete without some consideration being given to the cases and accessories used with them, and also to the minor variations in "get-up" and construction of the weapons themselves.

As stated by J. N. George, (*op. cit.*) the Second and Third Models are normally found in mahogany cases (sometimes with brass-bound corners), and an examination of Plates 23 and 24, will show the layout of compartments usually found, although small variations will be encountered. The First Model was ordinarily oak-cased, and the arrangement shown as Plate 23 is not strictly standard, since the revolver is a pocket weapon having a smaller case than normal.

The accessories found in the cases of these weapons varied little with any of the first three models; we traced no cased Fourth Model. The mould was a "W.D."-stamped, single-cavity, brass, tong bullet mould with an iron cut-off, having the bore number stamped upon it, and casting a Tranter-type projectile with a single grease-groove round it. A test specimen (most kindly provided by Major Dowell) from a stamped 60 bore mould cased with Second Model pistol No. 629, threw a 196-grain bullet, measuring ·635 in. long, with a diameter of ·430 in.

The powder flasks found in these cases varied considerably. So far as we could establish, the G. & J. Hawkesley planished tinned flask, with standing seam (see Plate 24 (*lower*)) was correct for that particular Third Model, and the preference for this "pear" shaped flask (over the more common "bag" shape found in most percussion revolver sets of this period) was often maintained in other Longspur sets. The flask shown with the fine cased Second Model on Plate 23 was probably associated, and we believe that the true layout should be for a flask similar in shape to the tinned Hawkesley flask (which would lie in the compartment now occupied by the nipple-key/screwdriver set), although a common-topped flask similar to Riling No. 1253 would be an acceptable alternative.

As for the rest of the accessories, oil bottle, wooden cleaning-rod, nipple-key, chamber and nipple cleaning tools, jags, etc., were standard for any cased revolver of this period, and probably supplied by Philip for each weapon ordered in a case. A complete set should, in the Second and Third Models, contain both a tin of lubricating compound and a tin of caps, whilst special lubricating trays and japanned or lacquered tin cartridge boxes are not unknown; a wooden lubricating tray can be seen in the bottom left-hand compartment of the Third Model on Plate 24.

The lids of the cases sometimes bore labels giving instructions for loading or cleaning, and that quoted by Dowell (*op. cit.* p. 35) is quite typical, being taken from the lid of cased Second Model pistol No. 629. The drawing accompanying these instructions featured a Longspur *rifle*, as well as the pistol, and although the hinged frame of the Second Model design was unlikely to withstand the vibration of a rifle-length barrel, on discharge, for an operational life of any real duration, a few specimens of such rifles may exist. We have never encountered one, but would expect it to be of the 60 or 120 calibres only.

Having regard to their manufacturing background, the details of these weapons can obviously vary quite considerably through the series, over and above any main Model variations. The chief differences to be observed (apart from those already noted) will be found in the occasional use of a sliding safety-bolt; the shape of the butt-cheeks, and their cross-pin escutcheons; the shape of the trigger guard; the types of rifling used; the omission of capping shields on some First Models; and an occasional disappearance from use (in First, Second and Third Models) of the additional see-saw cylinder-bolt, with its attendant shallow locking-slots in the nipple partition walls.

On the first of these differences, we observed that the true Third Model (i.e. that having a large-headed or butterfly screw to secure the barrel, as proposed in Joseph Bentley's Br. Pat. No. 960/1852) began at about No. 1100, but was preceded by a small number of Third Models in the 1000 numbering only, which used small-headed indexing screws, and had rather ineffectual sliding safety-catches mounted on their removable side-plates. We did not meet this latter refinement elsewhere.

As to the butt-cheeks and escutcheons, towards the end of the Third Model series, a completely chequered cheek was adopted in place of the original shape (which had an unchequered chamfer at the top), and diamond-shaped escutcheons were discarded in favour of an inverted "tear-drop" outline.

The trigger guard also was changed towards the end of the Third Model series; after revolver No. 1297, a bifurcation at the rear created a more graceful, curved form than the original solid guard, but this later reverted, finally, to a coarser form of the original shape, now having the tail-end externally anchored, and not pinned from inside the bottom butt-strap as in all previous models.

As to the rifling (apart from a revolver which appeared to have been misnumbered as No. 199, since it conformed to Second Model characteristics, and the repaired or converted No. 243), all First Models were rifled with sixteen fine grooves and a right twist. Second and Third Models (with two exceptions) and both embodiments of the Fourth Model were rifled with three broad grooves and a right twist; presumably, these had the "Minie Rifled Barrels" referred to in an advertisement shown as Fig. 40. Finally, two revolvers with the highest numbers (Nos. 6967 and 6974), and the repaired or converted First Model No. 243, had five-grooved rifling, again pitched to the right.

Single-action Revolvers

(2) *Philip Webley's single-action percussion revolvers*

Although Philip Webley continued production of his dead brother's single-action revolvers (and even improved the design; see Plate 40), he also produced a small number of such arms which owed nothing to the Longspur weapon. A specimen of one such revolver is shown on Plate 25, (*lower*) and W. C. Dowell (*op. cit.* Plate 11(b) & (c)) illustrated another version of it.

The pistol shown in Plate 25 is a handsomely mounted weapon, retailed by "Bentley's, Liverpool", and is unnumbered; beautifully cased, and highly engraved, it bears the legend "P. Webley's Patent" on its frame, and is five-chambered, Birmingham-proved, with a $5\frac{1}{2}$-in. barrel (in 48 bore) rifled with nine grooves, twisted to the right. The internal workmanship (in contrast to all Longspur revolvers) is deplorable.

Another such specimen, most courteously reported to us by Mr. P. S. McCarthy (of the Metropolitan Arms Co. Ltd., Mayfield), had been retailed by Crew, of Tetbury, and matched the dimensions set out above.

Other features, common to both specimens, were the Tranter-type rammer, with fixed head; thick nipple-partitions, which permitted entry of the hammer-nose into holes drilled therein, to lock the cylinder safely at rest; and the use of a barrel-indexing screw at the front of the barrel-lug, which passed into the end of the frame and had its head squared off to fit the nipple key, so that this tool might be used for turning it in and out.

Both weapons had one-piece butts, with cap boxes in the base, and their hammer-noses struck the nipples through a hole in the standing breech; both used a rear-sight notch on top of the frame, instead of a notched hammer-nose used in James Webley Longspur revolvers First, Second or Third Models.

As to patented features, we had no opportunity to examine the Crew weapon, but that illustrated had an "L"-shape pawl pivoted on the side of the hammer, the lower limb of which acted as a cylinder-bolt at full-cock. This device was claimed in Philip Webley's Br. Pat. No. 2127/1853, but the gas-seal cylinder from that patent was not used; however, this weapon did have the frame-mounted pawl spring claimed in Philip Webley's earlier Br. Pat. No. 305/1853.

Two more Philip Webley single-action revolvers illustrated by Dowell were numbered 122 and 134, and had been respectively retailed by "Capella, Newport", and "John Grainger, 106 King Street West, Toronto"; both weapons differed, in matters of detail, from the type illustrated in Plate 25, and described above.

On both specimens, the butt had a distinct prawl behind the hammer, and each trigger guard had a spur or finger-rest on the underside of it. A removable side plate (on the left) gave access to the single-action lock-mechanism, which embodied the frame-anchored pawl spring and "L"-shaped combined pawl/cylinder-bolt of Br. Pat. No. 2127/1853. The frames were engraved (on the left) "P. Webley's Patent No. . . .", and a peg for a loose, detachable rammer was screwed into the left side of the frame of No. 122, just forward of this engraving.

The peg on No. 122 served also to retain a sliding male pin projecting into the front of the frame, from the barrel-lug; if it was desired to replace the cylinder with another loaded one (instead of reloading *in situ* with the rammer), a Colt-type cross-bolt was knocked out of the barrel-lug, and the barrel could then be drawn forward and swivelled sideways on the male pin, to allow removal of the cylinder. This basic idea had come from Philip Webley's Br. Pat. No. 305/1853, but was restated in this slightly modified form (although not as a patent claim) in his patent No. 2127/1853. Revolver No. 134 had a permanently attached rammer, and did not embody any equivalent to the peg described above, as a means for preventing the male pin (on which the barrel could be pivoted) from being pulled forwards completely out of the frame. Both revolvers were Birmingham-proved. Their cylinder arbors were fitted with damper-springs and had reduced-diameter centre sections (as shown in Philip's earlier patent) to cut friction, and also to trap powder-fouling where it would do least harm.

Single-action Revolvers

(3) *Cartridge revolvers*

In September, 1863, the fabulous "gun-a-minute" market for Birmingham-made military rifles, sparked off by the American Civil War, came suddenly to an end. Little warning seems to have come to the gun trade generally, for a decline appeared in the proof-figures for August of that year, and thereafter the bottom simply fell out of the military arms market, as factors described in Chapter X destroyed American demand for such weapons.

In evidence before the British Association (in 1865), J. D. Goodman, Chairman of the Birmingham Small Arms Company, Ltd., recalled nostalgically that military barrels proved in the town in July, 1863 (and made up into complete arms in the following month) reached a total exceeding that for the *whole* of the last quarter of the same year. This had indeed been a boom period, the like of which the British gun trade was never to see again.

It must be assumed that P. Webley & Sons were hit as hard by this slump as any other military arms maker, for although their business was broadly-based as general suppliers to the trade, this general recession in the military market (and competition from the newly-formed Birmingham Small Arms Company, Ltd. for what business was left) must have closed the year rather bleakly for the Webley house.

Characteristically, however, other irons appear to have been warming in the fire during 1863, and disappearance of the old "James Webley" business, at about this time, probably heralded a decision to bring on additional revolvers as a part of the parent firm's activities.

One such weapon was undoubtedly a copy of the highly-successful Smith & Wesson breech-loading design, since (as earlier mentioned) P. Webley & Son advertised such pistols from 1864 onwards.

The Smith & Wesson design was a single-action, rim-fire revolver, made in ·22 or ·32 calibres, which had been first produced by D. B. Wesson and H. Smith (at Springfield, U.S.A.) during 1857; in it, the barrel could be tipped up on a hinge, and the cylinder then bodily removed for loading, or unloading by punching out the cartridge-cases against a rod mounted beneath the barrel.

Successful from their date of introduction, genuine Smith & Wesson revolvers were in short supply throughout the American Civil War, and there is little doubt that P. Webley & Son produced close copies of both ·22 (seven-chambered) and ·32 (six-chambered) versions during this period of shortage. A presentation arm in the possession of Messrs. Webley & Scott, Ltd. (successors to P. Webley & Son) is dated 1863, and closely resembles the pistol shown at the top of Plate 26, but the rarity of such arms argues that very few were made; Smith & Wesson began a serious export drive in 1867, and it seems probable that P. Webley and Son (who had introduced their own highly-successful double-action "Royal Irish Constabulary" revolver in that same year) did not thereafter try to compete with the American design, by merely copying it.

As mentioned in Chapter XVI, moreover, the Webley firm had earlier secured a Registered Design (No. 4634 of May 7th, 1864; see Fig. 41) for a powerful lever-rod cartridge-ejector for revolvers, and it duly introduced a hand-gun embodying this ejector. Such devices were actually a necessary adjunct to all pistols using early copper-cased rim-fire cartridges, for these swelled heavily on firing and required an ejector of real authority to unseat their cases, but not all manufacturers troubled to develop them.

The single-action revolver for which this ejector was apparently intended appears on Plate 26, and will be seen as a descendant of the old James Webley "Longspur" percussion revolver.

The specimen shown is in the Students' Room at H.M. Tower of London and was, curiously enough, not fitted with the Registered Design ejector, but with a simple rod ejector; another specimen, which we examined from the Buchanan Collection, certainly had the lever-ejector fitted to it, but simply could not be taken as a serious competitor to those excellent Tranter designs (in both single- and double-action variations) which were simultaneously available. As will be seen from Plate 26, the Webley revolver itself did have a certain grace and balance to it; however, when fitted with the heavy lever-ejector, both grace and balance are wanting.

A number of these revolvers were studied, and found to be of excellent quality and workmanship; the representation of their frames in Registered Design No. 4634 seems to confirm that they were introduced at about the beginning of 1864. Mr. A. N. Kennard (of H.M. Tower of London) has confirmed sight of a number of these pistols, amongst weapons surrendered to the police during recent arms amnesties, in which the characteristic thumb-spur on the hammer has been deliberately removed, and replaced by a coarser peg; one specimen arm, with this modification, rests in the Students' Room.

All revolvers examined were six-chambered (with chambers recessed for the cartridge-heads) and had 6-in. barrels, screwed into their solid frames, which were rifled with five grooves and a right twist; loading was effected through a rearward swinging loading-gate on the right. The lock work was basically that found in Tranter single-action pocket-sized weapons of similar vintage, where a powerful trigger spring caused the trigger to rock a see-saw cylinder-bolt out of engagement with the cylinder locking-notches, for the appropriate arc of hammer travel. Their short Birmingham-proved cylinders took ·440-in. or ·442-in. rim-fire cartridges, but might also have accepted ·430-in. Tranter rounds with safety; a removable side-plate on the left of the action-body, and the type of trigger guard fitted, were both interesting reversions to the old James Webley "Longspur" design, but a hammer-pivot screwed permanently into the frame was a departure from the method previously used for mounting the hammers of Webley single-action revolvers.

In addition to these heavy weapons, P. Webley & Son also produced single-action pocket arms for the ·30-in. rim-fire cartridge, which were closely modelled upon the equivalent Tranter design shown in Plate 31 (*top*). Fitted with stud triggers, but differing in manufacturing detail from Tranter pistols, these neat little weapons are extremely rare today.

We found no evidence that P. Webley & Son seriously pursued either of the solid-frame, single-action rim-fire designs described above. Both types of revolver were found as marked "Webley Patents" (or even simply "Patent"), but no relevant British patent could be traced, except for William Tranter's Br. Pat. 2067/1862; this had become void in July, 1865, and it therefore seems probable that P. Webley & Son used this patent-legend only after protection had passed into the public domain.

Self-cocking Revolvers

Although, as earlier mentioned, both of the Webley brothers secured British patents for self-cocking revolver designs, we were not able to trace specimens of such patented arms for examination and illustration. This failure is no authority for the view that arms were not made under the patents, but it can be stated with confidence that if either or both of the patents were worked, then it was upon a small scale only.

However, most collectors do recognize a particular kind of self-cocking percussion revolver by the title "Webley" (or "Webley-Bentley") and such arms will be considered here, as a matter of convenience. Although, for reasons made clear in Chapter XVI, there is little likelihood that either of the Webley brothers actually made complete revolvers of this type, we can be fairly certain that such arms were obtainable through either brother, albeit drawn, by them, from suppliers in the Birmingham gun trade at large. Such weapons should always be carefully distinguished from those true Bentley revolvers described in the Appendix (in relation to Joseph Bentley's Br. Pat. No. 960/1852) and illustrated at the top of Plate 33.

A typical self-cocking revolver of the type often called "Webley", or "Webley-type", or "Webley-Bentley", by collectors, is shown at the top of Plate 27, although this specimen actually lacks the hammer-nose safety protected by Joseph Bentley's patent of 1852.

These were coarsely-made weapons, commonly five-chambered (but occasionally with a six-chambered cylinder), and embodied a lock-mechanism barely distinguishable from that shown on Fig. 1 to have been used in pepperbox pistols; having horizontal nipples, their hammers had to be slightly modified in form (and a revolver-pawl fitted), but there was little to distinguish the mechanical principles of such a lock, from that shown by Stocker and Bentley (see Fig. 2) to have been within the public domain by 1839.

As a general rule, these weapons had malleable cast-iron frames, with integral butt-straps, and removable side-plates giving access to their lock mechanisms; upon occasion, the collector will encounter German silver frames, but these arms were really made for the cheaper end of the market, and such refinements were rare.

Many revolvers have Joseph Bentley's simple safety device embodied in the top of their hammers, and such a device is visible in Plate 33. This safety-catch was a lever pivoted in the top of the hammer, and formed with a tiny thumb-piece at its rear: if the trigger was lightly pressed to turn the hammer back off the nipples, the nose of this lever could be tipped upwards (by pressing upon the thumb-piece) until it rested against the frame. This lever then held the hammer clear of the nipples, and enabled the pistol to be carried safely; on pressing the trigger to fire the arm, however, the nose-heavy lever tipped down again as soon as it cleared the frame, and the hammer could then fall, without interference, when the sear released it.

The late J. N. George (*op. cit.* p. 215) stated that Bentley's name appeared "jointly with that of Webley upon several self-cocking revolver patents of the 1850/60 period", but there is no surviving Specification which supports that claim. Indeed, it would seem that Bentley's safety-catch was widely used by the Birmingham gun trade, with or without reference to the patentee, and that its inventor took no steps to enforce his patent rights against infringers.

As Plate 27 demonstrates, the collector will find revolvers of this "Webley-Bentley" type which do not embody Bentley's safety device at all; instead, a press-in spring safety-catch (engaging the hammer in the manner shown at the top of the Plate) was sometimes mounted upon the left side of the action body, and backed up by a conventional sliding safety-bolt.

Revolvers of the "Webley-Bentley" type were made in 54-, 80- and 90-, or (less commonly) 120-bore calibres; barrel lengths varied, but 6 in., $4\frac{3}{4}$ in. and $4\frac{1}{4}$ in. were, respectively, about standard for these weapons, and 80 bore was the most popular calibre. There was no common form of rifling for the type, and three-, five-, twelve- or fourteen-groove patterns have been encountered; in a few cases, the weapon was even smooth-bored.

As to the configuration of frame and barrel, a majority of such pistols were open-framed, as in Plate 27; however, their barrels were normally screwed on to the end of the arbor and secured by passing one screw (usually with flat "butterfly" head) through the barrel-lug and into the front of the frame, rather than by the cross-bolt visible in Plate 27. Less commonly (but still in some numbers), such arms were made with a top-strap between barrel and frame, and were assembled by a "wedge-frame" method described in the next section of this chapter, and shown in the middle weapon on Plate 27.

Nearly all of these pistols had rammers, and these were of two types; the majority of arms had a lever-rammer of Kerr-type mounted on the left of their barrels, but a few had the Colt-type rammer shown at the top of Plate 27 pivoted beneath them.

As to markings, the names of provincial retailers were engraved upon a handful of specimens examined, but the majority of weapons were not thus embellished. More common, apparently, was the practice of engraving a reference to patent protection upon such arms, and "Improved Patent" was one such legend, whilst "Patent No. . . ." or "London Patent No. . . ." was quite commonly used; in one case, we even encountered "By Her Majesty's Royal Letters Patent", engraved in James Webley style upon the top butt-strap. Such legends presumably referred to features of Joseph Bentley's Br. Pat. No. 960/1852 embodied in these arms, but it cannot be safely assumed either that the patentee had them engraved, or even that he approved the use of his devices.

A surprising proportion of these pistols were numbered. The lowest number found was 111 (upon an 80-bore, closed-frame weapon), and the highest number was 73200 (upon an open-frame, 80-bore revolver), but these numbers revealed no apparent connecting-link between their presence (or their magnitude), and, for example, the frame-configuration or type of rammer used in the pistol upon which they were engraved. It would be rash to suggest that all of these numbers were imaginary, being engraved upon the arms purely to suggest a large production, and without relation to the quantities actually produced; however, until thorough study of a very much larger number of revolvers has been made, we are reluctant either to style the numbers as "serial numbers" or to suggest a definite purpose behind their use.

In addition to the percussion revolvers described above, we encountered a number of arms using rim-, centre-, or even pin-fire metallic cartridges, and in all cases, the weapons conformed closely in dimension and appearance to their muzzle-loading counterparts. It was clear that some of these pistols had been manufactured as breech-loaders, from components made during the percussion era, but, in a few cases, it appeared that the revolvers had begun life as muzzle-loaders and been converted to use new cartridges. As a working assumption, we regard both types of breech-loading self-cocking revolver as falling just within the period of this book, like the double-action arms described in the closing section of this chapter.

Double-action Revolvers

In Chapter XVI, we detailed evidence to support the classification of a double-action percussion revolver as produced by P. Webley & Son, and suggested 1861 as the year in which such arms were introduced by that firm. Plate 27 shows two such revolvers, and it will be noted that each has a distinct type of frame construction; collectors call these pistols, respectively, "wedge-frame" and "solid-frame" Webley revolvers.

In the "wedge-frame" arm shown at centre on Plate 27, barrel and top frame-strap are forged as one piece; this unit is slipped on to the front of the arbor, and held there by a transverse wedge or key of Colt-type, which is passed through both barrel-lug and arbor to effect the union. This done, the top frame-strap butts against the standing breech (often in a transverse groove cut for it), and a stud on the front of the frame enters a corresponding hole in the back of the barrel-lug, to hold barrel in alignment with chamber. This method of construction was also used (see Plate 40) for the Fourth Model James Webley single-action revolver described earlier in this chapter.

On the pistol shown below in Plate 27, however, the barrel is still a separate component, but is here permanently screwed home into a solid one-piece frame of Adams's type. This method of construction was not an infringement of Robert Adams's English Patent No. 13527/1851, because the barrel was not forged integrally with the frame.

An examination of specimen arms suggested that far more wedge-frame revolvers were made than solid-frame, and that two butt-shapes were used in the more numerous arms. Major Dowell (*op. cit.* p. 37) stated that this wedge-frame version was made in 54-, 90-, 100- and 120-bore calibres, and that it was chambered for five or six shots; barrel lengths varied between 4½ in. and 6¼ in., and the malleable cast-iron frame had no removable inspection plate. Two types of rammer might be found (imitating the Colt or Kerr designs, as in those self-cocking revolvers earlier described), and a three-groove rifling was normally used; occasionally, arms were found to have a fourteen-grooved rifling, but even the three-groove pattern varied in size of groove from weapon to weapon.

The solid-framed revolvers were quite difficult to trace. A 38-bore model was reported (but not seen), and pistols of 54, 60, 80 and 120 bore were either examined or noted from Dowell (*op. cit.* Plates 18 and 20).

In this model, a rammer copied from Kerr's design seems invariably to have been fitted, and five chambers only were encountered; as in the case of wedge-frame arms, however, three-, five-, or multi-grooved rifling was used. The frame-construction also required a latch to anchor the arbor, and prevent it from falling out of the frame, and this latch was found either as a thumb-screw or as a spring-loaded stud.

Although retailers had not signed all of the double-action Webley revolvers traced, a majority of vendors had engraved their names upon these weapons. It was clear from the presence of such names as "Deane & Son", and "Robert Adams" that demand for such arms had to be met from gunmakers of considerable reputation, as well as from those of lesser renown, whilst the legends "P. Webley & Son, London", "P. Webley & Son, Makers", and "Webley & Sons Patent" (on three separate weapons) suggested that Philip Webley had opened a retail outlet of his own, during the service life of this design.

Whether in wedge- or solid-frame embodiments, only one lock-mechanism appeared to have been used in these arms; it is impossible to state that this design was always used, however, since both circumstance and good manners sometimes forbade the disembowelling of weapons to which we had access.

This lock-mechanism was one of four double-action lock-mechanisms shown in Joseph Bentley's Br. Pat. No. 2657/1857, and is shown in Fig. 43. As explained in Chapter XVI, P. Webley & Son appear to have secured an interest in this patent, during 1860, and it is interesting to note that the drawings to Bentley's Complete Specification illustrated a Colt-type rammer in a revolver of wedge-frame construction which was very similar in appearance to that shown on Plate 27.

As in the case of self-cocking arms, both wedge- and solid-framed revolvers were numbered; "Patent No. . . ." was one numbering style, whilst "London . . ." was also occasionally used. In these double-action weapons, however (unlike the self-cocking arms), it was in part possible to relate magnitude of number to one constructional detail.

Surprisingly enough, however, there was no point in the numbering at which one type of frame construction superceded the other; 1015 was the lowest number traced, and 36649 the highest, yet wedge- and solid-framed revolvers appeared (in no particular order) throughout the range of numbers between these two extremes. We noted also, that wedge-frame arms were made with two shapes of butt; normally that shown on Plate 27 was used, but pistols numbered 10676, 31312, and

36649 had a different configuration, with butts almost identical in shape to that used for Beaumont-Adams revolvers. (See Plate 13.) The one major and noticeable change was the disappearance of Colt-type rammers, in favour of the Kerr-type, after a 4000 numbering-range; this event may have signalled the expiry of Kerr's rammer patent, in 1862.

In all the circumstances, it seems safe to suggest that (unlike the numbering used on self-cocking arms) numbers engraved upon Webley double-action revolvers were true serial numbers. These numbers were probably inflated (in gun-trade style) by the addition of an initial digit or two, but for anyone with the key to their allocation, these numbers almost certainly recorded use of the lock-mechanism protected by Joseph Bentley's Br. Pat. No. 2657/1857, under the terms of the 1860 agreement described in Chapter XVI.

Reference to Fig. 43 will show that Bentley's lock-mechanism consisted of hammer (A), trigger (B), pawl (C), and sear lever (D). In thumb-cocking, (D) engaged suitable bents on (A), and was tripped by a shelf on (B), when the latter was pulled; a lever (E) hinged to the hammer, linked (A) and (B), by a tooth inserted into a slot through the trigger-nose, and dragged the trigger back to a suitable position for tripping (D), when the hammer was thumb-cocked. In self-cocking fire, the hinged lever (E) acted in a similar fashion to the sear described in Chapter V, as operating Robert Adams's lock-mechanism; in Bentley's case, however, the point of engagement for lifting the hammer was at the lower end of (E), where the tooth mentioned above (and a notch shown in Fig. 43) gave the trigger-nose purchase for its task.

It is worth noting also, that the cylinder-bolt pivoted on top of (B) was not patented by Bentley; Br. Pat. No. 2657/1857 had proposed the use of a combination trigger-spring and cylinder-bolt, along lines suggested in this patentee's earlier Br. Pat. No. 960/1852. The pivoted cylinder-bolt was a device claimed by Charles Reeves (a Birmingham revolver maker) in his own Br. Pat. No. 1623/1858; use of this bolt could have been another reason for numbering the revolvers, although no agreement relating to Reeves's patent was traced in the *Index of Assignees, etc.*, earlier mentioned, to support such a suggestion.

As in the case of Webley self-cocking arms, double-action revolvers (of both wedge- and solid-frame construction) were adapted to use metallic cartridges. In a few cases, solid-framed arms were cased-up with both percussion and breach-loading cylinders (permitting the owner a choice of ammunition to be used in them), whilst other pistols had been permanently converted to breech-loaders; such conversions were normally accomplished by the "Split-cylinder" method, in which loading was accomplished by removing the cylinder entirely from the

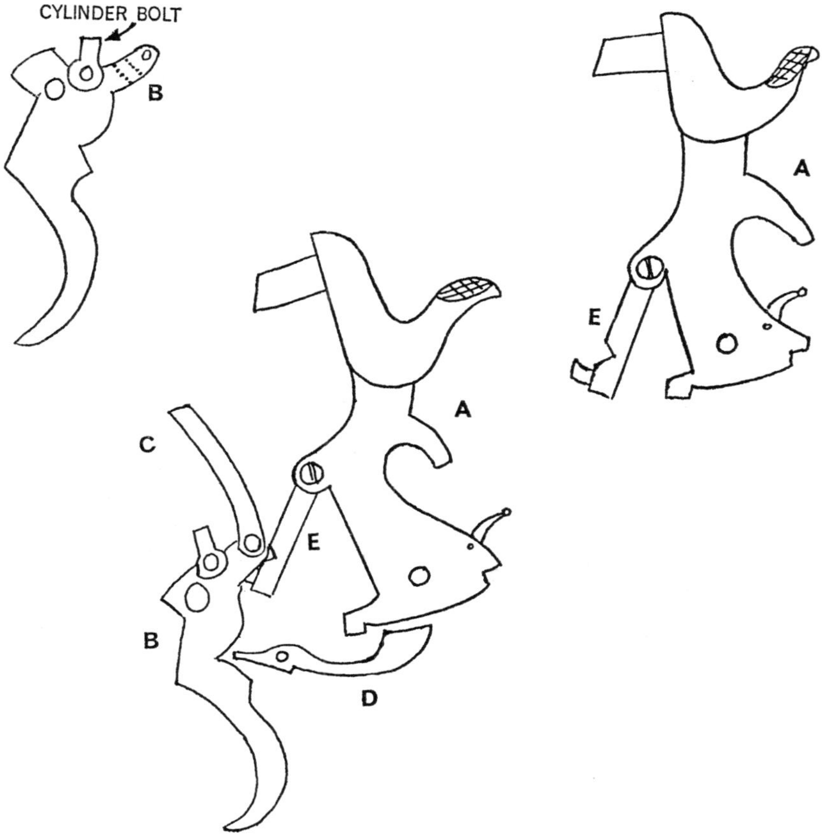

(A) Hammer; (B) Trigger, slotted through nose to accept tooth on (E); (C) Pawl; (D) Sear lever; (E) Hinged lever connecting hammer and trigger.

FIG. 43. *Joseph Bentley's double-action lock mechanism.*
(Br. Pat. No. 2657/1857)

frame, taking off from the cylinder a backing-plate (which contained firing-pins for each chamber) and inserting the cartridges. The cylinder was then reassembled, and replaced in the pistol.

In the light of P. Webley & Son's advertisement offering breech-loading revolvers in 1864 (which was noted in Chapter XVI), it seems likely that this firm was offering either or both of the metallic-cartridge revolvers described above, in the last two years of the period covered by this book.

CHAPTER XVIII

TRANTER REVOLVERS

IN the years between 1865 and 1885, William Tranter (a Birmingham gunmaker) was to achieve an almost dominant position within the British revolver-manufacturing industry. During the period of concern here, however, his progress towards that dominance was hampered by the existence of Robert Adams's English Patent No. 13527/1851, covering the principle of forming a revolver barrel and frame as one unit; until the expiry of Adams's patent (in 1865), therefore, two excellent Tranter percussion revolvers were deliberately developed under licence from that patentee, and their inventor assumes a lesser status in these pages, as a result.

William Tranter was reasonably well-to-do, but almost certainly served an apprenticeship in the Birmingham gun trade, like most of his rivals in that town; after setting up his own business, in Whittall Street, during the early 1840's, he was briefly in partnership with the Hollis brothers (famous Weaman Row gunmakers), between 1844 and 1848. It is clear that he maintained a good working relationship with his former partners, even after he had again set up business on his own account (at 13 St. Mary's Square), and "Hollis Brothers & Company" had been dissolved.

From St. Mary's Square, Tranter filed Registered Design No. 2054 (of October 13th, 1849) for a pepperbox lock-mechanism shown in Fig. 4, and he remained in those premises until a decade after 1865; in 1863, however, he completed construction of a new factory at Lichfield Road, Aston Cross, and carried on his business there, as well as at 13 St. Mary's Square.

William Tranter was successful both as gunmaker and as businessman for the whole of his life, but although he manufactured some excellent revolvers (both for himself, and for the gun trade), such arms never formed the entire base for his business, at any time. Revolvers were merely one item in a thriving concern, which supplied weapons of every kind to the gun trade, as it did to the British Government and to overseas markets.

Ingenious as Tranter's Registered Design of 1849 undoubtedly was, there is no evidence that he ever produced pepperboxes on this system, and it was not until 1853 that the first of his patents covering true revolvers was secured, and he properly entered the field of concern here.

In the years between 1853 and 1865, Tranter secured six British patents relating to revolvers, or to accessories for such arms, the claims of which can be conveniently summarized here.

British Patent No. 212, of January 28th, 1853.

Claims:

(1) Two self-cocking lock-mechanisms for revolvers, each embodying two triggers; pressure on the rear trigger (by the second finger) raised the hammer, whilst pulling the front trigger (with the first finger) discharged the piece.

(2) In association with (1), a spring safety-catch which prevented the hammer from striking the nipples, except when properly raised by the rear trigger. This was effected by attaching a flat spring to the left-hand side of the pistol frame, the tension of which constantly urged it to lie flat against that frame; at the top of the spring (and in alignment with the hammer-nose) was a stud projecting into the hammer-slot in the frame, *via* a hole in the frame-wall.

If the hammer was raised slightly (by pressing the cocking trigger), the spring snapped this stud into the hammer-slot, to separate nipples and hammer under all circumstances except those obtaining when the cocking trigger had been fully retracted. At this latter point in the lock-mechanism cycle, the pawl was thrust sufficiently high in the frame both to cam the spring safety-stud out of alignment with the hammer-nose, and to hold it clear. When pressure on the front firing trigger dropped the hammer, the pawl still held the spring safety-stud out of the hammer-slot in the frame, and the hammer could strike a nipple; release of the cocking trigger lowered the pawl (as the trigger-spring carried that trigger forward again), and permitted the spring safety-stud to be re-engaged by the firer, if he wished to use it.

Although similar to Robert Adams's press-in spring-safety in external appearance, Tranter's device was the more effective of the two, since it operated automatically if the firer cocked, or began to cock, his pistol, and then decided not to fire it. Adams's safety required a deliberate act by the user to set it, in similar circumstances, and was too readily disengaged from the hammer by a slight accidental movement of the latter.

(3) An "elongating socket" for the cylinders of revolvers. This was a tube, inserted in the arbor passage through the cylinder, which could be adjusted to take up any slop, or undue play fore and aft of the cylinder, in the frame-aperture.

Additional claims related to matters of no concern here, as to breech-loading and needle-fire guns, cartridges, and the manufacture of gun-barrels.

British Patent No. 2921, of December 16th, 1853.

Claims:

(1) A breech-loading gun, in which a set-screw mated breech to barrel.

(2) A single-trigger, hammerless, double-action revolving gun, in which the cylinder was propped up to the barrel by a similar set-screw to that used in (1).

(3) Lubricated bullets and waddings for revolvers and other firearms. Amongst these items, a one-piece lever-rammer and a bullet mould were illustrated and described, but not specifically claimed; presumably their presence in Tranter's Complete Specification offered some protection, even without specific claims of novelty for them.

(4) Bullets with a tapered hole formed through them.

British Patent No. 1913, of August 16th, 1856.

Claims:

(1) Two single-trigger, single-action revolver lock-mechanisms, with cylinder-bolt hinged in the top of the trigger.

(2) Three single-trigger, double-action revolver lock-mechanisms, all broadly based upon the principles shown in Fig. 45, i.e. that of linking hammer to trigger by a suitable hinged lever (for self-cocking action), and embodying a separate sear below the hammer (for single-action fire).

(3) Two double-trigger, double-action revolver lock-mechanisms. One of these was styled a "treble action" lock mechanism in the Specification, since the hammer could either be thumb-cocked and fired by the front trigger, or raised and fired by pulling the front trigger only, or raised by the rear trigger and then dropped by pressing the front trigger.

(4) Two lever-rammers for revolvers, pivoted "near the top square of the barrel" (i.e. above the centre-line of the arbor), which had jointed to them a swivelling slotted ram-head.

(5) A method for cutting out the lock-mechanism cavity in a revolver frame, by mounting the frame on a plate, and turning out the cavity on a lathe, instead of chipping and filing it out, by hand.

British Patent No. 2067, of July 19th, 1862.

This short-lived patent (see Appendix) had an unillustrated Specification, and the claims are not easily envisaged; it would appear, however, that protection was claimed for:

(1) A see-saw cylinder-bolt, operated by a stud on the hammer-breast.

(2) A removable inspection plate, giving access to the lock-mechanism of a revolver.

(3) A spring-loaded arbor latch.

(4) A lubricating ramrod for firearms.

(5) A form of rifling.

British Patent No. 1862, of July 27th, 1863.

Claims:

(1) A lever cartridge-ejector, for breech-loading revolvers.

(2) A rearward swinging loading-gate, for breech-loading revolvers.

(3) A method for attaching butt-cheeks to revolvers, without recourse to the conventional cross-pin.

(4) A cylinder-bolt operated by a "spring click" on the hammer-breast.

British Patent No. 1889, of July 20th, 1865.

Almost in passing (for the patent was primarily concerned with breech-loading sporting arms, having drop-down barrels), claim was laid to the combining of a detachable breech-piece with the cylinder of a revolver. This idea was, of course, a "split cylinder" conversion system for permitting use of metallic cartridges in arms originally made as muzzle-loaders (see Chapter XVII), and the patent also claimed a centre-fire cartridge for this purpose.

It is clear that Tranter began to produce percussion revolvers as soon as it was safe to do so under his British Patent No. 212/1853; the Complete Specification for this patent was filed on July 28th, 1853, but there can be no certainty that the patentee did not begin manufacture (as he was perfectly entitled to do), under protection of the Provisional Specification filed with his application for this patent, on January 28th, 1853.

Revolvers of the type produced under British Patent No. 212/1853 are shown on Plates 28 and 29, from which it will be seen that a double-trigger lock-mechanism was fitted, in which the long self-cocking trigger hung below the trigger guard (and was worked by the firer's middle finger), whilst the firing trigger was encased in the trigger guard and operated by the index finger.

This production lock-mechanism is shown in Fig. 44, from which the reader may see the principles of operation.

In Fig. 44, the long cocking-trigger (A) carries a sear (B), which lifts the hammer (C) in exactly the same manner as that used by

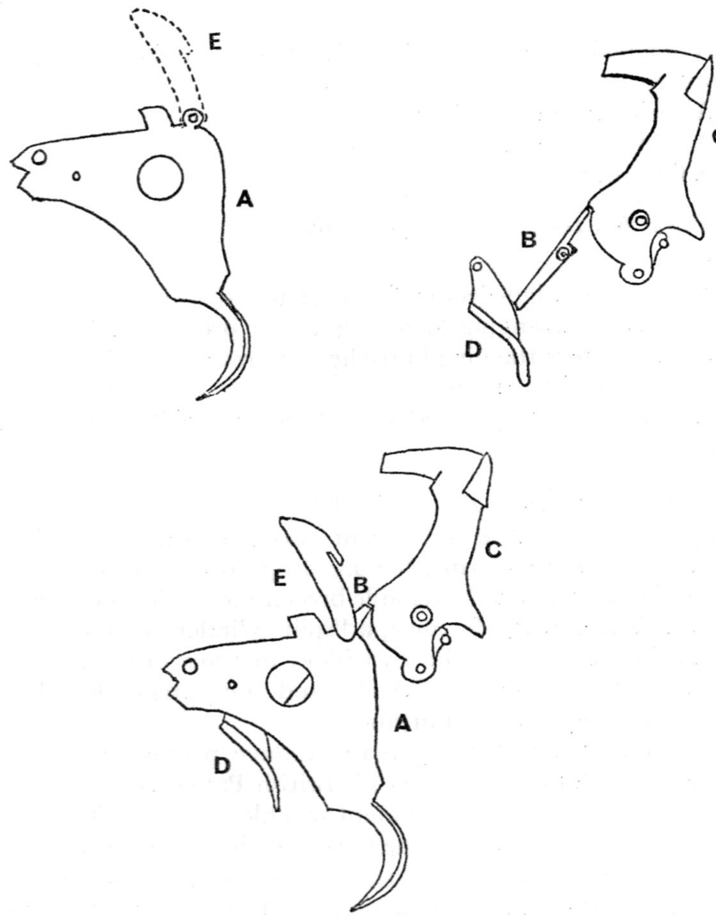

FIG. 44. *William Tranter's double-trigger lock mechanism.*

Robert Adams in his English Patent No. 13527/1851 (see Fig. 10); however, Tranter's hammer (C) is not formed in such a way as to push (B) out of its bent at full-cock, in Adams's fashion. Instead, (B) has a tail extending down through a slot in (A), which is engaged by the firing trigger (D) pivoted in the same slot; all that the firer therefore has to do is to cock (C) by pulling (A), take aim, then fire by pulling (D). The pawl (E) is pivoted on (A) in exactly the same manner as that for Adams's lock-mechanism, the pawl spring acting to thrust the sear (B) at all times towards the hammer bent, and a fixed lump on top of (A) engaging the nipple-partitions at full-cock as a cylinder-bolt.

The lock-mechanism shown in Fig. 44 was that illustrated in the Complete Specification to Tranter's British Patent No. 212/1853, save as to its use of a long cocking-trigger protruding below the trigger guard; in his Specification, the patentee had proposed to mount the triggers in tandem, so that both were encased by the trigger guard, and had described a long trigger only with reference to a minor variation in method for releasing the sear.

It is also clear, from the Complete Specification to British Patent No. 212/1853, that Tranter had originally conceived his pistol for a frame-construction that did not infringe Robert Adams's master-patent (English Patent No. 13527/1851) for the one-piece barrel and frame; thus both drawings and words to Tranter's Complete Specification indicate that the solid pistol frame was to be made as one component, whilst a separate barrel was to be screwed into that frame.

In the event, however, Tranter chose to use Adams's patented one-piece frame and barrel under licence, and to pay a royalty for each frame made under his licence.

This adoption of Adams's frame, by Tranter, presumably arose through the existing business relationship between the two men. As was explained in Chapter VII, many of Adams's self-cocking revolvers in a serial-range below about 13500R have frames stamped "WT" internally, and we suggested that Deane, Adams & Deane were drawing supplies of this main component from Tranter; given the existence of such a connection, it was only sensible for Tranter to use a frame for which he already had manufacturing facilities, rather than to develop his patent on a new two-piece frame/barrel assembly.

It is not impossible that Tranter had already reached a decision to use Adams's frame, when the Complete Specification to British Patent No. 212/1853 was filed; in negotiating licence terms with Robert Adams, there would be considerable advantage in showing an existing patent for a viable revolver design, capable of being manufactured without access to the one-piece frame, if necessary.

Modern collectors follow the classification proposed by J. N. George (*op. cit.* Plate XXV), by recognizing Tranter's double-trigger revolvers in "First", "Second", and "Third" Models. In this classification, "First Model" arms have a detached rammer of the type shown in Tranter's British Patent No. 2921/1853, which is carried separately from the pistol, and slipped onto a peg on the left side of the revolver frame for use: "Second Model" pistols have a detachable rammer with fixed head, which may be clipped on to the side of the weapon when not in use; "Third Model" arms have a permanently attached rammer, with swivelling head, pivoted on a screw in the left side of the frame, which embodies some principles from William Tranter's British Patent No. 1913/1856.

This system of classification carries an implication that the various models of Tranter's double-trigger revolver followed each other consecutively in point of time (without any overlap), and ignores the possibility that Tranter may have made earlier weapons with triggers encased in the trigger guard, in strict accordance with the Specification to British Patent No. 212/1853, or have made double-trigger revolvers (similar to those on Plate 28) without rammers of any kind, prior to filing the Specification to his British Patent No. 2921/1853. In point of fact, at least one pistol has survived which never had a rammer fitted to it, and which also embodies the second of the two lock-mechanisms claimed in British Patent No. 212/1853; if one such arm exists, it is reasonable to suppose that others were made (probably to use Robert Adams's thumb-loaded ammunition), and that they should strictly be accorded the "First Model" and "Second Model" titles. In addition, there can be no doubt that "First" and "Second" model revolvers (by J. N. George's classification) were made simultaneously, at one point, and sold side by side to customers, according to their preference for a particular type of detachable rammer.

Since this "Model" classification is now so firmly established, it will be used in this chapter to describe the same arms as were first mentioned above; the reader should, however, dismiss from his mind any conception of the "First Model" as having been the first produced by Tranter, or of the "Second Model" as an improved version of the "First Model" marketed to the exclusion of its forebear.

It is believed that the "First Model" appeared early in 1854 (after a minute production-run of earlier arms), and it is known to have been made in 38-, 50-, 54-, 80-, 90- and 120-bore calibres; specimens are quite rare, and 38-bore or 54-bore arms are the versions most commonly found. Like all Tranter percussion revolvers, these weapons were five-chambered, and barrel lengths varied from about 7½ in. (for the largest weapon) down to 4½ in. for pocket arms; examples appear at the top of Plate 28.

The "Second Model" could obviously have been made in all six of the calibres mentioned above, but we have only encountered arms in 38-, 54- and 80-bore calibres. The date of introduction is uncertain, but a unique arm shown below in Plate 28 suggests that this model was either introduced in 1856, or that the slightly modified frame-configuration visible in this Plate was introduced in that year; the pistol, being both unnumbered and engraved "Pattern 1856", suggests that Tranter himself was using sub-contractors for part of his revolver-manufacturing process, and that pattern arms were used to control dimensions and standards in the workshops of those contractors.

In Chapter VII, we were at pains to emphasize the use of "R"-suffixed serial numbers upon Robert Adams's patented revolver frame, to point out the appearance of "WT" and "HS" internal frame-stamps in certain of the Adams Model 1851 self-cocking revolvers sold by Deane, Adams & Deane (with the disappearance of such stamps after about serial number 13500R), and to list certain "Y"-suffixed serial numbers as appearing amongst the specimens of Adams's self-cocking revolver. In studying the serial numbers and markings used upon both the "First" and "Second" models of William Tranter's double-trigger revolvers, an obvious reflection of that maker's activities as a frame-manufacturer for Deane, Adams & Deane, was visible in the numbering and marking of his own patented arms.

A few "First Model" Tranter double-trigger revolvers were clearly built upon frames made and serially numbered for use by Deane, Adams & Deane, in their self-cocking Adams Model 1851 revolvers; we traced such Tranter arms numbered 12494R (38 bore), 13702R (see Plate 28, *top*), and 15590R (54 bore) in this category, and others are known. As Chapter VII explained, the "R"-suffixed serial numbers were apparently allocated to pistols made by (or for) Deane, Adams & Deane under Adams's English Patent No. 13527/1851, and it may be that Tranter used some frames already made by him for that firm, as a result of the change in frame-shape adopted by his principals after about 14000R.

Another serial-range was used upon both "First" and "Second" model Tranter double-trigger revolvers, and had the "Y"-suffix in a 20000 serial-range already noticed (in Chapter VII) as occasionally encountered upon Adams Model 1851 self-cocking pistols sold by Deane, Adams & Deane. In these Tranter arms (as in the "R"-suffixed serial-range described above) most frames were engraved "Adams's Patent No. . . .", and we traced serial numbers between 20148Y and 22746Y; an 80-bore example is shown in Plate 28 (*top*).

Finally, a third serial-range was established for use on these early Tranter double-trigger revolvers, which had a "T"-suffix, and apparently began at about 2000T. This serial-range (found upon no "First Model" pistols) embraced the majority of "Second Model" arms examined, and also carried over into the "Third Model" series later described; 6080T was the highest serial number definitely established as used for the purpose of these "Second" model arms, but we had three higher numbers (6733T; 14494T; 18564T) reported, and this may indicate that arms with detachable rammers were still available, on demand, throughout the life of the double-trigger design.

It can be seen from Plate 28, that the spring-safety device used upon all double-trigger revolvers had two points of anchorage, unlike that used in Robert Adams's self-cocking revolvers. Tranter's Specification to British Patent No. 212/1853 had proposed to secure the spring by one screw, and it is possible that the first arms made had this single-screw safety, and that it broke too often in use. We have never encountered such an arm, but Tranter broadsheets of the period illustrated a two-tined spring, and specifically stated that this was "made in a different manner", suggesting that readers were to be reassured about an earlier mistake in design.

Some of the earliest of these broadsheets do not mention the "elongating socket" for the cylinder, which was another specific claim in Tranter's British Patent No. 212/1853, and it was noticed that no arms in the "R"-suffixed serial-range mentioned above had this socket, that it did not appear in arms with "Y"-suffixed serials until after about 20400Y, but that it was used in all revolvers with "T"-suffixed serial numbers. When present, however, this elongating socket was not in a form specifically shown in the Specification to British Patent No. 212/1853; where the Specification had envisaged the use of locking and securing screws for the socket and cylinder-ratchet, a neater device was in fact produced, which was simply slotted so that it might be rotated, fore or aft, by using an appropriate key.

Another feature of the First and Second Model arms lay in the practice of numbering their double-trigger assembly in a series distinct from the frame serial number, and of stamping the weapon's calibre on the breast of its frame. Both practices ceased very early in the "Y"-suffixed serial-range, and it appears doubtful that any trigger was numbered above 2000; such numbers appeared in the centre of a small circular stamp reading "W. Tranter's Patent", and similar stamps were impressed upon both the detachable rammers and the bullet moulds sold with these early arms.

In general, a five-groove rifling, twisted to the right, was used in the First and Second Model arms, but First Model No. 20378Y (in the Darwent Collection) has oval-bored Lancaster rifling to suggest that other types of rifling may be encountered. A revolving rifle No. 5518T also dates with the First and Second Model pistols, if (as seems probable) rifles were numbered consecutively with the hand-guns and not allocated a serial-range, of their own, but all Tranter revolving long-arms are extremely rare, as a type.

Both "First" and "Second" model arms carried the names and addresses of a wide variety of retailers, both from the United Kingdom and the British colonies; one or two pistols had a Confederate provenance claimed for them, the name "A. B. Griswold & Co., New Orleans" appearing on one such arm.

The internal "WT" stamp continued to appear upon Tranter double-trigger revolvers, after it had been abandoned in Adams's self-cocking arms and replaced by a "D. A. & D." external stamp, but the "HS" internal marking on Tranter pistols (which we assume to have been that of "Hollis & Sheath", the successors to Tranter's associates in the late 1840's) disappeared during that series of arms with "Y"-suffixed serial numbers. Upon occasion, "First" and "Second" model arms were found to have the letter "H" externally impressed, beside the rammer-peg on their frame, and this may indicate that Hollis (or some other contractor having a name commencing with that letter) fitted the rammers to these arms.

Tranter's "Third Model" double-trigger revolvers (see Plate 30) apparently had no individual serial-range allocated to them, and were simply absorbed into the "T"-suffixed serial numbers mentioned above, from about 7000T onwards. The weapons shared this serial-range with the single-trigger Tranter arms next to be described, and the highest serial number traced upon a "Third Model" pistol was 22746T; amongst the higher serials, we also noted a 24-bore double-trigger rifle (retailed by "Challenger, Sydney") numbered 19967T.

The "Third Model" double-trigger revolvers were principally made as 54-bore arms (with 6-in. barrels) and 80-bore pistols with $4\frac{1}{2}$-in. barrels; less common today are 38-bore weapons (with $7\frac{1}{2}$-in. barrels and, occasionally, a detachable carbine butt), 120-bore pocket arms (with $3\frac{3}{4}$-in. barrels) and 54-bore arms with $4\frac{3}{4}$-in. barrels. All were rifled with five grooves, twisted to the right, and Birmingham proof-marks were usually present; reference to Adams's patent had been deleted from the frame engraving by the date of appearance for the "Third Model", and they were signed "W. Tranter's Patent" in an oval on the rammer and left frame, with "No. . . .T" on the right. As with preceding models, a wide range of retailers' names are found engraved upon the top frame-strap.

The date of introduction for this model is not known. As can be seen by comparing the specimen arm on Plate 29, with one illustrated as an advertisement in W. W. Greener's *Gunnery in 1858* (see Plate 30) the "Third Model" was clearly available by date of publication for this book. However, two dated presentation arms clearly suggest either that sales of this model moved very slowly, or else that Tranter did not start serious production until well after 1858; thus revolver No. 9295T (a 54-bore arm, retailed by Blanch, London) was the subject of a presentation in 1861, whilst No. 19882T (an 80-bore weapon, retailed by Cogswell & Harrison, London) bore an escutcheon dated 1865.

The "Third Model" was an excellent revolver, and some remarkable shooting has been done with it (in America) during recent years. As a result of this general excellence, the basic design of Tranter's double-trigger revolvers did not alter again. One pistol (No. 18167T) had a folding cocking-trigger, which could be turned up to lie about the underside of the trigger guard; since this was an 80-bore pocket arm, the refinement was most useful, but we have never encountered another weapon embodying it.

Before turning to the last of Tranter's percussion revolvers, it is as well to notice that the double-trigger arms were made in Belgium, and apparently under licence. Since the weapons embodied Robert Adams's patented frame, protection for which was recognized over there (see Chapter VII), the same Belgian manufacturers made Tranter arms, as had imitated Adams's self-cocking revolver; thus Dandoy, Pirlot Fréres, and A. Francotte all signed double-trigger Tranter pistols, in a serial-range between 6000 and 8000. Since these men were excellent and reputable gunmakers in their own right, these Belgian Tranter arms differ only in details from their British counterparts, and normally bear similar acknowledgements to Tranter and Adams, as those described in Chapter VII for the self-cocking Adams revolver.

As mentioned above, William Tranter also took out quite elaborate protection (as Br. Pat. No. 1913/1856) for a number of double-action revolver lock-mechanisms, and in due course manufactured pistols to use one of them.

Plate 29 shows a specimen of Tranter's single-trigger revolver, and it was made in 38-, 54-, 80- and 120-bore calibres, and upon the same frame as the equivalent "Third Model" double-trigger arms. Since the same frame was used, a single-trigger weapon and a double-trigger "Third Model", in the same calibre, have the same dimensions.

The lock-mechanism for the single-trigger revolver is shown in Fig. 45, from which it will be seen to consist of a hammer (A), lifter (B), trigger (C), pawl (D), and sear (E); three V-springs serve as main-, sear- and trigger-springs.

When (A) is cocked by the thumb, (E) engages bents under the hammer to hold it at full- or half-cock, as desired; (B) has a hook at the top of it, which engages a round slot cut and drilled across the hammer-

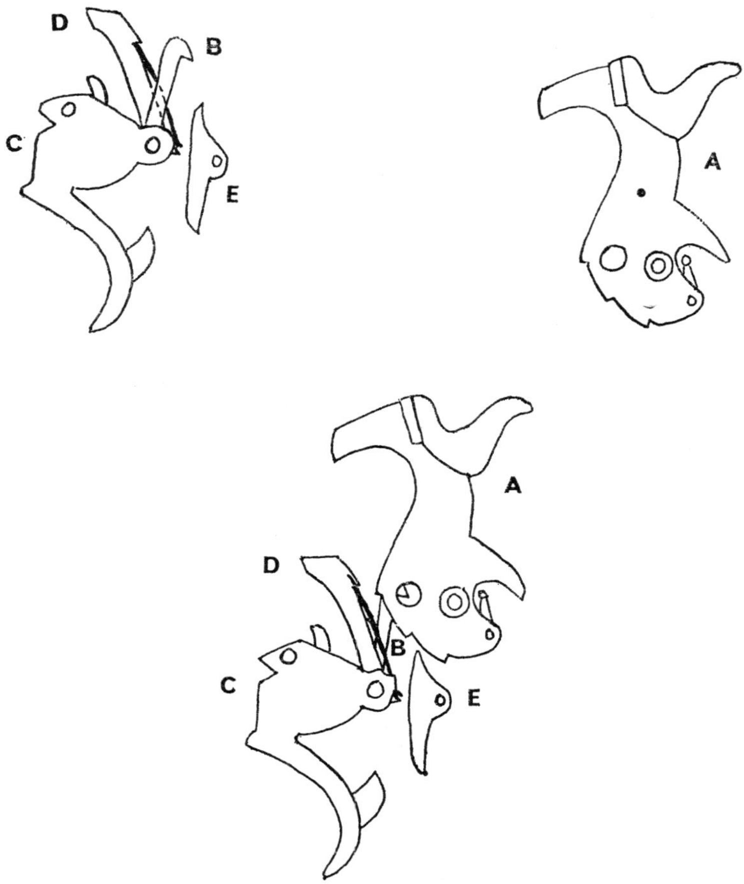

FIG. 45. *W. Tranter's single-trigger revolver lock mechanism (After Br. Pat. No. 1913/1856).*

breast, and, when the hammer is thumb-cocked, (B) pulls (C) back on its pivot, until the horn formed on the rear trigger-bow is almost touching (E). Pressure upon the trigger then disengages this sear, and the hammer falls.

In self-cocking fire, (B) lifts the hammer (A) in much the same fashion as in Robert Adams's lock, and eventually escapes from the slot in the hammer-breast, as the hammer tilts back; at this point, the horn on (C) is bearing against the sear (E), so that this cannot engage the hammer bents, and the hammer therefore falls free. The pawl (D) has a spring which engages a tail on (B) (in exactly the same manner as in the double-trigger arms), and this spring serves both to push the pawl forward against the cylinder ratchet, for rotation of the cylinder, whilst tilting the lifter (B) back towards engagement with the hammer. A lump on top of the trigger rises through the floor of the frame at full-cock, to engage a nipple-partition and bolt chamber in alignment with bore.

Although never officially adopted by the British armed forces, the double-action Tranter revolvers using this lock-mechanism were intended to compete with Beaumont-Adams weapons described in Chapter X, and were therefore made in 38-, 54-, 80- and 120-bore calibres, with barrel lengths of 7 in., 6 in. and 4½ in. each, respectively; the 80-bore pistol was not, in fact, matched by any equivalent Beaumont-Adams model.

In general, the single-trigger weapons were (apart from their lock-mechanisms) very similar in detail to "Third Model" double-trigger arms; the same rifling, rammer, and external markings were used, and the serial numbers, as later explained, in general marched with those of the less numerous "Third Model" double-trigger pistols. In two minor devices, however, the single-trigger arms departed from the well-tried details of the "Third Model", for two different types of safety-catch were used upon them, and variation occurred in the form of their arbor-latches.

As to safety devices, most single-trigger pistols had (in addition to a half-cock bent on the hammer) a pivoted catch similar in principle to that used on the Harding revolver described in Chapter XII; turning upon a screw low in the side of the frame, this catch (shaped like a flat gun-hammer) could be tipped forward to engage in holes drilled in the nipple-partitions, thus bolting the cylinder with nipples out of alignment with the hammer-nose. On a few arms, however, a sliding bolt (similar to that used in Beaumont-Adams revolvers) was fitted instead.

In Tranter's double-trigger arms, arbor-latches had followed the same form as that originally designed for Robert Adams's self-cocking revolver, which had a flat spring screwed into the forward right-hand end of the frame, carrying a peg (penetrating that frame) to engage a notch upon the arbor, and thus prevent the latter from falling out. In his single-trigger arms, Tranter first changed the shape of this spring to an "S"-shape, carrying a loose captive peg to engage the arbor, but towards the end of the service life of these arms, a sunken spring-loaded, rocking arm was fitted instead (after his Br. Pat. No. 2067/1862), which had to be tipped by the thumb to release the stud holding the arbor.

All serial numbers noted upon single-trigger arms were "T"-suffixed. In general, they marched with the serial numbers used upon "Third

Model" double-trigger arms, which began (it will be recalled) at about 7000T. An exceptional single-trigger 80-bore pistol (No. 2376T) is, however, actually numbered in a range normally found upon "Second Model" double-trigger arms only; in addition, Mr. Darwent (to whom the pistol belongs) has pointed out ("The Percussion Revolver of Lt.-Col. George James Ambrose", *Guns Review*, Vol. 5, No. 8) the existence of an Adams's self-cocking Model 1851 revolver serially numbered 2368T, and it is therefore obviously unsafe to offer any dogmatic statement about the purpose of, or course taken by, the serial numbers used upon Tranter arms.

It is enough, therefore, to say that the bulk of the single-trigger revolvers were found in the serial-range between 7010T and 22876T, and that the model was probably in intermittent production from about 1859 until 1870. We could find only six revolvers for which any clues as to an approximate date of issue could be suggested; thus pistol No. 8378T had the legend "R. Sawer. 25/2/59" pencilled on the case-lid label (when bought by one of us in 1953), weapons numbered 11372T and 11844T had a Confederate provenance (1861–65), revolver No. 15235T was presented in 1862, revolver No. 19586T had been carried through the Abyssinian operations of the British Army in 1869, and No. 21777T had been the subject of another presentation in 1867.

As to their markings, these single-trigger arms generally followed the practice used for the "Third Model" double-trigger series, and a variety of retailers (both at home and overseas) had their addresses engraved upon the top frame-straps; occasionally, the legend "W. Tranter Patentee" (as in "W. Tranter Patentee. Made for Henry Allport, Cork") was present, as well as the familiar "W. Tranter's Patent" stamps upon frame and rammer. In a majority of cases, too, the letters "HH" or "WH" were stamped minutely upon the left side of the frame, near the rammer (or on the lower butt tang) and may have been the initials of sub-contractors or out-workers who either fitted the rammers or finished the pistols.

Before turning to the breech-loading revolvers made by William Tranter, it is as well to mention that variant percussion arms by this maker can be encountered. Although we were unable to secure examples for illustration here, there are surviving specimens of Beaumont-Adams revolvers made by Tranter, and also of single-trigger Tranter pistols fitted with Kerr's rammer (see p. 140, *ante*), instead of the equally serviceable device shown in Plate 29, which was protected by Tranter's Br. Pat. No. 1913/1856, and fitted to the majority of the single-trigger arms. Beaumont-Adams revolvers made by Tranter are distinguished by the use of a separate barrel screwed into their frame, and also by the presence of a horizontal see-saw arbor-latch (of the kind described above) in place of the screw-in device used by the London Armoury Company, Ltd., and by other licensees under Adams's and Beaumont's patents, and described in Chapter VII. Finally, two specimens of the "treble action" percussion revolver (described above, in relation to Br. Pat. No. 1913/1856) have been traced, and others are known to exist. The identified arms were both 54-bore pistols, one (No. 10811T) having been retailed by B. Cogswell, of London, and the other (No. 21835T) by A. B. Griswold, of New Orleans, but 80-bore pistols were also made. Built upon the same frame as normal "Third Model" or single-trigger arms, these revolvers exhibit no obvious peculiarities, save for their unusual lock-mechanism and the occasional presence of Tranter's old spring-safety device upon them. The variant is now commonly called "Tranter's Export Model" by American collectors, but the majority of known specimens are signed by British retailers.

The date at which Tranter turned to manufacturing breech-loading revolvers is uncertain, but we are inclined to propose either late 1863, or early 1864, as the probable date for commencement. As earlier mentioned, it is known that a severe slump occurred in the Birmingham gun trade after September, 1863, and that Tranter then had a large new factory for which business had to be found. Although revolvers were only one of the lines available from that factory, it seems logical that Tranter's Br. Pat. Nos. 2067/1862 and 1862/1863 were worked, forthwith, to enable the patentee to offer novel arms under their protection.

Before dealing with pistols made under these patents, it should be emphasized that Tranter's muzzle-loading revolvers were not immediately superceded by them. As mentioned above, such percussion arms were made and sold until the 1870's, and contemporary pamphlets and trade-sheets often show the two types of arm as offered side by side.

For the buyer who feared that metallic cartridges would not always be available, Tranter followed the current practice of offering a revolver into which either a percussion or a breech-loading cylinder might be inserted; known as "Tranter's Patent Combination Revolver", this arm was a ·442 rim-fire or 54-bore percussion single-trigger, double-action weapon, based upon the new frame and barrel assembly under his 1862 patent, and using a lever cartridge ejector that could double as a rammer, at need, under the patent of 1863.

It is possible, too, that a few revolvers were permanently converted from percussion to breech-loading, in our period, under Tranter's Br. Pat. No. 1889/1865; in general, such conversions are more likely to date from the next decade (when even the most conservative user was satisfied that ammunition would always be available), but the possibility has to be admitted.

One such arm had been an 80-bore, single-trigger percussion revolver (No. 15235T), retailed by Parker Field & Sons, which was engraved "Presented by Capt. Field to the best shot in the 2nd C. Honble. Art. Coy. 1862. Won by Henry Parsons"; this weapon had later been converted (along lines described in Chapter XVII) in accordance with Tranter's patent of 1865, into a split-cylinder ·38-in. breech-loading centre-fire pistol.

A number of converted double-trigger Tranter revolvers have also survived, and H.M. Tower of London has several split-cylinder arms of that type. In this case, however, there is a much greater possibility of error in accepting a date for their conversion; towards the close of the century, a few target shooters had such work carried out, during a revival of interest in Tranter's original design of 1853, and there are also in existence double-trigger breech-loading revolvers which were built as such. When William Tranter retired, in 1885, his Aston Cross factory was leased to, and operated by, a company formed by George Kynoch, M.P.; as part of that company's operations, a few breech-loading, centre-fire Tranter revolvers were produced. It is possible, however, that conversion of one ·38 centre-fire weapon (No. 13347T, originally retailed by Parker Field & Sons) does actually date from the period of concern, and that other such specimens were produced.

Tranter revolvers designed and made as breech-loaders are found in two distinct models; the first model was a single-action, seven-chambered pocket pistol, made without trigger guard, whilst the second model was a double-action weapon with trigger guard; the majority of examples surviving today were in fact made after 1865, but, for reasons earlier stated, these two models fall within the ambit of these pages.

The single-action pocket arms were sold as "Tranter's Patent Breech Loading Revolver No. 1" (in ·230-in. rim-fire calibre), and as "Tranter's Breech Loading Revolver No. 2" (in ·320-in. rim-fire calibre); an excellent article by J. Darwent ("Tranter's No. 1 and No. 2 Revolvers", *Guns Review*, Vol. 6, No. 2) deals with these revolvers in considerably greater detail than space permits here.

A specimen arm shown on Plate 31 is typical of the single-action Tranter revolver as a type; the one-piece cast frame (in brass or bronze for the smaller pistol) had the detachable side inspection plate claimed in Br. Pat. No. 2067/1862, and a separate octagonal barrel of $2\frac{1}{2}$ in. (·230 in.) or $3\frac{7}{16}$ in. (·320 in.) in length (rifled with five grooves twisted to the right) was screwed home permanently into this component. Another design feature, giving a very smooth-working lock, was the use of a separate centre (screwed into the right wall of the frame) upon which the hammer rotated; the screw holding on the inspection plate was turned into this centre, and the same method of construction was also used in Tranter's double-action breech-loading arms.

All of the single-action Tranter revolvers were serially-numbered on the left side of the frame, this number being stamped in the centre of (or below) the familiar oval legend "W. Tranter's Patent". It is not clear why Tranter serially-numbered these weapons (or his double-action arms, for that matter) since they were almost certainly made by him, and no sub-licensing under the two short-lived first patents, covering his breech-loading arms, can have been necessary, but numbered they were; most specimens were traced in a serial-range between 1000 and 11000, but one or two pieces (made many years after the period of concern here) were numbered in a serial-range between 30000 and 38000.

During the course of production, one or two changes of detail occurred in these arms. The most important of these was the introduction of an alternative type of cylinder-bolt to that protected by Br. Pat. No. 2067/1862; the patent had broadly proposed the use of a see-saw cylinder-bolt (mounted in the frame, immediately below the cylinder, and engaging notches cut on the forward circumference of the latter), which had a spring-tempered tail that was operated by a projection on the hammer-breast. The principle was almost identical with that shown in Fig. 19, as used for Colt percussion arms of the period. A more simple cylinder-bolt was adopted as well, however, and used instead a strong trigger-spring, together with sufficient movement in the trigger, fore and aft, on cocking, to rock the bolt (which engaged a notch in the trigger) in and out of engagement with the cylinder. Curiously enough, it would appear that these two bolt mechanisms were used simultaneously during the whole service life of these little pistols, and no supercession, of one by the other, occurred; thus Darwent (*op. cit.*) shows pistol No. 3959 with the simplified bolt, 10047 with the patented bolt, and 10069 with the simplified device.

Another change of detail was the use of the rocking arbor-latch already described, which was protected by Br. Pat. No. 2067/1862 and is visible in Plate 31 (*top*); this was used on most of these arms, but a few pistols with very low serial numbers used a spring-loaded button latch. Similarly, although cartridge ejection was supposed to be done singly, through the loading-gate, by using a cleaning-rod, some cylinders had broad notches cut at the outer edges of the chamber, so that cases might be prised out by their rims.

These arms were handsome in appearance, and this was often

enhanced both by engraving and plating the frame (with gold or silver) in the manner shown at the top of Plate 31. In normal course, the butt-cheeks (which followed a design protected by Br. Pat. No. 1862/1863, and required no cross-screw to secure them) were of polished un-chequered walnut, but some chequered cheeks were fitted, and occasional specimens will be found with ivory butt-cheeks. Retailers from all over the United Kingdom signed and retailed both versions, the London firm of Cogswell & Harrison (then at 223/4 Strand, London) being particularly active; their manufacturer identified himself only by use of the stamp mentioned above, and by impressing the letters "WT" upon the breast of each frame.

It is quite possible that Tranter made these pistols by intermittent production-runs of each calibre, and only set up his machinery to produce them when stocks were low. Thus, in studying the serial numbers, it is noticeable that each block of 1000 numbers tends to be restricted to one calibre; we thus have ·230-in. arms in the 1000–1900 serial-range, ·32-in. arms in the 2000–4000 serial-range, and so on. One ·32-in. centre-fire revolver was reported (as No. 8365), but this may have been converted during its service life; there is, in fact, no evidence that Tranter produced these pocket arms (which were widely copied by other makers) in anything but the two rim-fire calibres mentioned above.

Tranter's double-action breech-loading revolvers of the period are best-known in the model shown on Plate 31, which was a six-chambered "Army" revolver in ·442-in. rim-fire calibre; however, smaller, five-shot, rim-fire models were also made in ·38 in. and ·32 in. calibres. Centre-fire revolvers on these models (or to an improved design patented in 1868) are quite as well known, but it is doubtful if Tranter produced any centre-fire versions under his patents of 1862 and 1863 before the year 1865 was sped.

The ·442-in. pistol had a 6¼-in. barrel, rifled with five grooves twisted to the right, and used the same double-action lock-mechanism (protected by Tranter's Br. Pat. No. 1913/1856) as had been chosen for the single-trigger percussion revolvers. The frame, however, had the barrel screwed into it, and had those design features already described above, for access to the lock-mechanism, and mounting the hammer. A lump on top of the trigger acted as a cylinder-bolt, engaging locking notches formed at the rear end of the cylinder, and a horizontal see-saw arbor-latch was again used. The powerful lever-ejector was based upon a design claimed in Tranter's Br. Pat. No. 1862/1863, but neither the improved cylinder-bolt nor the rearward-swinging loading-gate also claimed in that patent was ever used in these pistols. The markings followed those already described for single-action arms, and numerous retailers sold the weapon; it was never formally adopted for the British armed forces.

The smaller ·38-in. and ·32-in double-action rim-fire revolvers had, respectively, barrels of 5 in. and $3\frac{3}{4}$ in. in length, although other lengths may have been offered as well; the rifling was again five-grooved, twisted to the right, in a separate screwed-in barrel. Intended as pocket arms, these pistols were only five-chambered, had light frames, and dispensed with lever-ejectors, in favour of a simple rod screwed into the butt-cap.

Although using the same lock-mechanism as the larger pistol, these arms dispensed with the characteristic horn upon the trigger (see Fig. 45) and operated their sears by direct pressure from the rear of the trigger-bow. A small damper-spring was often mounted into the front of the frame aperture (to prevent accidental rotation of the cylinder), but a one-piece butt was fitted, as in the large revolver; in common with the latter (and with the single-action arms) these pocket revolvers had a loading-gate which opened to the right.

It would appear that all three models of Tranter's double-action rim-fire revolvers were numbered in a single serial-range, which began at about 4000; a serial number as high as 55966 has been reported upon a Tranter double-action rim-fire revolver, but models of concern here are not commonly found as numbered above the 12000 serial level.

As in the case of Tranter's single-action arms, of course, the majority of double-action rim-fire pistols were made after 1865.

APPENDIX

APPENDIX

English and British patents for revolving arms

CHAPTER II explained the basis for British patent protection, and the various periods during which an inventor might enjoy the fruits of his monopoly for making, using, and selling his invention. Part of the price that he paid for his rights to exclusive enjoyment lay in his duty to file a Complete Specification of his invention at the Patent Office, so that others might learn how to use the idea when his protection ended.

After 1852, an inventor might release his invention to the public at large before the full fourteen years of protection had expired, by failure to pay renewal fees after only three or seven years of exclusivity, and this fact was advertised in the *London Gazette*. It follows, therefore, that although the student may study details of revolving arms patents secured during our period (in the excellent *Abridgements of the Specifications relating to Firearms and Other Weapons, Ammunition, and Accoutrements*, Eyre & Spottiswoode, London, 1859 and 1870), yet to discover the exact length of time during which manufacture of any particular revolver was controlled by the inventor, or by his licensees, is often a matter for long and wearisome research.

Accordingly, we have assembled in this Appendix the date, number, name of patentee, title, and period of protection for *all* English and British revolving arms patents secured in our period; chain-guns, long-arms, pepperboxes and radial weapons are included with the conventional revolvers, and brief descriptive notes are added upon those arms for which space in our preceding Chapters could not be found.

In using this Appendix, the reader is referred to Chapter II for remarks upon "communicated" patents, and is reminded that where only six months of protection are shown, the patentee failed to file the Complete Specification mentioned above, and therefore lost his protection at the end of this short period of grace.

N.B. All Patents listed up to No. 14027 were protected in England, Wales and Berwick-on-Tweed under the stated patent number. Those Patents marked + were also protected in Jersey, Guernsey, Alderney, Sark and Man, whilst those marked * were protected in the "Colonies and Plantations overseas".

Patent Date	Number	Name of Patentee	Title	Expiry Date	Remarks
Nov. 24, 1818	4315*	E. H. Collier. *Partly communicated*	"Improvements in firearms of various descriptions; which improvement is also applicable to cannon."	Nov. 24, 1832	A revolver; see Chap. I.
Apr. 13, 1835	6826	I. Dodds	"Improvements in the construction of firearms", etc. Locks, revolvers, cannon with several barrels, boreing [*sic*] machinery.	Apr. 13, 1849	

Amongst the claims of an elaborate Complete Specification, was one for a single-action, four-shot, radial revolver shown in Fig. 46. Such arms had chambers bored as radii within a circular metal plate (like spokes in a wheel), and the latter might rotate upon either a vertical or a horizontal axis. Dodds (of Hoxley Iron Works, Tipton, Staffs) used the latter configuration, with four nipples at the starboard end of the axis on which his chambers turned; a convolute mainspring powered the hammer, and the chambers were rotated by jerking down an operating-lever beneath the latter. In a variant design, this lever was also used to cock the hammer.

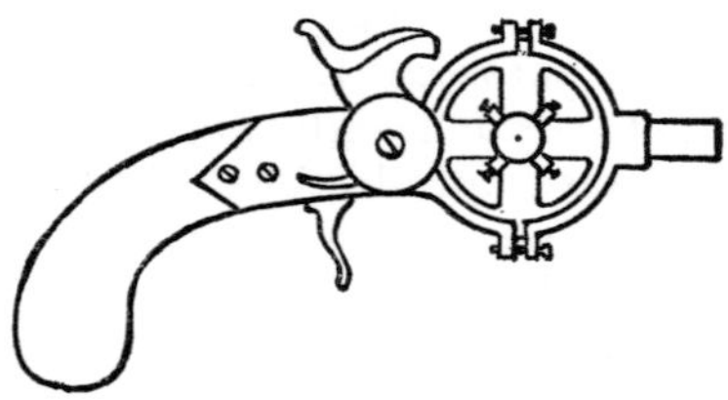

FIG. 46. *Isaac Dodds's Revolver.*
(After Eng. Pat. 6826/1835)

Patent Date	Number	Name of Patentee	Title	Expiry Date	Remarks
Oct. 22, 1835	6909	S. Colt	"Improvements applicable to fire-arms." Revolvers with single barrel.	Oct. 22, 1849	

This was the equivalent to Samuel Colt's U.S. Patent No. 9430X of February 25th, 1836, which protected the principal features of revolving arms known to modern collectors as "Paterson" Colt arms; the revolvers were distinctive, single-action pistols, with folding triggers, and their name springs from a "Patent Arms Manufacturing Company" (at Paterson, N.J.) which first promoted and manufactured them.

It is unlikely that any revolvers were made here, under the English patent, and it warrants little attention. Principal features of both the English and American patents were Colt's methods for rotating revolver cylinders (by a pawl linked to the hammer and engaging a ratchet), and for bolting them in alignment with the barrel, for discharge. In America, U.S. Patent No. 9430X became a master-patent, by which native revolver manufacturers were barred from using *any* mechanical cylinder-rotation system until 1857; in England, however, Colt never achieved this enviable position.

Patent Date	Number	Name of Patentee	Title	Expiry Date	Remarks
Jan. 19, 1837	7286	M. Poole. *A communication*	"Improvements in Ordnance and firearms."	Jan. 19, 1851	

Moses Poole (of Lincoln's Inn, London) was a patent agent, and identification of his principal, for this patent, remains uncertain. In all probability, it was W. Cochran (of New York City), who held an equivalent U.S. Patent No. 188, of April 28th, 1857; however, that patent was worked in America by C. B. Allen (of Springfield, Mass.), who may possibly have used Poole to secure this English protection for him.

The patent protected a single-action, under-hammer, radial percussion revolving rifle; unlike Dodds's design, the "turret" or magazine rotated upon a vertical axis, as shown in Plate 32, and was manually turned. A few pistols were made on this system, but it was principally used in long-arms.

A British version of the Cochran rifle is shown on Plate 32, and was made by James Wilkinson & Son, of Pall Mall, London; this was marked, on the top-strap, "15. Patent 5249", and it is suggested that the first number related to production under Eng. Pat. No. 7286 (presumably subject to royalties), whilst the second number placed this weapon in an overall numbering of Wilkinson firearms.

Patent Date	Number	Name of Patentee	Title	Expiry Date	Remarks
Apr. 9, 1839	8024*	G. Stocker and J. Bentley	"Improvements in guns, pistols, and other denomination of firearms." Lock. Revolvers.	Apr. 9, 1853	A pepperbox; see Chap. III, and Fig. 2.
Jan. 21, 1840	8347	J. R. Cooper	"Improvements in firearms and the balls used there-with."	Jan. 21, 1854	A pepperbox; see Chap. III.
Oct. 14, 1841	9119	M. Poole. *A communication*	"Improvements in firearms." Revolvers.	Oct. 14, 1855	

There is no record of the inventor who communicated this patent to Moses Poole; an elaborate Specification claimed both semi- and fully-automatic percussion revolving guns, operated by clockwork. The idea was a logical development to that in the Collier/Coolidge weapons described in Chapter I.

Patent Date	Number	Name of Patentee	Title	Expiry Date	Remarks
May 14, 1845	10667	C. J. Smith	"Improvements in artillery, guns, pistols, etc., and in apparatus to be used therewith." Primers, Revolvers, Flasks.	May 14, 1859	A pepperbox; see Chap. III and Fig. 7.
June 20, 1849	12668+*	S. Colt	"Improvements in firearms." Revolvers.	June 20, 1863	

In this patent, Colt illustrated revolvers of a type shown on Plate 16, which were those later manufactured at his London factory (see Chapter XIII); the Specification was acknowledged by the inventor, before an English consular official, in Paris, on October 27th, 1849. It was then forwarded to England for enrolment.

Patent Date	Number	Name of Patentee	Title	Expiry Date	Remarks
Sept. 20, 1849	12781	J. R. Cooper	"Improvements in firearms." Locks. Revolvers.	Sept. 20, 1863	A pepperbox; see Chap. III.
Feb. 24, 1851	13527	R. Adams. *Partly communicated*	"Improvements in rifles and other firearms." Rifles, locks, breech-loaders, revolvers.	Feb. 24, 1865	Adams's self-cocking revolver; see Chap. V *et. seq.*
Nov. 22, 1851	13823	S. Colt	"Certain improvements in firearms."	June 22, 1852	Patent lapsed, as no spec'n. was enrolled.
Mar. 3, 1852	13994+*	G. L. L. Kufahl	"Improvements in firearms." Needle-guns.	Mar. 3, 1866	

Needle-fire arms used combustible cartridges, with a percussion cap embedded just behind the bullet; a long needle (powered by any suitable lock-mechanism) passed through the powder-charge to reach this cap, and better ignition was claimed for such a system. Needle-guns had originated on the Continent some years before 1852, and British interest was aroused during the 1850's; some service trials were arranged for arms put forward by various designers. The system was never adopted here, although it had been taken up in Prussia as early as 1848.

Kufahl's patent was primarily concerned with conventional long-arms on the needle system, but it also claimed a self-cocking revolver using such cartridges; this weapon was later made on the Continent (by F. Dreyse, at Sommerda), but no English manufacturer appears to have been interested in it.

Patent Date	Number	Name of Patentee	Title	Expiry Date	Remarks
Mar. 20, 1852	14027	W. W. Richards	"Improvements in firearms and in the means used for discharging the same, also improvements in projectiles." Revolver, bullets, plugs of wood, etc.	Mar. 20, 1866	

William Westley Richards founded a famous firm of Birmingham gunmakers, and was almost unique in that town for having every facet of the trade carried on in his own factory, at the rear of No. 82 High Street, Birmingham. He secured a number of gun-patents (some of considerable utility), of which this patent was the first relating to revolvers. Under its protection, a number of frail five-shot self-cocking percussion revolvers similar to that visible in Plate 34 (*top*) were manufactured, although one version claimed in the Complete Specification (which had a side-striking hammer) is not known to exist today. Some 300 revolvers were made, but the arms are now extremely rare. Calibres of 36, 80, and 90 bore are known (with barrel lengths of 7 in., 4½ in., and 4 in., respectively), and a rammer described under Br. Pat. No. 993/1854 was usually fitted. The specimens observed have been numbered between 30 and 323, whilst both spring-loaded and gravitating automatic safety-bolts were used. The barrel-dismount system was another feature of these patent claims.

N.B. This concludes the listing of *English* revolver-patents secured prior to October 1st, 1852. A list of *British* patents follows, all of which were secured for terms up to fourteen years, under the Patent Law Amendment Act, 1852. Apart from granting protection for the United Kingdom, those post-1852 patents also covered such Colonies as were named in a warrant issued for sealing the actual letters patent. Any applicant could name specific Colonies in his petition for grant of protection, but such an extension of cover outside the U.K. would only be given where it was not unlawful to grant it.

Patent Date	Number	Name of Patentee	Title	Expiry Date	Remarks
Oct. 1, 1852	69	W. Moore and W. Harris	"An improvement in repeating pistols and rifles."	Oct. 1, 1855	

These patentees were Birmingham gunmakers, who protected a self-cocking revolver, in which the barrel was tipped up on a frame-hinge for removal of the cylinder; loading was accomplished by using a separate ball-seater, operated by a powerful screw. This frail weapon is rare now, and applied a reciprocal motion to its cylinder, *via* the lock-mechanism, to accomplish a gas-seal mating of chamber and barrel.

Patent Date	Number	Name of Patentee	Title	Expiry Date	Remarks
Oct. 1, 1852	125	T. Hunt	"Improvements in firearms."	Oct. 1, 1855	A pepperbox; see Chap. III.
Oct. 11, 1852	329	A. E. L. Bellford. *A communication*	"Improvements in the construction of revolving or repeating firearms."	Oct. 11, 1855	

Auguste Edouard Loradoux Bellford was a patent agent at 16 Castle Street, Holborn, London, and the protection was a British equivalent to H. S. North and C. D. Skinner's U.S. Patent No. 8982, of June 1st, 1852.

North & Skinner (of Middletown, and Haddam, Conn.) manufactured bar-hammer revolving guns and pistols, with reciprocating gas-seal cylinders; a vertically sliding crotch wedged cylinder to barrel (after first rotating the former), and was operated by a hinged trigger guard that also cocked the weapon. Disc primers were proposed, for igniting the charges, instead of percussion caps.

No specimens are known to have been made under this British patent, and only a few arms were produced in the U.S.: however, an interest in the British patent was registered at the British Patent Office (during 1853) in the names of "Skinner, Savage & North", and the basic gas-seal idea was later developed into a highly successful American military revolver. See also Br. Pat. No. 942/1860.

Patent Date	Number	Name of Patentee	Title	Expiry Date	Remarks
Oct. 30, 1852	583	R. A. Brooman. *A communication*	"Improvements in revolving firearms."	Oct. 30, 1855	

Richard Archibald Brooman was a British patent agent, in the firm of J. C. Robertson & Co., of 166 Fleet Street, London.

Although his principal was not named, the Specification clearly showed a famous American radial rifle, protected in the U.S. by P. W. Porter's U.S. Pat. No. 8210, of July 8th, 1851. This had its chambers turning on a horizontal axis, and was operated by an under-lever similar to that used on later Winchester rifles; this lever not only rotated the chambers, but also cocked a side-swinging hammer.

Designed to use an extremely dangerous powder-and-ball magazine for feeding the chambers, this Porter rifle did achieve some success in America, once divorced from that accessory, but it is doubtful if any arms were made, in England, under Brooman's patent.

Patent Date	Number	Name of Patentee	Title	Expiry Date	Remarks
Dec. 4, 1852	960	J. Bentley	"Improvements applicable to firearms."	Dec. 4, 1859	

In Chapter XVII, something was said of the principal feature in this patent, as they were applied in the manufacture of so-called "Webley-Bentley" self-cocking percussion revolvers. However, Joseph Bentley was a member of an extensive family, which both manufactured guns in Birmingham and sold them in Liverpool or London, and he did apply this patent to a small number of other self-cocking revolvers, of a distinctive type.

A specimen is shown on Plate 33 (*top*), and can be seen as a five-chambered arm, of fair quality, with Bentley's patented safety-catch mounted on top of the hammer; internally, the weapon has another patented feature, in using a flat trigger spring with a projecting lug on top, which intercepts the nipple-partitions to act as a cylinder-bolt. The lock-mechanism used in these revolvers was also claimed in this patent, but was actually little more than an improved version of the old pepperbox lock-mechanism shown in Fig. 1.

Although Bentley's self-cocking revolvers are quite distinctive when actually compared with "Webley-Bentley" arms, they are not always so readily identified in isolation. It is worth noting, therefore, that Bentley always produced his own revolvers without the removable side-plate so characteristic of the "Webley-Bentley" pistols, and also that where a lever-rammer was fitted to true Bentley revolvers it was always the "Colt-type" shown in Plate 33 (*lower*).

Bentley self-cocking arms normally have a malleable cast-iron frame, but German silver was sometimes used. These revolvers were principally offered in 54- and 80-bore calibres, with barrel lengths of 7 in. and 4¾ in. respectively; however, dimensions did vary quite markedly (as did the butt-shape) and 70-bore pistols with 7½-in. barrels are known. Serial numbers were used in a range between 1 and 850, and normally preceded by the words "Bentley's Patent No. . . ." or "Patent No. . . ."; from the two hundred range onwards, an "A" prefix was added to the serial number, and at least one revolver (numbered 80C) suggests that other serial-ranges may have been used.

A few provincial retailers engraved their names upon Bentley self-cocking revolvers, but more commonly the legend "J. Bentley & Sons, Liverpool" is found upon the barrel. Their period of production is also uncertain, but revolver numbered A.735 was dated August 1st, 1857, and fitted with a rammer protected by J. Bentley's Br. Pat. No. 768/1854 (*q.v.*).

Patent Date	Number	Name of Patentee	Title	Expiry Date	Remarks
Dec. 7, 1852	988	S. A. Goddard. (Provisional protection)	"Improvements in construction of pistols." Revolvers.	June 7, 1853	

Samuel Aspinwall Goddard was a Birmingham merchant and gun-manufacturer. Since only provisional patent protection was secured, it is not easy to grasp exactly what was claimed (without drawings) as being novel. However, a simple percussion revolver was described (which was apparently hand-rotated), and might be single- or multi-barrelled; the chambers were proposed to be aligned successively with the barrel or barrels, for discharge, by inserting the ramrod into one of them.

Patent Date	Number	Name of Patentee	Title	Expiry Date	Remarks
Jan. 12, 1853	86	E. Haslewood. *A communication*	"Improvements in firearms and projectiles."	Jan. 12, 1856	

Edward Haslewood (of Tufnel Park, Holloway) did not name the communicant for this patent. A mechanically-rotated, gas-seal radial revolver (with detachable carbine butt) was described by the Specification, in which specially-shaped bullets contributed to the gas-seal, and were finally seated, on loading, by forcing a metal hoop over the periphery of the rotating charge-carrier.

Patent Date	Number	Name of Patentee	Title	Expiry Date	Remarks
Jan. 28, 1853	212	W. Tranter	"Certain improvements in firearms."	Jan. 28, 1867	Tranter's double-trigger revolver; see Chap. XVIII.
Feb. 4, 1853	305	P. Webley	"Improvements in repeating pistols and other firearms."	Feb. 4, 1856	A revolver; see Chap. XVI.
Mar. 3, 1853	535	S. Colt	"Improvements in rotating breech firearms." Revolvers.	Mar. 3, 1867	

Colt claimed a method for cylinder rotation by cutting "long and short straight grooves connected by curved diverging grooves" around the periphery of a revolver cylinder. A sliding block, linked to the hammer and riding in the frame beneath the cylinder, engaged these grooves (by a projecting pin) and caused rotation of the latter as the block slid to-and-fro. The idea was not developed.

Patent Date	Number	Name of Patentee	Title	Expiry Date	Remarks
Mar. 3, 1853	538	S. Colt	"Improvements in rotating breech firearms." Revolvers.	Mar. 3, 1867	

Colonel Colt claimed in the Specification to this patent (1) cylinder rotation upon an arbor penetrating only part-way through the cylinder; (2) a screwed collar in the standing breech for taking up any fore-and-aft movement of the cylinder in the pistol-frame; (3) an improved lever-rammer; and (4) a hinged barrel, for which any slackness in union to the pistol-frame might be taken up by a pin in that frame.

Patent Date	Number	Name of Patentee	Title	Expiry Date	Remarks
Mar. 7, 1853	577	J. Hall and J. Crofts	"Improvements in revolving or repeating firearms."	Mar. 7, 1856	A pepperbox; see Chap. III.
Mar. 29, 1853	743	J. Webley	"Improvements in the construction of repeating or revolving and other pistols and firearms."	Mar. 29, 1856	The "Long Spur" Webley revolver; see Chap. XVII and Plate 22.
Apr. 26, 1853	966	I. B. Sheath	"Improvements in firearms." Revolvers.	Apr. 26, 1856	

Isaac Sheath was a Birmingham gunmaker, and associated with the Hollis family (mentioned in Chapter XVIII) of that town. Under this patent, a method for attaching a revolver barrel to the frame, by screws, was claimed, together with the use of a projection on the pawl of a self-cocking lock-mechanism, to serve as a cylinder-bolt, by intercepting the nipple-partitions. (The latter idea was also claimed by P. Webley's Br. Pat. No. 2127/1853 (*q.v.*), for single-action arms.) A five-chambered, 38-bore revolver, with 7½-in. barrel, is the only specimen arm that we have encountered; it was retailed by "Hollis & Sheath, Makers to H.M.H. Board of Ordnance", and marked "Sheath's Patent No. 59".

Patent Date	Number	Name of Patentee	Title	Expiry Date	Remarks
Apr. 29, 1853	1038	T. Pennell (Provisional protection)	"Improvements in the construction of revolving or repeating firearms," &c.	Oct. 29, 1853	

Thomas Pennell (a Birmingham gunmaker) claimed protection for the idea of completely encasing the lock-mechanism and hammer of a revolver; for a method of attaching a revolver barrel by an ingenious swivelling bolt; and for a detached rammer. He did not proceed with his application for patent protection, but did produce a small quantity of self-cocking "hammerless" revolvers on the lines described in his Provisional Specification.

Pennell's revolver is shown in Plate 33, and specimens are rare; he did not number his arms, and they were sold by retailers in the larger British cities. So far as we are aware, only six-shot, 54-bore revolvers were made, but calibres of 60 and 56 bore have been reported. Considerable skill and care was lavished upon the manufacture of these pistols, which vary occasionally in detail; thus barrel lengths of 6½ in. or 7½ in. have been encountered, and such refinements as German silver frames, belt hooks (for carrying the arms), press-button safety-catches, and butt-traps for percussion caps have also been observed.

Patent Date	Number	Name of Patentee	Title	Expiry Date	Remarks
May 26, 1853	1298	W. J. Harvey	"Improvements in firearms." Revolvers.	May 26, 1856	

William James Harvey was an Exeter gunmaker who is best remembered for the self-cocking revolvers that he made under his slightly later Br. Pat. No. 2602/1854 (*q.v.*).

This patent of 1853 protected and illustrated a "hammerless" self-cocking revolver, in which the firing-pin (or "discharging ram", as Harvey termed it) acted in a line parallel with the axis of the barrel. Very few arms were made under this protection, and specimens are characterized by the saw-handled butt visible in

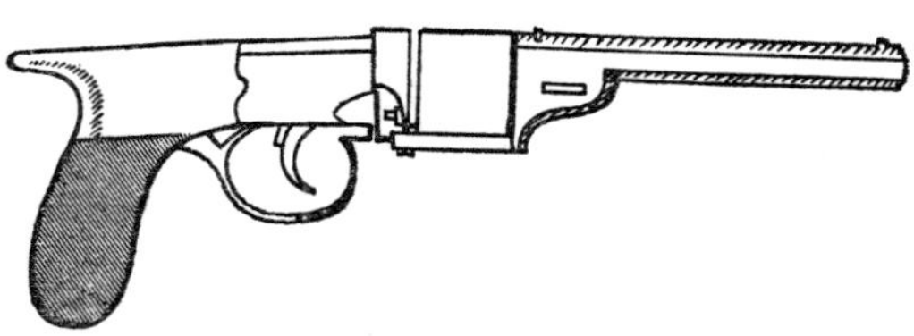

FIG. 47. *W. J. Harvey's First Model percussion revolver.*
(After Br. Pat. 1298/1853).

Fig. 47. The weapon from which this sketch was made is in 76-bore calibre (with a 5½-in. barrel) and numbered 3068; such a numbering probably placed the arm in a global total of all types of firearm made by Harvey, and did not relate solely to his revolving arms.

Patent Date	Number	Name of Patentee	Title	Expiry Date	Remarks
Sept. 14, 1853	2127	P. Webley	"Improvements in repeating pistols and other firearms."	Sept. 14, 1856	A single-action gas-seal revolver; see Chaps. XIV and XVII.
Sept. 20, 1853	2184	H. Needham	"Improvements in revolving firearms."	Sept. 20, 1856	

Henry Needham (of Wardour Street, London) was a member of a famous gun-making family. This patent covered (amongst other inventions) a self-cocking revolver not unlike the Westley-Richards weapon, shown on Plate 34, in appearance; novelty lay in designing the lock-mechanism so that only a light trigger-pull was needed to cock it, and in providing an anti-fouling tube around the arbor. In 1855, the patentee and Charles Lancaster (a famous barrel maker) registered a joint interest in the patent at the Patent Office, but no specimen arms are known to have survived.

Patent Date	Number	Name of Patentee	Title	Expiry Date	Remarks
Nov. 22, 1853	2712	R. Adams	"Improvements in firearms." Revolvers.	Nov. 22, 1856	Robert Adams's "hesitating" lock mechanism; see Chap. VII.
Dec. 16, 1853	2921	W. Tranter	"Improvements in firearms and in bullets and waddings to be used therewith." Breech-loaders. Revolvers. Grooved bullets.	Dec. 16, 1860	See Chap. XVIII.
Dec. 16, 1853	2924	T. Williams	"An improved revolving pistol."	Dec. 16, 1856	

This patent protected a single-action revolver lock-mechanism, in which a projection on the hammer engaged the nipple-partitions as a cylinder-bolt; an additional bolt was formed on top of the trigger, and a removable side-plate gave access to the lock mechanism. No specimen of Williams's revolver is known to have survived, but the Specification drawings showed a weapon of similar appearance to James Webley's "Longspur" arms. (See Plate 22.)

Patent Date	Number	Name of Patentee	Title	Expiry Date	Remarks
Apr. 4, 1854	768	J. Bentley	"Improvements in breech loading firearms." Revolvers.	Apr. 4, 1857	

This patent was primarily devoted to a double-trigger revolving rifle; the forward trigger (which was part of the hammer) was pressed forward to cock the weapon, whilst the rear trigger fired it. Features from Bentley's earlier Br. Pat. No. 960/1852 (*q.v.*) were mentioned as applicable to revolving the cylinder, and the rifles made under this protection are, though rare, extremely pleasant and effective to use.

In his Complete Specification, Bentley also claimed a "ramrod actuated by a screw", which is sometimes found embodied in the self-cocking revolvers made under his earlier patent, in place of a lever-rammer. In these screw rammers, a thread with a very quick pitch is cut around the actual ram, and a small rotatory motion of the thumb-piece visible at the top of Plate 33 moves the ram deep into any chamber aligned with it.

Patent Date	Number	Name of Patentee	Title	Expiry Date	Remarks
Apr. 27, 1854	955	J. H. Johnson. *A communication*	"Improvements in revolving firearms."	Apr. 27, 1861	

John Henry Johnson was a London patent agent, and did not name the inventor communicating this patent to him. Protection was secured for a six-chambered, single-action *breech-loading* pin-fire revolver (using metallic cartridges with integral ignition), which had actually been patented in France, twelve days earlier, by Eugene Lefaucheux. The latter was a famous Parisian gunmaker (whose father, Casimir, had shown breech-loading pin-fire pepperboxes at the Great Exhibition of London, 1851), and the French Navy adopted this pin-fire revolver, as the official issue weapon, two years later. Lefaucheux's design had a service life of more than twelve years, and specimens are quite rare today.

Patent Date	Number	Name of Patentee	Title	Expiry Date	Remarks
May 3, 1854	993	W. W. Richards	"Improvements in loading certain kinds of firearms." Ramrod for revolvers.	May 3, 1857	

This patent protected the rammer visible in Plate 34; here the ramrod slid horizontally in guides attached to the pistol barrel, and was worked to-and-fro by a thumb-operated pinion engaging a rack cut at one end of the ramrod. Use was principally confined to Westley Richards revolvers of the type protected by Eng. Pat. No. 14027/1852, but both Tranter's and Adams's self-cocking revolvers have been found as fitted with this device.

Patent Date	Number	Name of Patentee	Title	Expiry Date	Remarks
May 22, 1854	1136	H. S. Rogers (Provisional protection). *A communication*	"Improvements in firearms." Revolvers.	Nov. 22, 1854	

Henry Rogers was probably a patent agent (he gave his address as New Oxford Street, London), and named Eli Whitney as his principal. The latter was a famous American gun-manufacturer, who had pioneered the production of firearms with interchangeable parts, at Whitneyville, Ct.; this patent proposed the manufacture of revolver frames and butts in one piece, and outlined a lock-mechanism for use in them. In all probability, if a Complete Specification had been filed, it would have anticipated that of Whitney's U.S. Patent No. 11447 of August 1st, 1854.

Patent Date	Number	Name of Patentee	Title	Expiry Date	Remarks
Aug. 1, 1854	1694	W. E. Newton. *A communication*	"Improvements in the construction of repeating firearms."	Aug. 1, 1857	

William Edward Newton (of the "Office of Patents", Chancery Lane, London), was a professional patent agent, and did not name his principal in this application. The patent protected various clock-work revolvers, capable of discharging all of their chambers automatically.

Patent Date	Number	Name of Patentee	Title	Expiry Date	Remarks
Sept. 11, 1854	1976	J. Rigby	"Improvements in firearms and guns and in wadding to be used therewith."	Sept. 11, 1857	*Inter alia,* a revolver rammer; see Chap. VII.
Oct. 28, 1854	2297	E. Lindner	"Improvements in revolving breech firearms and magazine" [*sic*].	Oct. 28, 1861	

Edward Lindner, of New York, patented a revolving rifle, with tube magazines in the stock feeding the cylinder. A rod linked to the hammer rotated the cylinder as the hammer was cocked, by engaging grooves cut in the walls of the arbor passage. A preference for needle-fire cartridges was expressed, and one or two specimen arms are known to have survived.

Patent Date	Number	Name of Patentee	Title	Expiry Date	Remarks
Nov. 23, 1854	2473	C. Crickmay	"Improvements in single and repeating or revolving firearms and in the mode of attaching bayonets to breech loading firearms."	Nov. 23, 1857	

Charles Crickmay (of Lozells, Handsforth, Staffs.) was a gun-manufacturer, and here patented a frail gas-seal revolving gun, in which the only connection between barrel and stock was the arbor. A radially-projecting lever (behind the cylinder) was jerked down by the user both to revolve the cylinder and to cam it forward into a gas-seal upon the barrel.

Patent Date	Number	Name of Patentee	Title	Expiry Date	Remarks
Dec. 11, 1854	2602	W. J. Harvey	"Improvements in firearms when revolving barrels are used." Revolvers; lever ramrod; safety-bolt.	Dec. 11, 1861	

Effectively, this was a patent of improvement to Harvey's earlier Br. Pat. No. 1298/1853 (*q.v.*), and related to both revolving rifles and pistols. Basically, his improvement was only to fit a safety-device to the safety-catch on such arms, but an ingenious detachable rammer was also claimed; this device served as turnscrew and nipple key also, and was to be kept in a stock compartment when used with revolving rifles.

Six-chambered pistols of the type shown in Plate 34 (*lower*) were produced under this patent, and can be found in both 54- and 76-bore calibres; they are normally engraved "Harvey's Patent Breech Loading Revolving Pistol" on the top barrel-flat, with a serial number stamped between the second and third words. Serial numbers in the 3700–4000 range were the only numbers noted on this model, and their inventor appears to have had the arms made in Birmingham and to have been the only retail outlet for them. Barrel lengths vary between 4 in. and 5½ in., and have an unusual eight-grooved rifling, twisted to the left; the barrel itself is keyed on to the forward end of the arbor, like a Colt barrel.

Patent Date	Number	Name of Patentee	Title	Expiry Date	Remarks
Dec. 15, 1854	2645	R. Adams	"Improvements in firearms called revolvers."	Dec. 15, 1857	Rammer, and revolver details; see Chap. VII.
Dec. 21, 1854	2693	W. Greener	"Improvements in repeating military rifles, carbines and pistols and in cartridges to be used therewith." Revolvers. Breech-loaders. Cartridges.	Dec. 21, 1857	

This was the great William Greener's only excursion into the field of patents for revolving arms, and protected a solid-framed, gas-seal revolving rifle. The barrel slid to-and-fro in a stock-tube, and mated with stout cones protruding at the mouth of each chamber; a crank on the side of the stock moved the barrel, and no mention was made in the Complete Specification as to whether the cylinder was to be mechanically rotated, or turned by hand.

Patent Date	Number	Name of Patentee	Title	Expiry Date	Remarks
Feb. 20, 1855	374	F. B. E. Beaumont	"Improvements in firearms called revolvers." Lock.	Feb. 20, 1862	A revolver lock mechanism; see Chap. VIII and Fig. 14.
Mar. 3, 1855	479	T. W. Carter. *A communication*	"New and useful improvements in repeating firearms."	Mar. 3, 1858	

Timothy Walker Carter gave his address as at Massachusetts, U.S.A., and named Joshua Stevens (of the same State) as communicant of this patent.

The Specification claims covered a single-action percussion revolver, in which the trigger-pull (after dropping the hammer) was used to rotate the cylinder for the next shot; this method was practicable, because the charges were served by one nipple (mounted in the frame), and the hammer-nose did not therefore jam the cylinder on discharge. Instead of copper percussion caps, ignition was effected by a roll of paper caps (like those used in modern toy pistols) which were successively fed across the single nipple, each time that the hammer was cocked; the flame from each cap had to jump a gap between standing breech and cylinder, to penetrate pin-hole ignition passages drilled through the rear of the chambers.

This type of "tape primer" ignition had been patented by Doctor Edward Maynard (as U.S. Patent No. 4208, dated September 22nd, 1845), of Washington, D.C., and enjoyed considerable popularity in the United States. Arms using this ignition were made by the Massachusetts Arms Company, of Chicopee Falls, Mass. (amongst others), and Joshua Stevens was a designer with that Company; at the time of filing this British patent, the Company was entirely barred (with all other American manufacturers) from making mechanically-rotated revolvers in America. This situation had arisen through a successful patent infringement action brought against the Company by Colt; the Colonel alleged that Leavitt revolvers made by the Company, using bevel gears to rotate their cylinders (under Edwin Wesson's U.S. Patent No. 6669, of August 28th, 1849), were an infringement of his own U.S. Patent No. 9430X of October 22nd, 1835, and the Circuit Court of the U.S., District of Massachusetts, upheld Colt's contention. The Massachusetts Arms Company paid heavy compensation to Colt, and desisted from the manufacture of Wesson's revolver until 1857, when Colt's patent expired and American revolver manufacturers became free to take up mechanically-rotated cylinders again. Stevens's Specification to this patent shows, incidentally, cylinder rotation by a Colt-type pawl, and not by Wesson's bevel gears; no British-made specimens of Stevens's revolver are known today, and the commercial purpose in this patent is not clear.

Patent Date	Number	Name of Patentee	Title	Expiry Date	Remarks
Apr. 3, 1855	760	J. Brazier	"Improvements in revolving or repeating firearms." Ramrods.	Apr. 3, 1858	A revolver rammer; see Chap. VIII and Fig. 15.
Apr. 24, 1855	911	W. W. Richards	"Improvements in repeating or revolving firearms." Locks, etc.	Apr. 24, 1858	

This patentee (see also Eng. Pat. No. 14027/1852 and Br. Pat. No. 993/1854) now protected a six-shot, single-action, percussion revolver, which used a simple lock-mechanism, like a conventional shot-gun or rifle lock of the day, mounted on a plate formed in the frame; both pawl and cylinder-bolt were mounted upon the tumbler. Such arms are even more rare than W. W. Richards's self-cocking pistols, but are believed to have been made in 52- and 60-bore calibres, having 6-in. barrels rifled with twenty-one fine grooves; serial numbers below 100 are known, and the rammer protected by this patentee's Br. Pat. No. 993/1854 was fitted.

Patent Date	Number	Name of Patentee	Title	Expiry Date	Remarks
May 24, 1855	1175	S. E. Robbins. *Partly communicated*	"New and useful improvements in firearms." Revolvers.	May 24, 1858	A pepperbox; see Chap. III.
June 13, 1855	1346	F. C. Bakewell. *A communication*	"Improvements in rotating breech firearms."	June 13, 1858	

Frederick Collier Bakewell (of 6 Haverstock Terrace, Hampstead) named Josiah Ells, of America, as communicating this patent to him, and was probably a British patent agent; in 1857, he assigned the patent jointly to Ells and to James Maslin Cooper, of Pittsburg, Penn., U.S.A.

The patent claimed a method for mounting a revolver cylinder, and an ingenious "hesitating" self-cocking lock-mechanism. No British-made specimens of this weapon are known, but Cooper later formed the "Cooper Firearms Co." (at Frankford, Pa.), which manufactured five-shot, double-action percussion revolvers, similar in appearance to the "Navy" Colt pistol, from 1863 until 1869; no less than five U.S. patents were cited upon the barrels of these arms, and it is probable that Bakewell's British protection was the equivalent to J. Ells's U.S. Patent No. 10812 of April 25th, 1854, which was amongst those five patents.

Patent Date	Number	Name of Patentee	Title	Expiry Date	Remarks
June 14, 1855	1359	J. Enouy	"The means of 'Removal' of every Rotary or Revolving 'Barrel or Cylinder' containing Chambers from all Revolver Pistols, Guns, and Firearms and the 'Substitution' in their Place by another and other 'Barrels or Cylinders' in succession."	June 14, 1858	

Joseph Enouy (residing at No. 31 Denbigh Place, Pimlico) patented the idea for increasing the potential fire-power of revolvers by attaching two or more cylinders to them, and in such a manner that an empty cylinder might readily be replaced by a loaded one, at need; basically, he envisaged mounting the cylinders rotatably in a framework that could itself be rotated to bring them successively in position behind the barrel.

Two versions of Enouy arms have survived, one being considerably more rare than the other. A six-cylindered, thirty-six shot transition-type revolver (see Chapter IV) can be seen in a number of collections, and is known to American collectors as the "Ferris Wheel" model; less well known is a possibly unique twin-cylindered rifle (No. 301, in the Museum of Artillery, Turin) marked "Joseph Enouy London Patent Multiplying No. 1", illustrated as Plates 628/9 in H. L. Blackmore's monumental *Guns & Rifles of the World* (B. T. Batsford, Ltd., 1965). Both types of arm were clearly envisaged in the Complete Specification to this patent.

Patent Date	Number	Name of Patentee	Title	Expiry Date	Remarks
July 11, 1855	1552	T. G. Treeby	"Improvements in revolving firearms and cannon."	July 11, 1862	

Thomas Gardener Treeby claimed the use of an endless chain of chambers (flexibly connected into a circle), in place of a solid cylinder, drilled to form chambers; the chain was to be rotated behind a single fixed barrel. A rifled version was quite energetically promoted by the inventor; see remarks to Br. Pats. Nos. 2629/1855 and 1306/1858.

Patent Date	Number	Name of Patentee	Title	Expiry Date	Remarks
July 28, 1855	1722	J. Kerr	"Improvements in revolver firearms." Lever ramrod.	July 28, 1862	A revolver rammer; see Chap. X and Plate 13.
Aug. 16, 1855	1857	T. Williams. (Provisional protection)	"Improvements in breech loading firearms and in the mode or method of loading the same."	Feb. 16, 1856	

Thomas Williams (a Liverpool gunmaker) here patented a revolver rammer connected to a hinged trigger guard; teeth cut on the ram were engaged by a pinion formed about the trigger-guard pivot, and the ram was thus forced into a chamber as the guard was turned forward. This idea was later used by Belgian gunmakers (in a Colt-type revolver), and also by the American firm, Allen & Wheelock, of Worcester, Mass., manufacturing six-shot percussion revolvers under Ethan Allen's U.S. Pats. Nos. 16367 (of January 13th, 1857), 18836 (of December 15th, 1857), and 21400 (of September 7th, 1858). British revolvers embodying Williams's patented rammer have been reported, but no specimen could be traced for illustration here.

Patent Date	Number	Name of Patentee	Title	Expiry Date	Remarks
Sept. 6, 1855	2018	C. Pryse and P. Cashmore	"Improvements in repeating fire-arms." Locks, ram-rod, safety-bolt.	Sept. 6, 1858	See Chapters XIV and XV.
Sept. 13, 1855	2069	J. Blissett, (Provisional protection)	"Improvements in the construction of revolving chamber firearms." Ramrods.	Mar. 13, 1856	

Something has been said of this rammer (in Chapter XIV), and the lack of any illustration to Blissett's Provisional Specification will probably prevent any final decision as to whether or not his rammer was used in gas-seal revolvers of the "Lang" type.

The patentee stated of his device that the lever employed was "of the second order", the fulcrum being attached to the barrel-lug. Near this fulcrum, the rammer-lever was to be formed with a projecting lug, to which was to be connected a short link, which was also jointed to the actual ram. This ram slid in a hole through the barrel-lug, and the action of the link allowed the ram always to preserve its parallel position (*vis-à-vis* the bore) as the lever moved on its pivot in the action of ramming home charges.

Patent Date	Number	Name of Patentee	Title	Expiry Date	Remarks
Nov. 7, 1855	2497	C. Hanson	"Improvements in firearms." Revolvers.	Nov. 7, 1858	

Charles Hanson (a Pimlico gunsmith) proposed use of a single-action lock mechanism, in a solid-framed revolving percussion gun; a grip-safety prevented accidental discharge, and a lever-rammer was also shown. Use of these ideas in pistols was claimed, but no gun or pistol on Hanson's system has yet been identified.

Patent Date	Number	Name of Patentee	Title	Expiry Date	Remarks
Nov. 21, 1855	2629	T. W. G. Treeby (Provisional protection)	"Improvements in revolving firearms."	May 21, 1856	

This patent improved the chain gun described under Treeby's Br. Pat. No. 1552/1855, by proposing to slide the barrel of his weapon, to-and-fro, to effect a gas-seal with the chambers; the patent also claimed a single-shot breech-loading rifle or gun, and a cap-magazine feeding direct to the nipples. H. L. Blackmore (*British Military Firearms*, Herbert Jenkins, 1961) records that reciprocating motion was given to the barrel, in production arms, by connecting barrel to frame by a tapered steel sleeve; this sleeve had a handle to rotate it, and move the barrel, and on some rifles the sleeve also cocked the hammer. On most Treeby guns surviving today, however, a separate cocking-piece is fitted to the single-action lock-mechanism as shown in Plate 37 (*top*). (See remarks to Br. Pat. 1306/1858, as to tests on Treeby guns.)

Patent Date	Number	Name of Patentee	Title	Expiry Date	Remarks
Jan. 21, 1856	161	G. A. Blittkowski	"Improvements in repeating firearms." Self-charging apparatus.	Jan. 21, 1859	

Gustav Adolph Blittkowski (a New Yorker) proposed a side-hammer, gas-seal, percussion, revolving rifle, in which reciprocating movement of the cylinder successively rammed home charges (from a tube magazine in the stock) into its chambers; the simple act of cocking the hammer, therefore, was to feed a cartridge into a chamber, ram it home, feed a cap to a nipple, and rotate the cylinder; a needle-gun variant was claimed.

Patent Date	Number	Name of Patentee	Title	Expiry Date	Remarks
Feb. 8, 1856	337	T. Restell	"Improvements in breech-loading and revolving firearms and in cartridges."	Feb. 8, 1859	

Thomas Restell (a chronometer maker in the New Kent Road, London) secured a number of firearms patents, one or two of which were moderately successful. In this patent (which is not known to have been developed) a hammerless revolving gun was proposed, for use with rimless metallic centre-fire cartridges; a straight-pull trigger operated the self-cocking lock-mechanism, and a gas-seal was effected by mating each chamber, successively, to a vulcanized rubber washer on the breech of the barrel.

Later in this year, an interest in the patent was recorded at the British Patent Office, in the names of "Brooman, Clement and Mount"; Brooman was probably the patent agent already mentioned in relation to Br. Pat. No. 583/1852 (*q.v.*), but the identity of his partners is still unknown.

Patent Date	Number	Name of Patentee	Title	Expiry Date	Remarks
Feb. 13, 1856	370	W. E. Newton. *A communication*	"Improvements in the construction of firearms." Breech-loaders. Revolvers.	Feb. 13, 1859	

W. E. Newton (see Br. Pat. No. 1694/1854) acted here for Simon French Stanton, of America, although he did not specifically name him. A radial revolver gun was proposed, which had the charge chambers hanging by hinges from a plate revolving upon a vertical axis. The chambers were successively offered up to the barrel for firing, and freedom from accidental discharge of other chambers was claimed for the arrangement; Stanton himself secured U.S. Patent No. 14780 of April 29th, 1856, for this weapon.

Patent Date	Number	Name of Patentee	Title	Expiry Date	Remarks
Mar. 15, 1856	627	J. Rice and W. Rice	"Improvements in breech-loading repeating guns and rifles."	Mar. 15, 1859	

James Rice (a surgeon, at Foley Place, London) and William Rice (a Lieutenant in the H.E.I.C.S.) patented an unusual centre-hammer, percussion revolving gun, complete with telescopic sight.

The easiest way to envisage this weapon is to picture a percussion rifle with an old-fashioned rolling-pin mounted transversely through the breech; chambers were drilled across the "rolling-pin" (but at right angles to one another), and on the periphery of the "rolling-pin" was cut a spiral groove or thread. This thread was so pitched that manual rotation (by a handle on the right end of the "rolling-pin") would successively bring chambers into alignment with the barrel for firing, as the "rolling-pin" travelled across the weapon, to the right. Balancing such a rifle, so as to make use of the telescopic sight, must have been extremely difficult.

Patent Date	Number	Name of Patentee	Title	Expiry Date	Remarks
May 17, 1856	1177	C. C. Tevis. (Provisional protection)	"An improved revolver."	Nov. 17, 1856	

Lieutenant-Colonel Charles Carroll Tevis (of Paris) protected a ring-trigger, self-cocking percussion revolver, with a two-piece cylinder from which a nippled backing-plate was to be removed, for loading the chambers from their rear. This plate had coned plugs fitting into each chamber (to stop gas-leak, on discharge), and the barrel was vertically pivoted; Tevis's revolver must have been incredibly dangerous to fire, and no specimen of it is known today.

Patent Date	Number	Name of Patentee	Title	Expiry Date	Remarks
July 8, 1856	1608	A. V. Newton. *A communication*	"Improvements in repeating firearms."	July 8, 1863	

Alfred Vincent Newton shared quarters with W. E. Newton (see Br. Pat. No. 1694/1854) as a professional patent agent. He did not name the communicant of this patent, but the Complete Specification was a British equivalent to that in Henry Simeon North's U.S. Pat. No. 15144 of June 17th, 1856; North registered an interest in the British patent (at the Patent Office), during 1857.

A ring-trigger, self-cocking, gas-seal, percussion revolver was claimed, in which the ring-trigger cocked the hammer, and the weapon was then discharged by pulling a smaller trigger on front of the ring.

The American patent above referred to was but one of several, leading eventually to the production of an efficient gas-seal revolver, that was sold in thousands by the Savage Repeating Firearms Corporation, during the American Civil War. (See also remarks to Br. Pat. No. 329/1852.)

Patent Date	Number	Name of Patentee	Title	Expiry Date	Remarks
Aug. 16, 1856	1913	W. Tranter	"Improvements in firearms." Locks.	Aug. 16, 1870	A double-action revolver; see Chap. XVIII.
Aug. 30, 1856	2021	H. Conant	"A new and useful improvement in firearms." Breech-loader.	Aug. 30, 1863	

Hezekiah Conant (of Connecticut, U.S.A.) was primarily interested in protecting a metallic gas-seal breech ring for Sharps (and other) breech-loading rifles; however, application of the idea to revolvers was claimed.

PLATE 33. RARE BRITISH PERCUSSION REVOLVERS

(*Top*) Joseph Bentley's self-cocking percussion revolver, in 56 bore, retailed by "John Calvert. Leeds". Serially-numbered A502, this pistol has the hammer-safety claimed in Bentley's Br. Pat. No. 960/1852 and his No. 768/1854 protected the screw-rammer. *Andrews Collection.*

(*Lower*) Joseph Bentley's double-action percussion revolver, in 74 bore calibre, retailed by "J. Bentley & Sons. Liverpool". This pistol was made under Bentley's Br. Pat. No. 2657/1857, and has not the lock-mechanism shown in Figure 43, but one of the variants claimed in his Specification. Serially-numbered 39. *J. B. Bell Collection.*

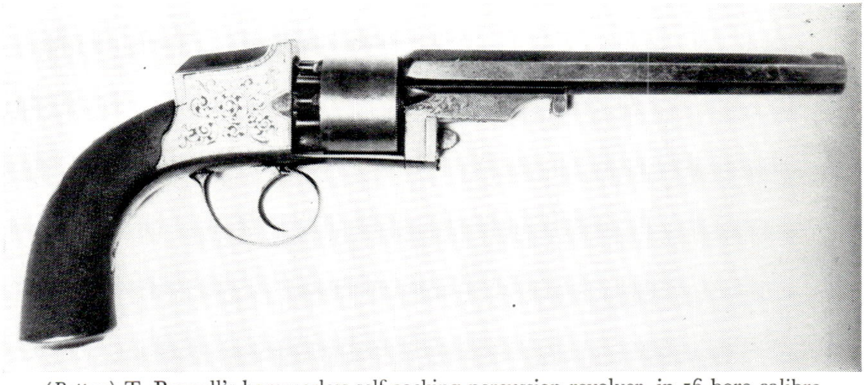

(*Bottom*) T. Pennell's hammerless self-cocking percussion revolver, in 56 bore calibre, retailed by "H. Adkin, Bradford". Made under Pennell's Br. Pat. No. 1038/1853. Detachable rammer missing. *Private Collection.*

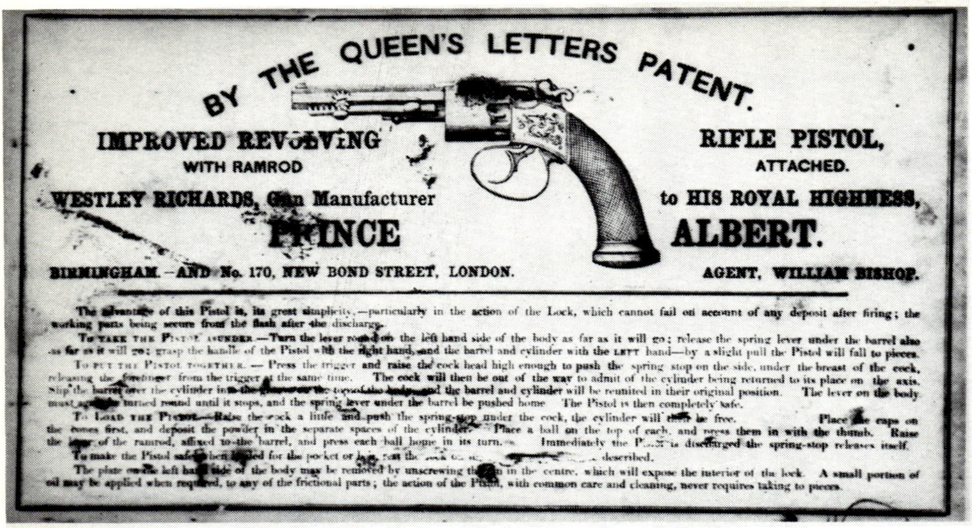

PLATE 34. HAMMERLESS SELF-COCKING PERCUSSION REVOLVERS

(*Top*) Case-lid label for Westley Richards' self-cocking percussion revolver. (After Br. Pats. 14,027/1852 and 993/1854.) *P. A. Bedford Collection.*

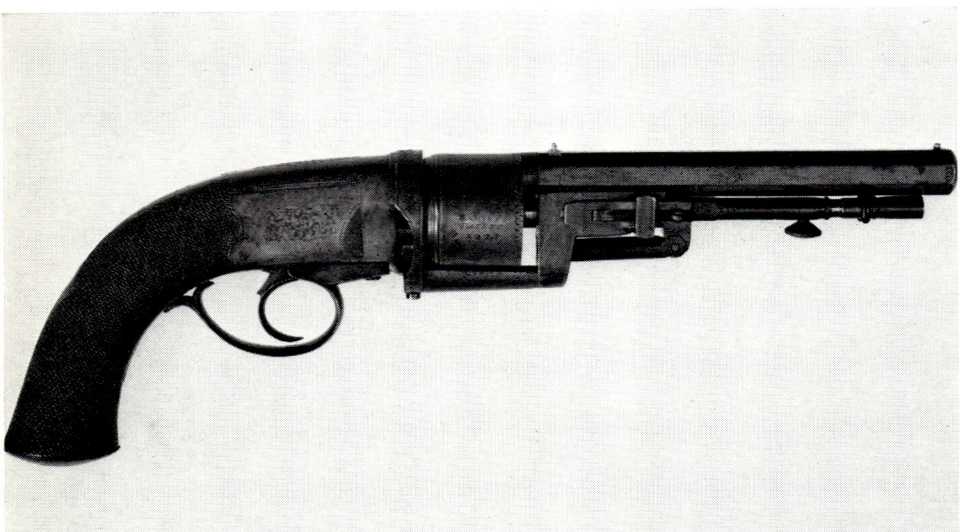

(*Lower*) W. J. Harvey's second model hammerless percussion revolver, to Br. Pat. 2602/1854. In 54 bore calibre, and serially-numbered 3922. *Crown Copyright. By permission of the Master of the Armouries.*

PLATE 35. LANG-TYPE REVOLVING RIFLE

This six-shot single-action gas-seal percussion rifle is in 24 bore calibre, and has a $26\frac{1}{2}''$ barrel, rifled with twenty grooves. All parts are numbered "1", and the frame is marked only "New Patent Revolving Rifle – London". *Photograph by Ray Riling.*

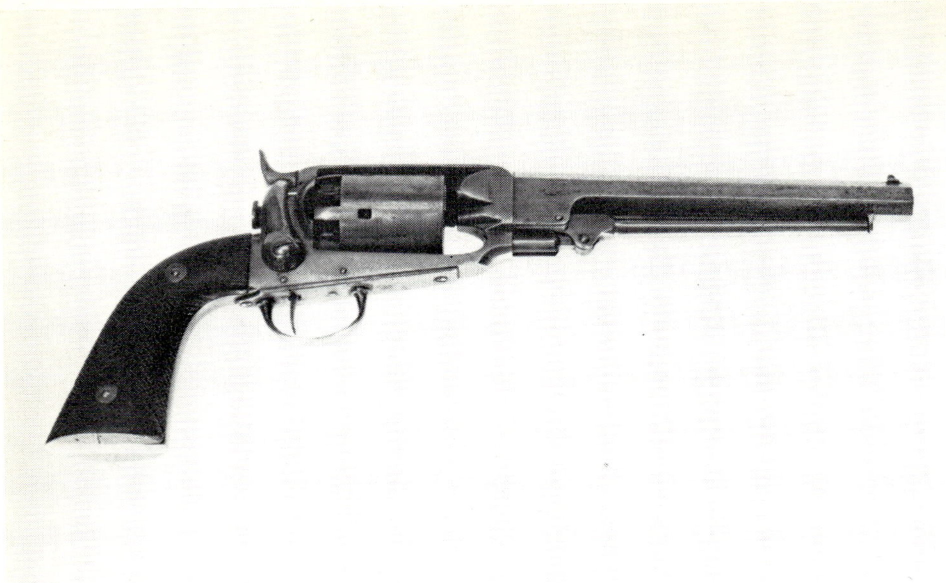

PLATE 36. AMERICAN PERCUSSION REVOLVERS
(*Top*) B. F. Joslyn's five-shot ·44 calibre single-action percussion revolver. After U.S. Pat. 20160 of 1858. *McCarthy Collection.*

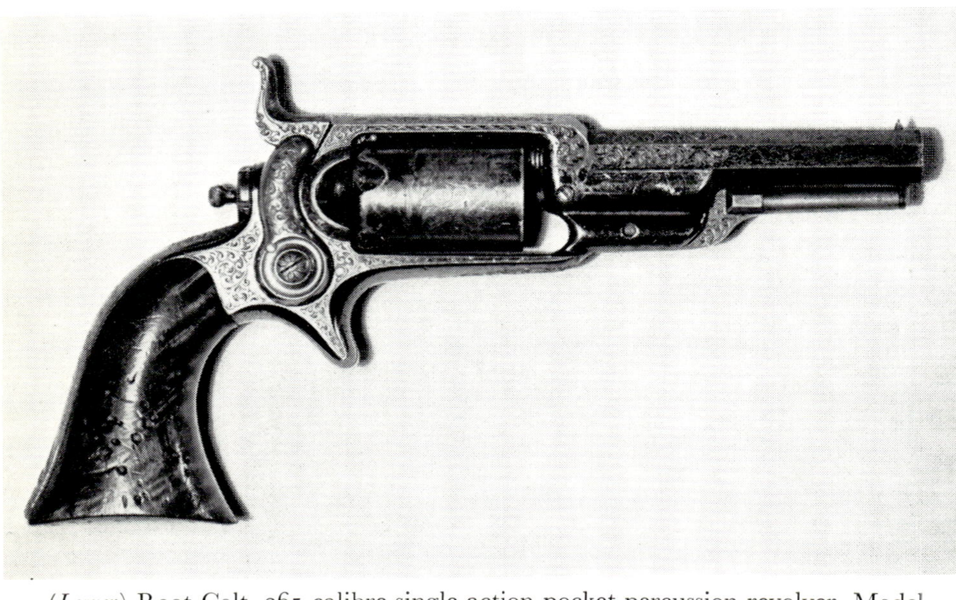

(*Lower*) Root-Colt ·265 calibre single-action pocket percussion revolver, Model of 1857. Serially-numbered 882, and engraved "George Peabody. From the Inventor, Sam'l Colt". Peabody (an American business man, in London) apparently received the gift on December 27th, 1856. *Hugh E. Hayes Collection.*

PLATE 37. UNUSUAL PERCUSSION ARMS
(*Top*) T. W. G. Treeby's single-action percussion chain gun. After Br. Pats. 1552/1855 and 2629/1855. *Private Collection.*

(*Lower*) C. E. Wallis's double-loaded percussion revolver, made under Br. Pat. 264/1864. *J. B. Bell Collection.*

ESTABLISHED ON OLD LONDON BRIDGE, A.D. 1700.

GEORGE AND JOHN DEANE,

WHOLESALE AND EXPORT

IRONMONGERS, CUTLERS, JEWELLERS, AND HARDWAREMEN

STOVE, RANGE, AND FENDER MANUFACTURERS.

SADDLERS, AND GUN MAKERS,

AND GENERAL FURNISHING WAREHOUSEMEN

OPENING TO THE MONUMENT,

46, KING WILLIAM STREET, LONDON BRIDGE.

CU[illegible]

[illegible] STOVES.

[illegible]ore than Fou[illegible]ousand of these stoves were sold during the last winter. They have the most unqualified recommendation of the first medical practitioners, because, while they yield a genial heat, they, at the same time carry off all noxious gases, and ventilate the apartments which they warm. They are sold in various sizes for chambers, libraries, and conservatories, from 19s. and upwards. They are also manufactured with two, three, and four cylinders for halls, chapels, or churches.

Prospectuses and testimonials may be had and the stove seen in operation daily.

IRON BEDSTEADS.

It seems almost superfluous to say any thing in behalf of Iron Bedsteads. Every one must admire their neatness; and their cheapness and durability are strong and universal recommendations. More than all, their cleanliness (in that vermin will not lodge in them, and that they may be thoroughly washed as often as you please without damage,) is a peculiarity which every careful and experienced housekeeper will know how to appreciate. Their portability makes them invaluable to the Traveller, the Emigrant, and the Soldier. And as George and John Deane are themselves manufacturers of these articles, they are enabled to offer them to their friends at prices especially low.

A [illegible]carefully [illegible]ted Stock [illegible] Harness a[illegible] Saddlery, [illegible] for private use and exportation, warranted of Deane's manufacture.

GEORGE AND JOHN DEANE'S

LONDON BRIDGE,

GUN AND PISTOL MANUFACTORY,

30, KING WILLIAM STREET, LONDON BRIDGE.

The rapid increase of their gun business rendering separate premises necessary, G. and J. Deane have Removed their Stock, workshops, &c. to the warehouse lately occupied by Messrs. Groocock and Co., Gunmakers, 30, King William-street, nearly opposite to Deane's warehouses and manufactories, No. 46, but nearer the statue of King William the Fourth. In addition to their usual stock, and that which they have purchased of Messrs. Groocock and Co.—all offered at prices unusually low—G. and J. Deane have some very exquisite specimens of their Registered Detonating Gun, and other manufactures. Their stock of pistols and shooting tackle is complete, and every article they sell is warranted.

Guns and Pistols of every description, manufactured on the Premises, and warranted of the best workmanship.

PLATE 38. "MESSRS. GEORGE & JOHN DEANE"
Advertisement from Kelly's "Post Office Directory for London, 1846" (pp. 2062/63). *Reproduced by permission of the Publishers. Copyright reserved.*

PLATE 39. Advertisement from Kelly's *London Directory*, for 1858, p. 2469.
Reproduced by kind permission of the Publishers. Copyright reserved.

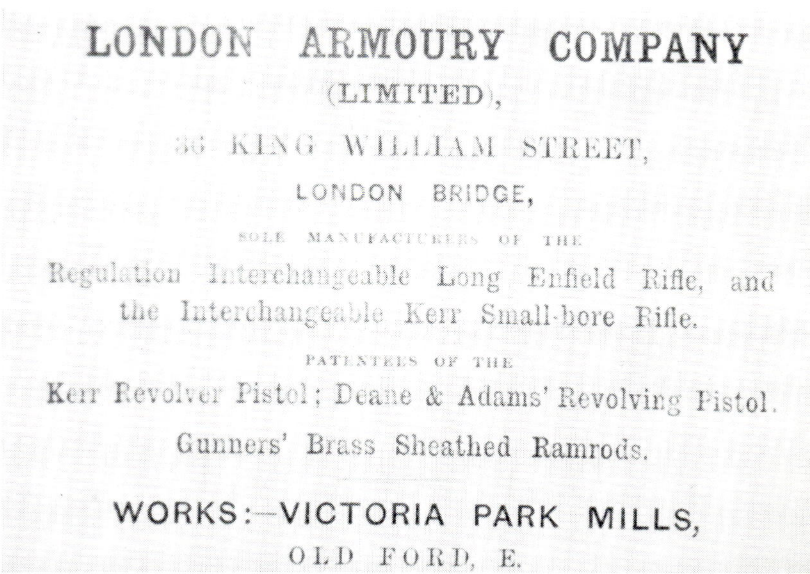

PLATE 39. Advertisement from Captain Heaton's "Notes on Rifle Shooting". London. 1864.

6 *ADVERTISEMENTS.*

PHILIP WEBLEY,

84, WEAMAN STREET, BIRMINGHAM,

PRESENT CONTRACTOR TO THE HON. BOARD OF ORDNANCE

PATENTEE OF SAFETY REVOLVING PISTOLS.

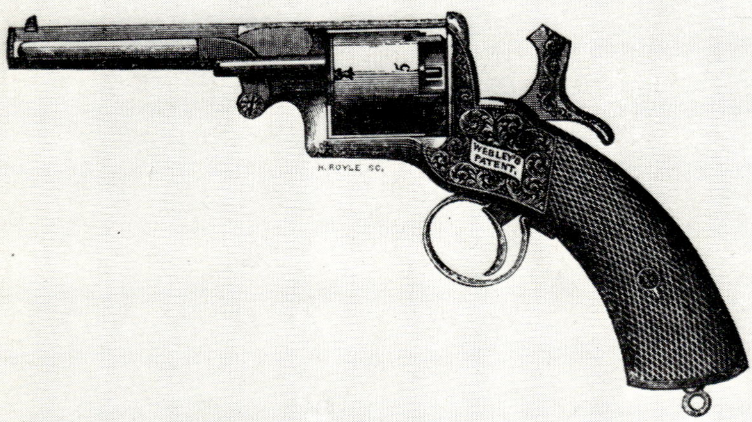

P. WEBLEY respectfully informs the public, that he is prepared to supply in any quantity his

PATENT REVOLVING PISTOLS,

which he can confidently recommend, as they embrace all latest improvements with the greatest possible simplicity of construction, and are pronounced by men of experience, both civil and military, to be most efficient weapons.

PLATE 40. Advertisement from "Gunnery in 1858" (W. Greener), featuring a Fourth Model Wesley "Longspur" percussion revolver.

Patent Date	Number	Name of Patentee	Title	Expiry Date	Remarks
Sept. 12, 1856	2134	J. T. Pitman. *A communication*	"Improvements in repeating firearms." Lock.	Sept. 12, 1859	

John Talbot Pitman filed application from the "American & European Patent Offices" (at 67 Gracechurch St., London), but did not name the communicant. A hammerless, self-cocking revolver was claimed, in which the lock-mechanism rebounded the hammer to full-cock automatically, after each shot; a device was used to take all tension from the mainspring, except when the hammer was falling.

Patent Date	Number	Name of Patentee	Title	Expiry Date	Remarks
Dec. 3, 1856	2868	H. Genhart	"Improvements in firearms and ordnance and of the cartridges or projectiles to be used therewith." Breech-loaders.	Dec. 3, 1859	

John Deane (*op. cit.*) described this inventor as an "officer of Swiss artillery", but a namesake had a gun-making business at 42 Rue aux Chevaux, Liège, and produced revolving arms of the type covered by this patent.

Genhart ten-shot pistols and rifles had radially-bored charge-carriers, turning upon a vertical axis, and were hand-rotated; a side-lever permitted rotation (by withdrawing the barrel from a gas-seal in the chamber), and unbolting of the charge-carrier was accomplished through a button release. A cartridge with integral priming was used, and fresh loaded charge-carriers could be readily fitted into the weapons.

Patent Date	Number	Name of Patentee	Title	Expiry Date	Remarks
Feb. 7, 1857	351	C. Crickmay	"Improvements in breech-loading guns or pistols," etc. Breech-loaders.	Feb. 7, 1860	

Charles Crickmay (a gun-manufacturer, at Guildford Place, Lower Kennington Lane, London) here improved a bolt-action, breech-loading under-hammer gun, which might be designed to use cartridges with a normal percussion cap embodied in them, or rounds with a special detonating tail. The cartridges were carried in a "revolving collar" embracing the breech, and dropped by gravity into the gun, when the breech-bolt was open; this cartridge-carrier might be manually or mechanically rotated, and was claimed as an improvement to an earlier patent by this inventor (No. 1499 of June 20th, 1853) which had no revolving features.

Patent Date	Number	Name of Patentee	Title	Expiry Date	Remarks
Apr. 2, 1857	921	A. V. Newton. *A communication*	"Improvements in repeating firearms."	Apr. 2, 1860	

A. V. Newton (see Br. Pat. No. 1608/1856) did not name the inventor communicating this invention to him, but (later in 1857) an interest in the patent was registered at the British Patent Office, in the name of William Montgomery Storm, a noted inventor of breech-loading rifles, who may therefore have been Newton's principal. A self-cocking revolver with "hesitating" lock-mechanism was claimed as suitable for use by horsemen, and had a metal casque surrounding its cylinder. A slot cut through the hammer permitted sight to be taken, and the mainspring also served as a pawl spring. Another unusual feature was that the cylinder had an integral arbor formed at its rear end, with a gear wheel mounted upon it for effecting rotation.

Patent Date	Number	Name of Patentee	Title	Expiry Date	Remarks
Apr. 18, 1857	1100	G. Jahn. (Provisional protection)	"Improvements in revolver firearms."	Oct. 18, 1857	

From 43 Bedford Row, London, Gustav Jahn sought provisional protection for an under-hammer, radial, needle-fire revolver, in which the hammer struck back towards the firer; a revolving magazine fed cartridges to the chambers.

Patent Date	Number	Name of Patentee	Title	Expiry Date	Remarks
Apr. 30, 1857	1213	H. Ball. (Provisional protection)	"Improvements in repeating and other firearms."	Oct. 30, 1857	

Henry Ball (a Birmingham gunmaker) filed only a short Provisional Specification relating to a lock-mechanism, a method for attaching a barrel to a stock, a lever ramrod, and "a kind of shield provided with a firearm". He is better remembered for his Br. Pat. No. 235/1858 (*q.v.*).

Patent Date	Number	Name of Patentee	Title	Expiry Date	Remarks
June 4, 1857	1574	W. Harding. (Provisional protection)	"An improvement in pistol holsters."	Dec. 4, 1857	

William Harding was well known for the revolvers made under his Br. Pats. Nos. 669/1858 (*q.v.*) and 1159/1858; see Chapter XII.

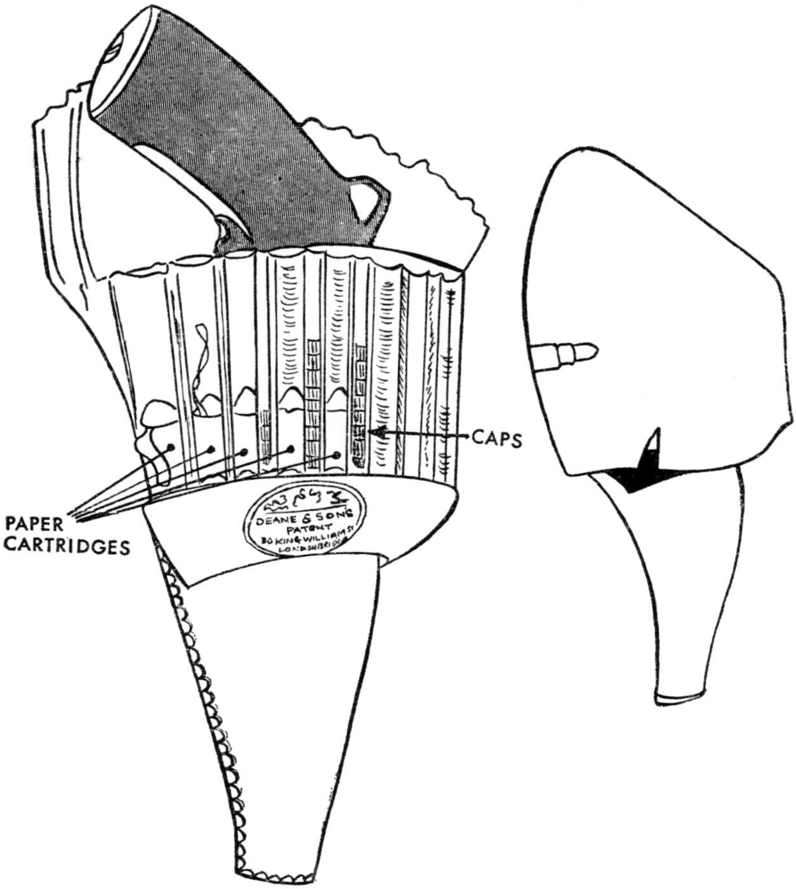

FIG. 48. *Harding's holster magazine.*
(After Br. Pat. No. 1574/1857)

In this patent, however, Harding proposed to attach a cartridge and cap magazine to a revolver holster. One surviving specimen is sketched as Fig. 48, from which it will be seen that a fan-shaped cluster of eight open-sided japanned tin tubes held the cartridges, and three smaller tubes (soldered between the larger tubes) held caps; the wearer of such a holster had only to open the flap, and slide his ammunition up the tube, and into his fingers. Specimens are rare, and the problem of weight must have defeated interest by promoter and public alike; even a 120-bore pocket revolver weighed some four pounds in Harding's holster, when accompanied by the twenty-four cartridges for which the latter was designed.

Patent Date	Number	Name of Patentee	Title	Expiry Date	Remarks
Oct. 17, 1857	2657	J. Bentley	"Improvements in firearms." Locks. Revolvers.	Oct. 17, 1860	

Use of a lock-mechanism from this patent (in double-action percussion revolvers made by P. Webley & Son) has already been described in Chapter XVII; however, as in the case of his Br. Pat. No. 960/1852 (*q.v.*), Joseph Bentley also made a few revolvers, on his own account, under this patent of 1857.

Double-action revolvers of the type shown in Plate 33 (*lower*) were made in 74-bore calibre, and had 5-in. barrels, rifled with three grooves twisted to the right; on the specimen illustrated, a circular metal plate was attached to the barrel-lug, and covered all but one of the chambers (in a manner somewhat similar to that shown in Plate 2), presumably to prevent the balls from being jarred out of the chambers on discharge. The lock-mechanism (which we have never encountered in any other type of revolver) was illustrated as Fig. 1 in Bentley's patent Specification, and is quite different from that used in the Webley arms described in Chapter XVII. A noteworthy detail is the use of a long, flat trigger spring (bearing the cylinder-bolt at one end), which was illustrated in all the lock-mechanisms claimed in Bentley's patent, but never used in the Webley version shown in Fig. 43.

No serial number above 50 has so far been noted on these arms, and the pistol illustrated is numbered 39. Their octagonal barrels were normally marked "J. Bentley & Sons, Liverpool", on the top-flat, and "Bentley's Patent No. . . ." appeared on the frame; a few similar self-cocking revolvers were also made.

Patent Date	Number	Name of Patentee	Title	Expiry Date	Remarks
Oct. 22, 1857	2690	C. Reeves	"Revolving firearms." Locks.	Oct. 22, 1860	

Charles Reeves was a Birmingham gunmaker, and supplier to the gun trade at large, who took out three British patents relating to percussion revolvers; see also Br. Pat. Nos. 3156/1857 and 1623/1858. The double-action lock-mechanism claimed in this patent is not often encountered, the patentee apparently preferring a mechanism covered by the later of his patents (*q.v.*).

Patent Date	Number	Name of Patentee	Title	Expiry Date	Remarks
Nov. 7, 1857	2824	J. Adams	"Improvements in revolver firearms." Locks. Ramrods.	Nov. 7, 1871	

John Adams (of Queen's Road, Dalston, Middlesex) was Robert Adams's brother, and made a considerable contribution to the improvement of British revolvers after the years of concern here. His percussion revolvers are rare, and, when encountered, commonly embody features both from this patent, and his later Br. Pat. No. 1758/1861 (*q.v.*). The interest in John Adams's patent of 1857 lies not in the claimed use of a firing-pin on percussion revolvers, or the sliding safety-bolt shown, or even in his rather complex double-action lock-mechanism, but in the patentee's grasp of possible improvements to existing designs. Thus his Complete Specification claimed (1) designing the lock-mechanism so that the frame cavity for it might be machined out by circular cutters, (2) retention of the arbor by the rammer, and (3) a two-part barrel/frame construction which (without infringing his brother's master patent for the one-piece revolver frame) permitted manufacture of a solid-framed revolver by simplified methods.

Patent Date	Number	Name of Patentee	Title	Expiry Date	Remarks
Nov. 17, 1857	2882	G. T. Bousfield. *A communication*	"Improvements in firearms, and in detonating compounds to be used therewith."	Nov. 17, 1860	

George Tomlinson Bousfield was a London patent agent, and did not reveal the identity of his principal. This patent claimed various methods for discharging firearms by use of a granular detonating compound, instead of the conventional copper cap, and is of interest also as illustrating (in the Complete Specification) a nipple-less Colt revolver, adapted to use that compound.

Patent Date	Number	Name of Patentee	Title	Expiry Date	Remarks
Dec. 24, 1857	3156	C. Reeves	"Improvements in repeating or revolving firearms."	Dec. 24, 1860	

In this patent, Reeves (see Br. Pat. No. 2690/1857) protected the basic lock-mechanism used in that model of his percussion revolvers most likely to be encountered; the patentee's Br. Pat. No. 1623/1858 refined this patent, and the note thereon should be studied.

Reeves here claimed only the double-action lock-mechanism, a "jointed lever loading rod", and an arbor latch.

Patent Date	Number	Name of Patentee	Title	Expiry Date	Remarks
Feb. 8, 1858	235	H. Ball	"Improvements in repeating and other firearms, a portion of which improvements may be applied to ordnance."	Feb. 8, 1861	

This patent is probably best remembered for protecting an "anti-garrotter" firearm, which was attached to a belt, and could be fired at any attacker behind the wearer. However, Ball (see Br. Pat. No. 1213/1857) also claimed, in considerable detail, double-action lock-mechanisms for percussion revolvers, a method for connecting a revolver-barrel to the frame, and a lever-rammer. Specimen arms are extremely rare, but an interest in the patent was registered at the Patent Office (during 1858) by a man named Lloyd.

Patent Date	Number	Name of Patentee	Title	Expiry Date	Remarks
Mar. 29, 1858	669	W. Harding	"Improvements in revolver firearms and in apparatus for manufacturing projectiles."	Mar. 29, 1861	

A complete revolver was claimed in this patent, but essentially it contributed only a rammer used in most of the Deane-Harding revolvers described in Chapter XII.

Patent Date	Number	Name of Patentee	Title	Expiry Date	Remarks
May 24, 1858	1159	W. Harding	"Improvements in revolver firearms."	May 24, 1865	A revolver; see Chap. XII.
June 9, 1858	1306	T. W. G. Treeby. (Provisional protection)	"Improvements in revolving firearms and cannon and cartridges."	Dec. 9, 1858	

In an unillustrated Provisional Specification, Treeby proposed improvements to his chain-guns (see Br. Pats. Nos. 1552/1855 and 2629/1855), as to the hinging of the chambers, a rammer, a cartridge, and the use of separate strikers for each chamber.

According to a booklet published in the following year, Treeby succeeded in arranging a public trial, during June, 1854, in the presence of newspaper reporters, for his chain-gun, and for a single-shot breech-loading rifle protected by his Br. Pat. No. 2310/1858; the latter had no revolving features. On July 15th of that year, he also demonstrated his chain-gun at the Small Arms School, Hythe, before General Hay (First Instructor of Musketry), Colonel Wilford (Second Instructor) and other army officers; one specimen target is shown in Treeby's booklet, for a burst of 30 shots (in 1 min. 20 sec.) fired at 200 yards range, and with $\frac{3}{4}$ dr. charges; this card showed a creditable pattern, but no further official interest was shown. The booklet gave Treeby's address as being 1 Westbourne Terrace Villas, Westbourne Square, Paddington, but a contemporary calling card has this address struck out, in favour of No. 12 Blomfield Terrace.

Patent Date	Number	Name of Patentee	Title	Expiry Date	Remarks
July 15, 1858	1593	R. Brazier	"Improvements in repeating firearms."	July 15, 1861	

Richard Brazier was a member of the famous firm of Wolverhampton gun-lock manufacturers, already noticed in Chapters VII and VIII.

In this patent he claimed a double-action percussion revolver and a lever-rammer for such arms. The lock-mechanism of the former required a finesse in manufacture probably only available to its inventor, and specimens of the revolver are very rare; the rammer had a superficial resemblance to that claimed in J. Brazier's Br. Pat. No. 760/1855 (see Fig. 15), but was of genuine invention.

Patent Date	Number	Name of Patentee	Title	Expiry Date	Remarks
July 19, 1858	1623	C. Reeves	"Improvements in repeating firearms."	July 19, 1861	

This was a patent of improvement to Reeves's Br. Pat. No. 3156/1857 (*q.v.*), for a double-action revolver lock-mechanism, and related to self-cocking operation of that lock-mechanism; it also claimed a quick-thread arbor release. Reeves's percussion revolvers are quite rare, but normally embody features of these two later patents; only five-shot 54-bore pistols have been noted, which as a rule are serially-numbered in a range (between 500 and 2000) which does not accurately reflect total production. Most weapons are engraved with the name of their maker (with the legend "Reeves Patent" on the left of their frame, and the serial number on the right), but may be encountered in revolver cases bearing the labels of retailers. A 6½-in. barrel (rifled with five grooves, twisted to the right) is screwed into a solid frame, and an inspection plate is fitted on the left of that frame. The plate is sometimes dovetailed into position (and has to be slid out of place), whilst a safety-bolt may be found on the right of the action. An exceptional pistol (in the Giggal Collection) is marked "Thomas Sylven, London" and has a serial number ("London Patent No. 39372") normally found on those Webley percussion revolvers described in Chapter XVII. Superficially similar in appearance to those arms described above, this weapon lacks any externally visible feature claimed in Reeves's Br. Pat. No. 1623/1858, and may actually have been made under his preceding patent (No. 3156/1857).

Patent Date	Number	Name of Patentee	Title	Expiry Date	Remarks
July 20, 1858	1634	T. Bailey	"Improvements in repeating firearms."	July 20, 1865	

Thomas Bailey (formerly of New Orleans, U.S.A.) filed his application from 17 Great Russell Street, Birmingham, and his occupation is unknown. The patent protected a double-action, side-hammer percussion revolver with a solid frame constructed in two pieces; the cylinder turned on centres, and not upon an arbor, and considerable ingenuity was expended in so encasing the lock-mechanism that penetration of fouling into it was virtually impossible. Despite the long period for which Bailey's patent was maintained, specimens of his revolver are extremely rare; one observed weapon was a five-shot pocket revolver (with 5-in. barrel) in 120-bore calibre, which had been retailed by the Birmingham firm of Swinburn & Son.

Patent Date	Number	Name of Patentee	Title	Expiry Date	Remarks
Sept. 4, 1858	2009	A. V. Newton. *A communication* (Provisional protection)	"An improvement in firearms."	Mar. 4, 1859	

A. V. Newton (see Br. Pat. No. 921/1857) did not name the communicant of this patent; it protected a detachable carbine butt for pistols and revolvers, which rested against two pins protruding from the weapon frame, and was held by a vertical screw.

Patent Date	Number	Name of Patentee	Title	Expiry Date	Remarks
Nov. 1, 1858	2436	W. Palmer	"Improvements in firearms."	Nov. 1, 1861	

William Palmer (of New York) may have been a patent agent, or some kind of entrepreneur; his name appears upon one or two firearms patents, in circumstances making it unlikely that he was the true inventor. Here, protection was secured upon two solid-framed, self-cocking percussion revolvers, with "hesitating" lock-mechanisms. No British-made specimen arms are known, but American equivalents may exist.

Patent Date	Number	Name of Patentee	Title	Expiry Date	Remarks
Dec. 17, 1858	2896	J. Kerr	"Improvements in the construction of revolving firearms."	Dec. 17, 1872	A revolver; see Chap. X.
Dec. 30, 1858	2990	R. A. Brooman. *A communication from L. F. Devisme*	"Improvements in revolving firearms."	Dec. 30, 1861	

Brooman (see Br. Pat. No. 583/1852) here acted for the Parisian gunmaker, L. F. Devisme, to protect a hinged-frame, single-action, breech-loading revolver; the inventor did not, in fact, seriously develop this design until after the Paris Exhibition of 1867. However, the Complete Specification showed clearly all features of Devisme's revolver as eventually produced, and even proposed the use of a plate cartridge-extractor for extracting all cartridge-cases in one motion.

Patent Date	Number	Name of Patentee	Title	Expiry Date	Remarks
Jan. 26, 1859	242	J. Kerr	"Improvements in the construction of revolving firearms."	Jan. 26, 1866	A revolver; see Chap. X.
July 8, 1859	1622	F. A. Le Mat	"Improvements in the construction of revolving or repeating firearms."	July 8, 1862	

François Alexandre Le Mat was an American physician, from Louisiana, and here patented the idea of a revolver capable of firing two different kinds of charge, at the user's election; Le Mat's method was to mount the ball-firing cylinder of a single-action revolver upon a single large-calibre barrel for discharging buck-shot. A moveable hammer-nose permitted the firer to select cylinder, or shot-barrel, by a flick of the thumb. In this patent, Le Mat proposed to use a sliding rammer for loading the cylinder, and a jointed rammer for charging the shot-barrel; on production versions of his revolvers, however, a jointed rammer serves the chambers, and has a removable ramrod for the shot-barrel inserted into its lever. (See also remarks to Br. Pat. No. 1081/1862.)

Patent Date	Number	Name of Patentee	Title	Expiry Date	Remarks
July 29, 1859	1764	A. V. Newton. *A communication from J. Walch*	"Certain improvements in that class of firearms known as revolvers and in bullets for the same."	July 29, 1862	

A. V. Newton (see Br. Pat. No. 921/1857) acted in this patent for J. Walch, an American inventor, who had secured the equivalent U.S. Pat. No. 22905 of February 8th, 1859, for a single-action revolver with unusually long chambers, each of which held two charges; these superimposed loads were successively fired by two hammers, and twelve-shot revolvers on this system were made by the "Walch Firearms Co." of New York. It is not known if manufacture under the British patent was achieved, but an interest in this patent was registered (in 1860), at the British Patent Office, in the names of "Bray, Cole, Dauchy & Lindsay".

Patent Date	Number	Name of Patentee	Title	Expiry Date	Remarks
Aug. 22, 1859	1918	H. Spratt	"Improvements in revolving pistols and other firearms."	Aug. 22, 1862	

Henry Spratt was a Captain, in the Royal Marines, and proposed to speed up the loading of percussion revolvers by using spare loaded cylinders. This idea was a throw-back to a practice envisaged when Robert Adams first introduced his rammer-less self-cocking revolver; within its limitations, however, Spratt's system was well-conceived. His Complete Specification illustrated a Beaumont-Adams revolver, with frame and cylinder modified so that the latter might be removed or replaced with the minimum of care or attention, and proposed the fitting of metal caps to spare cylinders carried loaded on the user's person.

Patent Date	Number	Name of Patentee	Title	Expiry Date	Remarks
Oct. 5, 1859	2263	A. V. Newton. *A communication from Messrs. Perrin and Delmas*	"Improvements in revolving firearms."	Oct. 5, 1866	

A. V. Newton (see Br. Pat. No. 921/1857) here acted for Messrs. Perrin and Delmas, of Paris, to patent a gate-loaded, self-cocking centre-fire revolver, and its cartridge. The lock-mechanism was ludicrously similar to that used in British self-cocking pepperboxes or percussion revolvers of the day, but the weapon was unusual in having barrel and frame formed as one unit, *without* a top-strap connecting barrel to frame. The cartridge here protected (which had an extremely thick rim) long outlived this revolver, and was in use a decade later; the pistol was, however, bought for trial by the Federal Army, during the American Civil War.

Patent Date	Number	Name of Patentee	Title	Expiry Date	Remarks
Feb. 22, 1860	474	C. Greves. (Provisional protection)	"Improvements in firearms."	Aug. 22, 1860.	

Charles Greves, of Birmingham, did not file drawings with his Provisional Specification. So far as his written description can be grasped, he proposed to form the barrel, lock-case, and butt of a revolver in one piece, and to use the trigger of it as a cylinder-bolt also.

Patent Date	Number	Name of Patentee	Title	Expiry Date	Remarks
Mar. 15, 1860	693	G. Sturrock. (Provisional protection)	"Improvements in revolving firearms."	Sept. 15, 1860	

Staff Sergeant George Sturrock, R.E., envisaged a double-barrelled percussion revolver with two sets of (7) chambers in concentric rows. No specimen arm is known to have survived.

Patent Date	Number	Name of Patentee	Title	Expiry Date	Remarks
Mar. 15, 1860	694	W. N. Norman and J. Lubé	"Improvements in guns and other firearms and in cartridges to be used therewith."		*Provisional protection refused*

William Nicholas Norman, Gentleman, and John Lubé, Esquire, were domiciled in Dublin; amongst various claims, they sought protection for three ideas accepted for Br. Pat. No. 899/1860, later described. Reasons for the Comptroller of Patents' refusal to grant this particular application are not known.

Patent Date	Number	Name of Patentee	Title	Expiry Date	Remarks
Apr. 7, 1860	880	W. Clark. *A communication from Starr Arms Co.*	"Improvements in firearms."	Apr. 7, 1867	

William Clark was a British patent agent, in practice at 55 Chancery Lane, London; his principal was the Starr Arms Company (of Yonkers and Binghampton, N.Y., U.S.A.), which developed the revolver patents of E. T. Starr between 1858 and 1867, and registered its interest in this patent (at the British Patent Office) during 1860. The patent protected a percussion revolver, with a two-part barrel/frame assembly, and an excellent double-action lock-mechanism for it; ·44 and ·36 versions were made in quantity under U.S. patents, during the American Civil War, but there is no evidence that this British patent was worked over here.

Patent Date	Number	Name of Patentee	Title	Expiry Date	Remarks
Apr. 10, 1860	899	J. Rigby and W. N. Norman	"Improvements in guns and other firearms and in cartridges to be used therewith."	Apr. 10, 1874	

John Rigby (a famous Dublin gunmaker) joined with W. N. Norman to patent the ideas for which provisional protection had been refused as Br. Pat. No. 694/1860 (*q.v.*), and something has been said, in Chapter II, of the application to revolving arms of certain features from this patent.

As to revolvers, claim was laid to stocking the pistol so that the line of resistance for the firer's hand was along the line of recoil of the weapon; for using the hinged barrel as a rammer-lever; for a thumb-operated safety-catch; and for a centre-hammer, self-cocking lock-mechanism.

Patent Date	Number	Name of Patentee	Title	Expiry Date	Remarks
Apr. 14, 1860	942	A. V. Newton. *A communication from E. Savage* (Provisional protection)	"Improvements in rotating breech firearms."	Oct. 14, 1860	

A. V. Newton (see Br. Pat. No. 921/1857) here acted for Edward Savage of Cromwell, Conn., U.S.A.

Protection was sought for improvements to a toggle-joint used to effect the reciprocal cylinder motion in the gas-seal revolver protected by the A. V. Newton/ H. S. North's Br. Pat. No. 1608/1856 (*q.v.*). Claim was also laid to the use of thimbles or bushings at the mouth of the chambers, which were to be driven into the barrel, as the weapon was fired, to effect a gas-seal.

This British patent, if pursued, would presumably have had a similar Complete Specification to Savage and North's U.S. Patent No. 28331 of May 15th, 1860, which partly protected the gas-seal revolver, mentioned in our remarks upon Br. Pat. No. 1608/1856 (*q.v.*).

Patent Date	Number	Name of Patentee	Title	Expiry Date	Remarks
May 31, 1861	1362	F. Tolhausen. *A communication from M. Javelle*	"Improvements in revolving firearms."	May 31, 1864	

Frederick Tolhausen was a civil engineer and patent agent, at 35 Boulevart Bonne Nouvelle, Paris; he acted, in this patent, for Michel Javelle, of 54 Rue Beaubrun, St. Etienne, France. The communicant's occupation was not stated, but his address suggests that he was a gunmaker.

Protection was secured for a breech-loading, double-action centre-fire revolver, which used cartridges with a leaden or zinc base, and had an ingenious tip-down barrel, by means of which a discharged cylinder might readily be replaced with a loaded one. Specimen arms are not common in the United Kingdom.

Patent Date	Number	Name of Patentee	Title	Expiry Date	Remarks
July 12, 1861	1758	J. Adams	"Improvements in revolving firearms and in cartridges for the same."	July 12, 1868	

John Adams (see Br. Pat. No. 2824/1857) here improved upon his first revolver design, by providing for barrel and frame to be formed as one component for attachment (by screws) to the lock-case; such an improvement was, once again, not an infringement of his brother's master-patent (Eng. Pat. No. 13527/1851) for the complete barrel, frame and lock-case formed as one integral component.

In addition, Adams's new pistol was designed for optional use with a cylinder for "cartridges made on the Lefaucheux principle", or with a percussion cylinder substituted; specimen arms are not known today, but this patent did form the basis for post-1865 patents, under which were produced a centre-fire breech-loading Army revolver that was adopted by the British armed forces, in 1867.

Patent Date	Number	Name of Patentee	Title	Expiry Date	Remarks
Nov. 6, 1861	2785	G. Davies. *A communication from B. F. Joslyn* (Provisional protection)	"Improvements in firearms and ordnance."	May 6, 1862	

George Davies was a civil engineer and patent agent, at 1 Serle Street, Lincoln's Inn, London (and 28 St. Enoch Square, Glasgow), who acted for Benjamin Franklin Joslyn, of Stonington, Conn., U.S.A. The latter was an American firearms designer, with a viable U.S. patent (No. 20160 of May 4th, 1858) already to his credit, for a single-action, side-hammer percussion revolver of the type shown in Plate 36 (*top*); 1,100 weapons made under that patent were purchased by the Federal Army during the American Civil War.

Joslyn became a prolific patentee after 1863, but this provisional British protection was for a revolver not so far identified in an English collection; of Colt-type, the cylinder of the arm was to be rotated by use of a miniature clutch, and a rack-operated ramrod (adjustable as to length of stroke) was also claimed.

Patent Date	Number	Name of Patentee	Title	Expiry Date	Remarks
Mar. 8, 1862	626	J. Deane (Jun.)	"Improvements in revolving firearms."	Mar. 8, 1865	A revolver; see Chap. XII, and Plate 19 (*lower*).
Apr. 15, 1862	1081	F. A. Le Mat and C. F. Girard	"Improvements in the construction of revolving and repeating firearms, part of which invention is also applicable to other arms."	Apr. 15, 1869	

Dr. F. A. Le Mat (see Br. Pat. No. 1622/1859) became a Confederate sympathizer during the American Civil War, and came to Europe to arrange manufacture of his "grape shot" revolvers, for use by the South; he became associated with Charles Frederic Girard, of Paris, in this venture, and their efforts are dealt with, in a detail impossible here, by W. A. Albaugh, H. Benet and E. N. Simmons (*Confederate Handguns*. Riling and Lentz. Philadelphia, 1963), to whose excellent work the reader is commended.

In their British patent, the two men protected a mainspring mount and hammer linkage for Le Mat's single-action revolvers, a cylinder-rotation system, and a cylinder-bolt; the designs were sound enough, but their abominable execution by Continental manufacturers led the two men to order rectification work upon the new arms, in England. Satisfied with that work, the partners had a quantity of arms made here (possibly by the Birmingham Small Arms Company, Ltd.), and about 1,000 revolvers eventually reached the Confederacy, before its defeat and the dissolution of the Le Mat and Girard partnership. During 1865 an attempt was made to register a joint interest in this patent (and in both partners' names) at the British Patent Office, but the entry was later cancelled.

Le Mat carbines and "Baby" revolvers are occasionally encountered, but the average collector is only likely to meet percussion arms ante-dating 1865, in one model. This is a nine-shot revolver of about ·400 calibre, with a shot barrel of about 18 bore; the rifled barrel is about 6¾ in. long, and the shot barrel about 5 in. in length. In a serial-range observed between 252 and 564, we noted the top barrel-legend on these pistols as reading "Col. Le Mat's Patent"; in a range observed between 1195 and 1845, the legend was "Col. Le Mat Bte s.g.d.g. Paris": in a third range observed between 2003 and 2412, the legend became "SYST^ME^ LE MAT BTE S.G.D.G. PARIS": and in a concluding range (for percussion arms) observed between 8016 and 8974, we found the words "Le Mat & Girard's Patent London". Upon occasion, the names of British or French retailers were also engraved upon the shot barrel.

Le Mat pin-fire/percussion and centre-fire revolvers can also be encountered, but date from a period after 1865.

Patent Date	Number	Name of Patentee	Title	Expiry Date	Remarks
May 27, 1862	1589	G. H. Sanborn. *A communication from G. Gates* (Provisional protection)	"Improvements in revolving firearms."	Oct. 27, 1862	

George Henry Sanborn (formerly of Boston, Mass., U.S.A.) acted from 100 Fleet Street, London, on behalf of George Gates, of Worcester, Mass., U.S.A. Provisional protection was secured for a breech-loading revolver, with tip-up barrel, of the Smith & Wesson type described in Chapter XVII. It is not clear (from the unillustrated Provisional Specification) exactly what novel features were to have been embodied in that basic design, but it is possible that Gates, in turn, was acting for Lucius W. Pond, of Worcester, Mass. The latter secured U.S. Patent No. 35623, of June 17th, 1862, for a revolver meeting the general description in Gates's Specification, which was later ruled to be an infringement of a patent, held by Smith & Wesson, for the breech-loading cylinder.

Patent Date	Number	Name of Patentee	Title	Expiry Date	Remarks
May 27, 1862	1592	W. Palmer	"Improvements in revolving firearms."	May 27, 1865	

William Palmer (see Br. Pat. No. 2436/1858) here patented magazine arms, in which a rotating carrier received loaded and capped steel chargers from a hopper; turning a crank-handle or wheel brought each one up (to mate with the breech), fired it, and bore it away to an ejection port.

The patent is included here because the Complete Specification claimed this design as suitable for pistol, gun, or "swivel"; in fact, the patent was probably part of the protection upon an unsuccessful American crank-operated machine-gun (called the "Union Repeating Gun"), in which Palmer is known to have been interested at about this time.

Patent Date	Number	Name of Patentee	Title	Expiry Date	Remarks
July 19, 1862	2067	W. Tranter	"Improvements in firearms."	July 19, 1865	A revolver; see Chap. XVIII.
Aug. 22, 1862	2351	D. Moore	"Improvements in revolving firearms."	Aug. 22, 1865	

Daniel Moore (of Brooklyn, King's County, New York, U.S.A.) here patented a single-action, breech-loading revolver, in which barrel and cylinder tipped sideways (on a pivot in front of the frame) for loading or removing the copper-cased, rim-fire cartridges that the weapon was designed to use. Like Pond (mentioned in relation to Br. Pat. No. 1589/1862), Moore fell foul of Smith & Wesson over use of a bored-through cylinder in America, during the currency of a patent in which that partnership held an interest. The patent (U.S. Pat. No. 12648, of April 3rd, 1855, in the name of Rollin White) prevented Moore from manufacturing his revolver, in America, until 1869, but although White's patent had no force in the United Kingdom, there is no evidence that Moore manufactured under his British patent, whilst thus balked in America.

Patent Date	Number	Name of Patentee	Title	Expiry Date	Remarks
Nov. 7, 1862	3012	A. V. Newton. *A communication from W. A. Shepard* (Provisional protection)	"Improvements in repeating firearms."	May 7, 1863	

A. V. Newton (see Br. Pat. No. 921/1857) here acted for Albert Shepard, of New York; protection was sought for revolving pistols, rifles, or carbines (using metallic cartridges) in which one spring served both sear and hammer. The cylinder-bolt was attached to (and projected from) the hammer, and entered locking recesses in the cylinder, when it fell. No American patent on these lines is known to have been taken out by Shepard, and it is assumed that he was acting for somebody else.

Patent Date	Number	Name of Patentee	Title	Expiry Date	Remarks
Dec. 26, 1862	3452	W. Clark. *A communication from P. A. Mathieu and J. F. Gevelot*	"Improvements in firearms."	Dec. 26, 1869	

William Clark (see Br. Pat. No. 880/1860) here acted for Philippe Antoine Mathieu, Proprietor, and Jules Felix Gevelot, Manufacturer, of 29 Boulevart St. Martin, Paris. Revolvers to be attached to the underside of guns or rifles were described, claimed, and illustrated in the Complete Specification, together with means for attaching and firing the same; these revolvers were breech-loading, and although not stated to be so, were (from the drawings) probably intended to use pin-fire cartridges. No specimen arm is known to have survived.

Patent Date	Number	Name of Patentee	Title	Expiry Date	Remarks
Mar. 5, 1863	628	W. Clark. *A communication from A. Guerriero*	"Improvements in firearms."	Mar. 5, 1866	

William Clark (see Br. Pat. No. 880/1860) here acted for a Genoese inventor, Alexandre Guerriero; the latter sought to improve certain defects in the basic Lefaucheux pin-fire revolver (see Br. Pat. No. 955/1854), as to attachment of the barrel, construction of the frame, and prevention of discharge if chamber and barrel were not properly aligned. Specimen arms may have survived on the Continent, but no pin-fire revolver is known, in any major British collection, to embody Guerriero's improvements as patented; an interest in the patent was entered at the British Patent Office (during 1864) in the names of "Cauvin, Pavese, Williamson & Weiland".

Patent Date	Number	Name of Patentee	Title	Expiry Date	Remarks
Apr. 9, 1863	902	A. V. Newton. *A communication from J. C. Campbell*	"An improved construction of offensive weapon."	Apr. 9, 1866	

A. V. Newton (see Br. Pat. No. 1608/1856) here acted for James Cameron Campbell, of New York. The patent proposed to mount a large pepperbox pistol around the shaft of a lance, with the barrels aligned along that shaft; manual rotation of the barrels automatically discharged them in succession.

Patent Date	Number	Name of Patentee	Title	Expiry Date	Remarks
May 19, 1863	1251	J. H. Johnson. *A communication from A. Pollak*	"Improvements in breech loading firearms."	May 19, 1866	

John Henry Johnson (see Br. Pat. No. 955/1854) here acted for Anthony Pollak, of Washington, D.C.; amongst various ideas, the latter claimed a mechanically-rotated, centre-hammer revolving gun for use with rim-fire cartridges. The cylinder was only long enough to contain the cartridge cases, and moved longitudinally (with the lock-mechanism) to thrust each exposed bullet, in turn, into the breech-end of the barrel. If this gas-seal had not been effected, the arm could not be discharged.

Patent Date	Number	Name of Patentee	Title	Expiry Date	Remarks
July 23, 1863	1844	G. Davies. *A communication from C. Sharps*	"An improvement in revolving firearms."	July 23, 1866	

G. Davies (see Br. Pat. No. 2785/1861) here acted for Christian Sharps, a famous American rifle designer, and for a patent that the latter never developed in the United Kingdom. Sharps proposed to use a detachable or swivelling cap, attached to the rear of breech-loading revolver cylinders, which had to be removed or turned before the metallic cartridges could be inserted. In America, this type of cylinder was under study, as not being an infringement of the Rollin White patent described in our remarks upon Br. Pat. No. 2351/1862 (*q.v.*); in the United Kingdom, there was no equivalent restriction upon using bored-through cylinders, and therefore no interest in Sharps's devices.

Patent Date	Number	Name of Patentee	Title	Expiry Date	Remarks
July 27, 1863	1862	W. Tranter	"Improvements in breech loading and other revolving firearms."	July 27, 1866	A revolver; see Chap. XVIII.
Sept. 12, 1863	2248	C. E. Wallis. (Provisional protection)	"Improvements in revolving firearms."	Mar. 12, 1864	

Charles Edwin Wallis, a Londoner, proposed use of two superimposed charges in each chamber of a percussion revolver; two nipples (or a "compound nipple with two channels") were to serve each chamber. Unlike the Walch revolver (see Br. Pat. No. 1764/1859) Wallis proposed to use only one hammer in his arms. (See also his Br. Pat. No. 624/1864.)

Patent Date	Number	Name of Patentee	Title	Expiry Date	Remarks
Sept. 12, 1863	2250	W. Clark. *A communication from T. J. Vail*	"Improvements in revolving firearms."	Sept. 12, 1866	

William Clark (see Br. Pat. No. 880/1860) here acted for Thomas Jefferson Vail, of Hartford, Conn., U.S.A.; it seems probable that Vail himself was an American patent agent, since an equivalent U.S. Patent No. 37693 of February 17th, 1863, had been secured by John C. Howe, of Worcester, Mass., U.S.A. Vail registered an interest in the British patent (at the Patent Office), during 1864, and this probably recorded an assignment of the patent to him, by Clark, upon completion of all formalities of grant.

A single-action, hinged-frame, breech-loading revolver was both claimed and illustrated in the Clark/Vail Complete Specification, together with various details of the barrel-latch, lock-mechanism, and frame construction. In years after 1865, features of Howe's patent were to be used by Smith & Wesson, of Springfield, Mass., in a range of magnificent hinged-frame, single-action, breech-loading revolvers; there is no evidence, however, that any arms were manufactured in England under this Clark/Vail patent.

Patent Date	Number	Name of Patentee	Title	Expiry Date	Remarks
Oct. 2, 1863	2420	G. T. Bousfield. *A communication from S. Wood* (Provisional protection)	"Improvements in revolver firearms."	Apr. 2, 1864	

G. T. Bousfield (see Br. Pat. No. 2882/1857) here acted from Loughborough Park, Brixton, for Stephen Wood, of Cornwall, New York. The latter claimed a double-action revolver lock-mechanism, in terms suggesting that the cylinder was to rotate after pulling the trigger, when the hammer was at full-cock; no specimen arm has been identified. (See also Br. Pats. Nos. 760/1864 and 309/1865.)

Patent Date	Number	Name of Patentee	Title	Expiry Date	Remarks
Nov. 7, 1863	2768	J. K. Hoyt. *A communication from B. F. Joslyn*	"Improvements in revolving firearms."	Nov. 7, 1866	

Jehiel Keeler Hoyt (of London and New York) was presumably a patent agent, and acted here for Benjamin Franklyn Joslyn, of Stonington, Conn., U.S.A. (see Br. Pat. No. 2785/1861). Joslyn claimed the use of a spring-loaded recoil plate in a breech-loading revolver (to prevent jamming of the cylinder, by bulging cartridge heads), and a cylinder that could be swung laterally out of the frame; in that position, a rod (hinged to the arbor) could be used to eject the metallic cartridge-cases, one by one.

Essentially, Joslyn's ideas were to be embodied in U.S. Patent No. 41117 of January 5th, 1864, by H. A. Briggs and S. S. Hopkins, of Norwich, Conn.; a six-shot single-action, rim-fire revolver (in ·38 calibre) was made under that patent, by the Bacon Manufacturing Company, of Norwich, Conn. No British arms made under Joslyn's patent have been identified.

Patent Date	Number	Name of Patentee	Title	Expiry Date	Remarks
Jan. 8, 1864	50	R. Adams	"Improvements in the construction of breech-loading firearms."		*Provisional protection refused*

This was Robert Adams's last appearance as an applicant for patent protection on revolving arms; he claimed a breach-loading revolver (with a moveable plate behind the cylinder), which could be converted to use a percussion cylinder. One subsidiary claim echoed his master-patent (Eng. Pat. No. 13527/1851) and claimed the "barrel-stock and lock frame and trigger guard" to be made of one piece of metal. The reasons for refusal of even provisional patent protection for these ideas are not known.

Patent Date	Number	Name of Patentee	Title	Expiry Date	Remarks
Feb. 25, 1864	469	B. Burton. (Provisional protection)	"Improvements in breech-loading and revolving firearms and cannon and metallic cartridges."	Aug. 25, 1864	

Bethel Burton (of Brooklyn, N.Y.) was primarily concerned with a breech-loading rifle design, and with cannon. However, he also claimed a peculiar chambering for revolvers; this was bored from the front, but did not extend clear through the cylinder. Instead, a hole of smaller diameter was drilled through the rear wall, and a front-loading cartridge (with a suitable percussion tail formed on it) was loaded into the chamber from the front. A suitable slit in the chamber wall permitted the hammer-nose to strike the tail of the cartridge, and to fire it.

Such cartridges were claimed to be superior to rim-fire rounds, as easier to prime and as not sticking in use, but Burton's system was really of interest only to those American manufacturers balked (by Smith & Wesson) from using conventional breech-loading cylinders. (See remarks to Br. Pat. No. 2351/1862.)

Patent Date	Number	Name of Patentee	Title	Expiry Date	Remarks
Mar. 12, 1864	624	C. E. Wallis	"Improvements in revolving firearms."	Mar. 12, 1867	

Wallis (see Br. Pat. No. 2248/1863) here filed application from Millman Street, Bedford Row, Holborn, London, for refinements to the basic idea claimed in his first patent. The double-loaded chambers of his revolvers were to be fitted with two rows of vertical nipples, and fired by a long bar-hammer (of the kind shown on Plate 37 (*lower*)), which had a sliding component for firing the rear charge, after the front loads had been cleared.

Two models of Wallis's revolvers are known. The first (of transition-type) follows closely the drawings to this patent; the second, however, has a set of horizontal nipples at the rear of the cylinder. Specimens of either model are extremely rare, but it is known that a man named Richardson registered an interest in this patent (during 1865) at the Patent Office, and may have been the Irish gunmaker, in business at Cork in those times.

Patent Date	Number	Name of Patentee	Title	Expiry Date	Remarks
Mar. 26, 1864	760	G. T. Bousfield. *A communication from S. Wood* (Provisional protection)	"Improvements in revolver firearms."	Sept. 26, 1864	

Bousfield (see Br. Pat. No. 2420/1863) again acted for S. W. Wood, and here secured provisional protection for a "hesitating" revolver lock-mechanism.

Patent Date	Number	Name of Patentee	Title	Expiry Date	Remarks
June 14, 1864	1464	C. Armbruster	'Breech-loading and other small firearms."		*Provisional protection refused*

Charles Armbruster, a gunmaker, at 8 Vernon Place, Bloomsbury, London, claimed that "by altering the breech part of a revolver or other small firearms, and using the bottom igniting cartridges, I add to the number of shots, economize the ammunition, diminish the weight of firearms and render it safe against ordinary accidents". If, for some reason, the applicant was unwilling to enlarge upon these curious claims (and, more important, how they were to be effected), the Patent Office can hardly have been blamed for refusing protection.

Patent Date	Number	Name of Patentee	Title	Expiry Date	Remarks
July 16, 1864	1785	A. Wyley	"Improvements in breech loading firearms, including revolvers and ordnance."	July 16, 1867	

Andrew Wyley proposed to connect the "breech-case" of a firearm to its barrel by a screw, having portions of both the internal and external threads cut away; this screw could thus be slid into position, and then turned to engage the surviving areas of its thread. Application to revolvers was claimed (and would have been useful to latch any hinged barrel), but the inventor's interest was primarily directed to long arms and artillery.

Patent Date	Number	Name of Patentee	Title	Expiry Date	Remarks
Aug. 19, 1864	2059	B. Burton	"Improvements in breech loading and revolving firearms and cannon and metallic cartridges."	Aug. 19, 1867	

Bethel Burton (see Br. Pat. No. 469/1864) again claimed the use in revolving arms of cartridges embodying their own ignition, amongst other inventions relating to conventional firearms.

Patent Date	Number	Name of Patentee	Title	Expiry Date	Remarks
Sept. 28, 1864	2383	J. Jongen	"Improvements in the construction of firearms."	Sept. 28, 1867	

Jean Jongen (a Liège gun-manufacturer) claimed an idea for drilling the chambers of a percussion revolver from end to end, and then closing them at the breech, by screwed blocks carrying their nipples; he also proposed to place a "ram pin" between the nipples and the hammer. This latter idea had been anticipated in John Adams's Br. Pat. No. 2824/1858.

Patent Date	Number	Name of Patentee	Title	Expiry Date	Remarks
Oct. 12, 1864	2513	I. Williams. *A communication from A. Micheloni* (Provisional protection)	"An improved revolver."	Apr. 12, 1865	

Isaiah Williams (a London patent agent) acted here for Ambrogio Micheloni, of Italy; the latter proposed to embody a revolver in the handle and guard of a sabre. Three years later (at the Paris Exhibition) a number of makers offered revolving arms of this type, and considerable interest in them was aroused; it is not known if Micheloni benefited from that interest.

Patent Date	Number	Name of Patentee	Title	Expiry Date	Remarks
Feb. 4, 1865	309	S. W. Wood	"Certain improvements in revolving firearms and cartridges for the same."	Feb. 4, 1872	

S. W. Wood (see Br. Pat. Nos. 2420/1863 and 760/1864) did not here bring his earlier efforts to fruition, but patented a front-loading revolver (using metallic cartridges) which enjoyed a minor commercial success in America. Essentially, however, the weapon and its cartridge were only desirable at all, as a means for avoiding a Smith & Wesson-controlled patent (for the bored-through revolver cylinder) mentioned in relation to Br. Pat. No. 2351/1862 (*q.v.*), and there is no evidence that arms and ammunition were made under Wood's British patent.

Patent Date	Number	Name of Patentee	Title	Expiry Date	Remarks
Mar. 9, 1865	659	W. Clark. *A communication from S. A. Noël and F. Gueury*	"Improvements in revolving firearms."	Mar. 9, 1872	

William Clark (see Br. Pat. No. 880/1860) here acted for Sebastien Amedee Noël and François Gueury, of France, to patent a weapon known, on the Continent, as the "Rotovolver". This was a self-cocking radial revolver, with a side-swinging hammer, and proved both compact and reliable in use; despite the long maintenance of this patent, there is no evidence that arms were manufactured in England, under it.

Patent Date	Number	Name of Patentee	Title	Expiry Date	Remarks
Apr. 12, 1865	1046	T. J. Mayall	"Improvements in firearms and in cartridges to be used therewith and with other firearms."	Apr. 12, 1872	

Thomas Jefferson Mayall (of Roxbury, Mass., and Red Lion Court, London) patented a breech-loading revolver, which could be used with a percussion cylinder, and which had the butt hinged to the action-body; by releasing the butt, a fresh cylinder could be rapidly inserted in place of an empty one. An ingenious rammer-cum-cartridge ejector (operated by rack and pinion) was also claimed. No specimen arm of British manufacture is known to have survived.

Patent Date	Number	Name of Patentee	Title	Expiry Date	Remarks
July 20, 1865	1889	W. Tranter	"Improvements in firearms and in cartridges for the same."	July 20, 1879	A revolver cylinder; see Chap. XVIII.
Aug. 4, 1865	2027	H. A. Bonneville. *A communication from P. Drivon and C. J. Biron* (Provisional protection)	"Improvements in revolver pistols."	Feb. 4, 1866	

Henri Adrien Bonneville was a patent agent (with offices in Paris and London), and here acted for Pierre Drivon and Claude Joseph Biron, both of Belleville, France. The protection had a more important French equivalent patent, and claimed a revolver lock-mechanism and a cylinder-rotation device.

Patent Date	Number	Name of Patentee	Title	Expiry Date	Remarks
Aug. 4, 1865	2030	T. W. Webley. (Provisional protection)	"Improvements in breech loading firearms and in revolving firearms and in cartridges."	Feb. 4, 1866	A revolver; see Chap. XVII.
Aug. 7, 1865	2050	W. C. Dodge	"Improvements in firearms."	Aug. 7, 1868	

William Castle Dodge (of Washington, D.C.) was a former official of the U.S. Patent Office, and a firearms designer of stature. By this patent, he protected the basic idea of plate- or star-shaped simultaneous cartridge extractors for revolvers, adaptable to almost any such weapon with a barrel that could be tipped or swivelled away from the frame. Smith & Wesson came to terms with Dodge to use the U.S. equivalent of this British protection (U.S. Pat. No. 45912, of January 17th, 1865), and developed their magnificent hinged-frame, self-extracting revolvers on the basis of it, in the years after 1865. A later British patent by Smith & Wesson's patent agent, W. R. Lake (which was secured as Br. Pat. No. 1510, of May 17th, 1869) must have been in preparation when Dodge's patent fell due for renewal, and may account for the early expiry date noted above.

Patent Date	Number	Name of Patentee	Title	Expiry Date	Remarks
Aug. 15, 1865	2108	J. Broun. (Provisional protection)	"Improvements in revolving firearms, in projectiles and cartridges."	Feb. 15, 1866	

James Broun (a Scottish barrister) claimed a revolver in which the charges were to be pushed successively forward out of their chambers, by a "self-acting bolt", and fired in the barrel.

Patent Date	Number	Name of Patentee	Title	Expiry Date	Remarks
Aug. 28, 1865	2210	P. Polain	"Improvements in breech loading revolvers."	Aug. 28, 1868	

Prosper Polain, of Liège, patented a curious pin-fire revolver, which is very rare today. A composite cylinder was proposed, in which chamber tubes could be drawn forward simultaneously, to reveal ports in the cylinder wall. Loading was to be effected through these ports, and the empty cases might fall from them, when the pistol was opened; an ejector rod was providently fitted.

Patent Date	Number	Name of Patentee	Title	Expiry Date	Remarks
Dec. 16, 1865	3258	A. V. Newton. *A communication from S. Crispin*	"Improvements in breech loading firearms and rotating breech cylinder pistols and cartridges to be used therewith."	Dec. 16, 1868	

A. V. Newton (see Br. Pat. No. 1608/1856) here acted for Silas Crispin, of New York. A hideously dangerous metallic cartridge was protected, which had the primer in an annular ring around its case; the cylinder of a revolver was formed in two halves, and mated over each end of the rounds, the hammer striking over the rear half, to reach this annular primer. A gas-tight joint was claimed as one virtue for this ammunition.

INDEX

Numbers in **heavy type** refer to illustrations.